# Catalytic Oxidation

## *Principles and Applications*

Netherlands Institute for Catalysis Research

# Catalytic Oxidation

## Principles and Applications


Editors

**R.A. Sheldon**
Delft Univ. of Technology

**R.A. van Santen**
Eindhoven Univ. of Technology



World Scientific
Singapore • New Jersey • London • Hong Kong

*Published by*

World Scientific Publishing Co. Pte. Ltd.

P O Box 128, Farrer Road, Singapore 9128

*USA office:* Suite 1B, 1060 Main Street, River Edge, NJ 07661

*UK office:* 57 Shelton Street, Covent Garden, London WC2H 9HE

**CATALYTIC OXIDATION: PRINCIPLES AND APPLICATIONS**



ISBN 981-02-2186-X

Printed in Singapore.

# Preface

Selective catalytic oxidation is a key technology for converting oil and natural gas-derived feedstocks to a wide variety of bulk chemicals. Moreover, in the wake of increasingly stringent environmental legislation, attention is being focused on the development of 'greener' processes, such as catalytic oxidation, for the manufacture of fine chemicals. A thorough understanding of fundamental mechanistic pathways of catalytic oxidations is of paramount importance for the improvement of existing processes and the development of new ones. Hence, the substantial interest, in both industrial and academic laboratories, in the mechanisms of oxidation catalysis.

Compared to hydrogenations, for example, which are mechanistically relatively straightforward processes, oxidations are enormously complex. For starters, molecular oxygen, in contrast to hydrogen, reacts with most organic molecules even in the absence of a catalyst. Hence, a thorough knowledge of these free radical chain processes (autoxidations) is an essential ingredient for understanding oxidation catalysis.

Catalytic oxidations with molecular oxygen (dioxygen) can be divided into two types: heterogeneous, gas phase and homogeneous, liquid phase processes. Researchers generally affiliate themselves with one or the other group and the two tribes speak largely different languages. Consequently, there is little cross-fertilization of ideas and, moreover, a third type-heterogeneous catalysis in the liquid phase- tends to suffer from lack of attention by either group.

A major aim of this book is to provide a sound mechanistic basis for understanding catalytic oxidation processes, which should be useful to reseachers in the field, irrespective of their tribal affiliation. It is based on a course on oxidation catalysis held in Rolduc, The Netherlands, in June 1994, under the auspices of the Dutch Research School in Catalysis (NIOK). The course was given by international authorities from industry and academia in the fields of both gas and liquid phase oxidations. It was targeted mainly at postgraduate research students wanting to acquaint themselves with the basic principles and industrial applications of oxidation catalysis. Moreover, it was hoped that the course would foster a synergistic cross-fertilization of concepts and ideas between the aforementioned groups. The participants had backgrounds ranging from surface science to organic synthesis.

The opening chapter (Sheldon) consists of an introductory overview of the subject, in which different processes to particular products are compared and the reader is introduced to mechanistic aspects. This is followed by a chapter (Haber) dealing with the elementary mechanisms of hydrocarbon oxidations on metal oxide surfaces and emphasizing the role of electrophilic versus nucleophilic oxygen species. Chapter three (Vedrine) contains a further description of the general features of oxidations on metal oxides, including multicomponent systems. The role of Mars -van Krevelen type mechanisms is emphasized. Chapter four (van Santen) continues with a detailed description of the mechanistic features of two industriallly important gas phase processes: the oxidation of ethylene to ethylene oxide and vinyl acetate, over silver and palladium-based catalysts, respectively. The elementary reaction steps taking place on the metal surface form the basis for this discussion,

Chapter five (Schmidt and Huff) focuses on the high-temperature oxidation of small molecules on noble metal catalysts. Four industrially important processes are discussed: ammonia oxidation to nitric acid, methane oxidation to syngas and ammoxidation to hydrogen cyanide, and oxidative dehydrogenation of ethane. A consideration of mass and heat transfer effects in conjunction with reaction kinetics is shown to be fundamental to understanding these processes. Chapter six (Marin), continuing in the same vein, elaborates the reaction pathways involved in the high-temperature oxidative coupling of methane. The consequences of the interplay between chemical kinetics and mass transfer for the selectivity to ethane and ethylene are highlighted. Chapter seven (van Veen) strikes a different note by discussing the principles and prospects of fuel cells, in which the basic reaction is equivalent to the combustion of hydrogen.

The second part of the book is devoted to liquid phase oxidation processes. It opens in chapter eight (Sheldon) with a review of the basic principles of free radical chain autoxidations. The intricate mechanism of the Amoco process for the catalytic oxidation of p-xylene to terephthalic acid is discussed in detail. Chapter nine (Sheldon) continues with an elaboration of the various types of heterogeneous catalysts for liquid phase oxidations. Particular emphasis is placed on molecular sieve catalysts containing redox metal ions incorporated in the framework or metal complexes encapsulated in the micropores (ship-in-the-bottle catalysts).

Chapter ten (Moiseev) elucidates the mechanistic roles of σ and π complexes in palladium-catalyzed oxidations of olefins. The oxidative acetoxylation of ethylene and propylene, to vinyl acetate and allyl acetate, respectively, catalyzed by giant palladium clusters is discussed. The latter can be considered as bridging the gap between homogeneous palladium complexes and a palladium metal surface. Chapter eleven (Sheldon) is devoted to the application of catalytic oxidations in fine chemicals synthesis. The characteristics of fine versus bulk chemical manufacture and catalytic oxidation versus oxygen transfer are explained. It also includes a section on catalytic asymmetric oxidation. This part of the book concludes with a chapter (van Veen) on selective electrochemical oxidations, encompassing both ex-situ and in-situ generation of metal oxidants. Attention is also focused an new developments in electrode materials and solid polymer electrolyte cells.

The third part of the book, chapter thirteen (Mills, Harold and Lerou), returns to the subject of heterogeneous gas-phase processes with a comprehensive overview of catalytic reactor technology. The advantages and limitations of existing and emerging catalytic reactors are discussed.

The book contains more than 600 literature references and a thoroughy cross-referenced index. Hopefully it will be widely used by aspiring researchers in this fascinating and economically important field. Finally, the editors would like to express their sincere thanks to their friends and colleagues who have contributed chapters to this book and Frank Sheldon for preparing the index.

R.A.Sheldon

R.A. van Santen

# Contents

# CATALYTIC OXIDATIONS: AN OVERVIEW

R.A. SHELDON
*Laboratory for Organic Chemistry and Catalysis,*
*Delft University of Technology, Julianalaan 136,*
*2600 GA Delft, The Netherlands*

## ABSTRACT

**Processes for the manufacture of industrial chemicals by catalytic oxidation of feedstocks derived from oil or natural gas are reviewed. Both heterogeneous, gas phase and homogeneous liquid phase processes are discussed and different processes to particular products, such as acetic acid, propylene oxide, phenol and cyclohexane, are compared. Recent trends in chemicals manufacture are also outlined. Finally, the reader is introduced to mechanistic aspects of metal oxidations.**

## 1. Introduction

The controlled partial oxidation of hydrocarbons, comprising alkanes, alkenes and aromatics, is the single most important technology for the conversion of oil- and natural gas-based feedstocks to industrial organic chemicals[1-3]. For economic reasons, these processes predominantly involve the use of molecular oxygen (dioxygen) as the primary oxidant. Their success depends largely on the use of metal catalysts to promote both the rate of reaction and the selectivity to partial oxidation products. Both gas phase and liquid phase oxidations, employing heterogeneous and homogeneous catalysts, respectively, are practiced industrially. (Table 1). Moreover, the pressure of increasingly stringent environmental regulation is stimulating the deployment of catalytic oxidation in the manufacture of fine chemicals (see chapter 11). Traditionally, the production of many fine chemicals has involved oxidations with stoichiometric quantities of, for example, permanganate or dichromate, leading to the concomitant generation of large amounts of inorganic salt - containing effluent. Currently there is considerable pressure, therefore, to replace these antiquated technologies by cleaner, catalytic alternatives[4-8].

## 2. Homogeneous Catalysis / Liquid phase

Several important liquid phase processes were developed during the 1950's and 1960's. Examples include the Wacker process for ethylene oxidation to acetaldehyde, the Celanese process for n-butane oxidation to acetic acid and the Amoco/Mid-Century process for the production of terephthalic acid from p-xylene (figure 1).

Although, as noted above, catalytic oxidation is the most favorable technology for the manufacture of many industrial chemicals this is not always the case. The most important process for the production of acetic acid, for example, is the rhodium-catalyzed carbonylation of methanol developed by Monsanto[9] (see figure 2). This process has the advantage of high selectivity (99%) coupled with cheap raw materials (methanol and carbon monoxide).

Table 1. Catalytic Oxidation Processes.

| Product | Primary raw materials | Volume [a] ($10^6$ tons) | Oxidant/ Process |
|---|---|---|---|
| Styrene | Benzene / ethylene | 5.0 | None/G |
| Terephthalic acid | *p*-Xylene | 3.9 | $O_2$ /L |
| Formaldehyde | Methanol | 3.8 | $O_2$ /G |
| Ethylene oxide | Ethylene | 2.9 | $O_2$ /G |
| Phenol | a. Benzene / propylene<br>b. Toluene | 1.9 | $O_2$ /L |
| Acetic acid | a. n-Butane<br>b. Ethylene | 1.8 | $O_2$ /L |
| Propylene oxide | Propylene | 1.4 | $RO_2H$ /L |
| Acrylonitrile | Propylene | 1.4 | $O_2$ /G |
| Vinyl acetate | Ethylene | 1.3 | $O_2$ /L;G |
| Acetone | Propylene | 1.2 | $O_2$ /L |
| Benzoic acid | Toluene | 1.0 | $O_2$ /L |
| Adipic acid | Benzene | 0.9 | $O_2$ /L |
| Caprolactam | Benzene | 0.7 | $O_2$ /L |
| Phthalic anhydride | *o*-Xylene | 0.7 | $O_2$ /G |
| Methyl methacrylate | Isobutene | 0.5 | $O_2$ /G |
| Acrylic acid | Propylene | 0.5 | $O_2$ /G |
| Methyl ethyl ketone | 1-Butene | 0.3 | $O_2$ /L |
| Maleic anhydride | n-Butane | 0.3 | $O_2$ /G |

a. USA, 1993 b. L=liquid phase; G=gas phase c. Acetic acid predominantly made via methanol carbonylation

### 3. Heterogeneous Catalysis / Gas Phase

Several important gas phase oxidation processes are outlined in figure 3. These processes were developed in the 1950's or earlier, although process improvements are still being made, e.g. in maleic anhydride manufacture from n-butane (see chapter 12).

The manufacture of methylmethacrylate (MMA) is an interesting case in point for comparing different technologies (see figure 4). The classical process involves the methanolysis of acetone cyanohydrin, has an atom utilization[5-8] of 29%, and produces 2.5 kg. of ammonium bisulfate per kg of MMA. More recently Asahi and Mitsubishi have developed alternative processes[10] based on the gas phase, catalytic oxidation of isobutene to methacrolein analogous to the oxidation of propylene to acrolein (see figure 3). Alternatively, methacrylic acid can also be prepared in a two-step, oxidation/dehydrogenation of isobutyraldehyde. The latter is a byproduct of propylene hydroformylation to n-butyraldehyde. Routes have also been developed based on ethylene carbonylation or hydroformylation[10] (see figure 4). Finally, Shell workers [11] have reported an elegant one-step synthesis of MMA by methoxycarbonylation of methylacetylene, the latter being available as a byproduct of naphtha cracking.

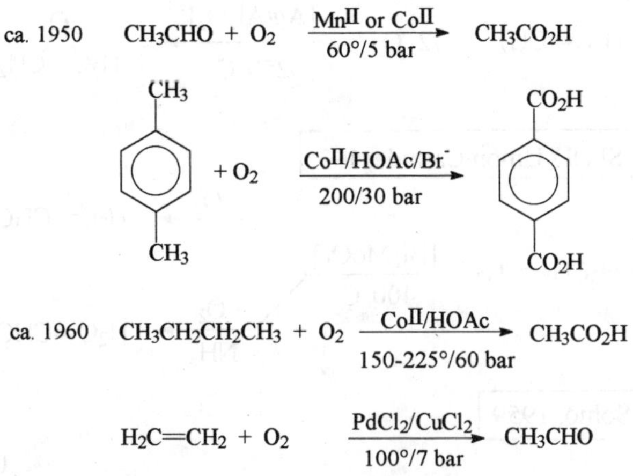


Figure 1. Liquid phase Oxidation Processes.

## 4. Propylene Oxide Manufacture: the Selectivity Problem

Table 1 contains one example of an industrial process, the manufacture of propylene oxide, that is notable in two respects : it involves the use of an alkyl hydroperoxide as the primary oxidant and, in one variant of the process at least, a heterogeneous catalyst in the liquid phase. As noted above, (figure 3) ethylene oxide is produced by gas phase oxidation of ethylene over a silver catalyst. Unfortunately, oxidation of propylene under the same conditions is unselective, due to competing oxidation of the olefinic double bond and the allylic C-H bonds.

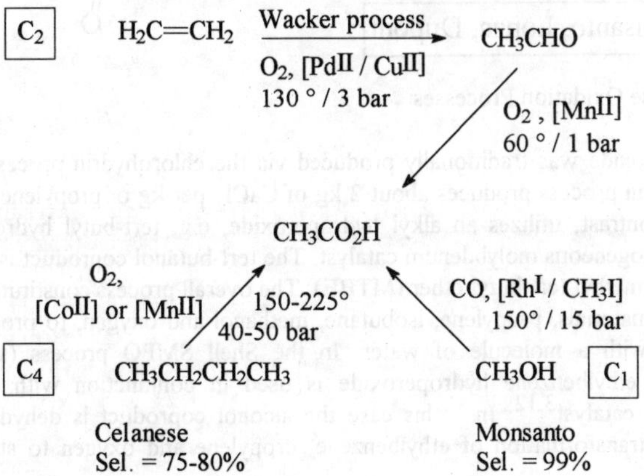


Figure 2. Acetic Acid Manufacturing Processes.

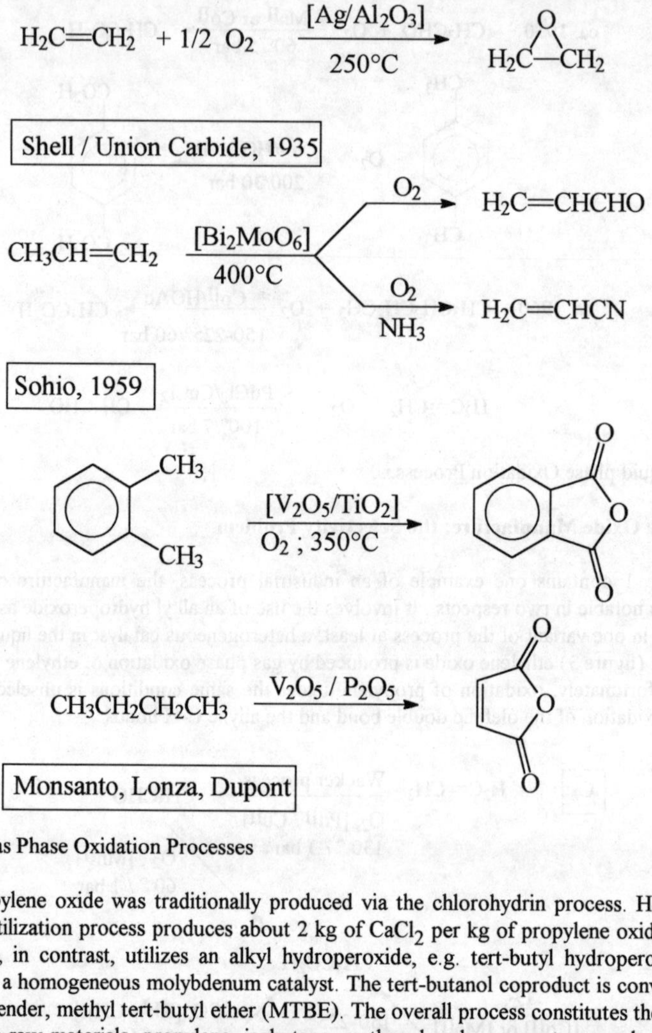


Figure 3. Gas Phase Oxidation Processes

Propylene oxide was traditionally produced via the chlorohydrin process. However, this low atom utilization process produces about 2 kg of $CaCl_2$ per kg of propylene oxide. The Arco process[3,12], in contrast, utilizes an alkyl hydroperoxide, e.g. tert-butyl hydroperoxide, in the presence of a homogeneous molybdenum catalyst. The tert-butanol coproduct is converted to the gasoline extender, methyl tert-butyl ether (MTBE). The overall process constitutes the conversion of the basic raw materials, propylene, isobutane, methanol and oxygen, to propylene oxide and MTBE together with a molecule of water. In the Shell SMPO process (styrene monomer propylene oxide) ethylbenzene hydroperoxide is used in conjunction with a heterogeneous titanium(IV)/silica catalyst[3,12]. In this case the alcohol coproduct is dehydrated to styrene, giving an overall transformation of ethylbenzene, propylene and oxygen to styrene, propylene oxide and water.

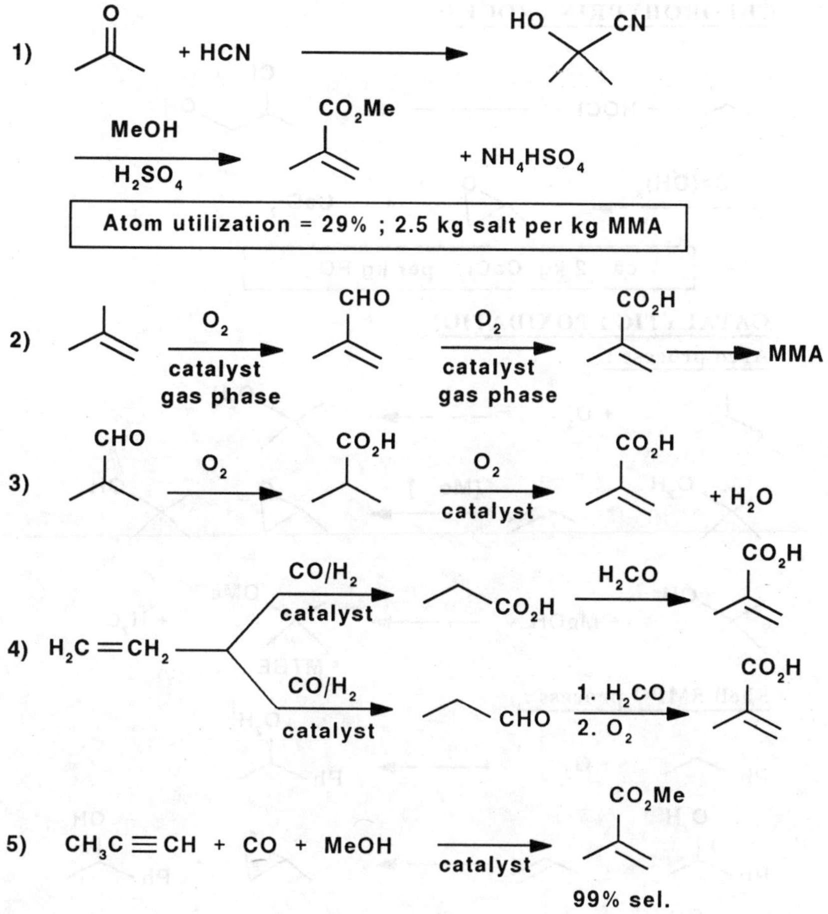


Figure 4. Alternative Routes to Methylmethacrylate.

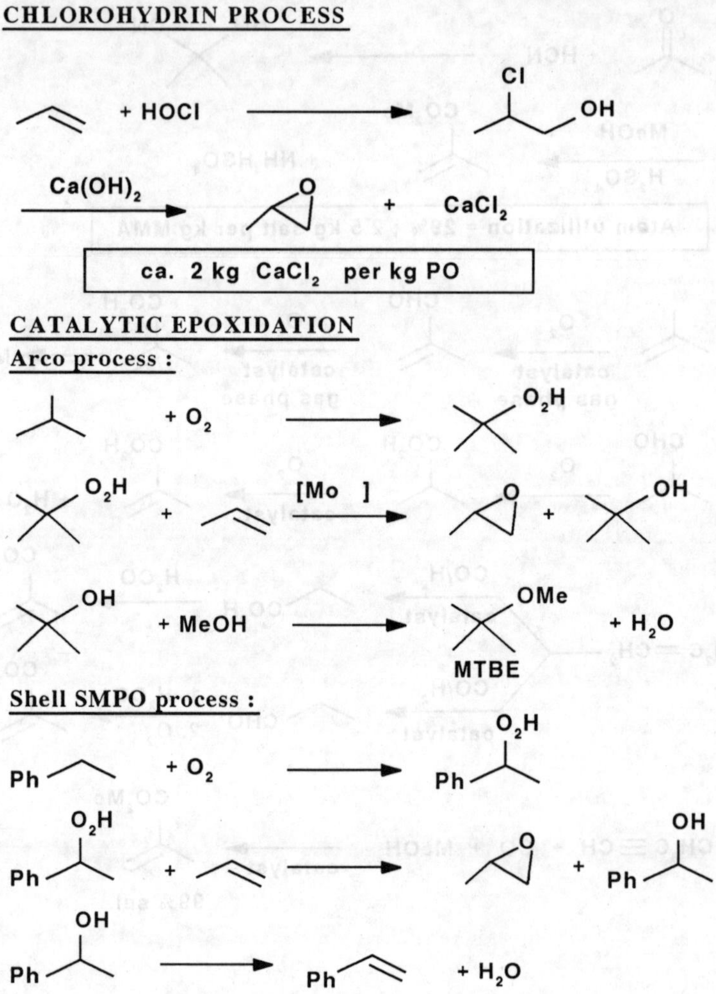


Figure 5. Routes to propylene oxide.

## 5. Manufacture of Aromatic Acids

As noted above, phthalic acid (anhydride) is produced by gas phase oxidation of o-xylene over a vanadium oxide catalyst. In contrast most other aromatic carboxylic acids are produced by metal-catalyzed autoxidation of the corresponding toluene in the liquid phase (reaction 1).

$$ArCH_3 + 1^1/_2O_2 \xrightarrow{\text{Catalyst}} ArCO_2H + H_2O \qquad (1)$$

Traditionally many aromatic carboxylic were produced by nitric acid oxidation of the corresponding toluene or by side chain chlorination followed by hydrolysis. Unfortunately, these processes suffer from serious drawbacks: large amounts of inorganic salt-containing effluent and chloro(nitro) compounds as impurities. Catalytic oxidation, on the other hand, is a high atom utilization, low salt process with no chloro(nitro) compounds as impurities.

Benzoic acid, for example, is produced commercially by the cobalt-catalyzed autoxidation of toluene (see chapter 8). The most important aromatic carboxylic acid is, however, terephthalic acid (see Table 1). It is produced by autoxidation of p-xylene in acetic acid in the presence of a cobalt catalyst and a promotor : either acetaldehyde (Eastman-Kodak and Toray processes) or bromide ion (Amoco/MC process)[13]. The two processes are compared in Table 2. The mechanism of the bromide ion promoted catalysis is discussed in chapter 9.

Table 2. Terephthalic Acid Manufacture.

| | **Cooxidation** (Eastman Kodak/ Toray) | **Bromide-mediated** (Amoco-MC) |
|---|---|---|
| Catalyst | $Co(OAc)_2$ | $Co(OAc)_2$/ $Mn(OAc)_2$ |
| Solvent | HOAc | HOAc |
| Promotor | $CH_3CHO$ | $Br^-$ |
| Temp. (°C) | 100-140 | 195 |
| Pressure (bar) | 30 | 20 |
| Conv./ Sel. (%) | >95/ >95 | >95/ >95 |
| Advantage | Less corrosion | Lower [catalyst] |
| Disadvantage | Coproduct HOAc (0.21 $^t/_t$) | Corrosive (Ti-lined reactor) |

## 6. Phenol Manufacture: Benzene versus Toluene as primary building block

The two industrial routes to phenol are outlined in figure 6. The benzene-based cumene process is more selective but this tends to be offset by the lower price of toluene and the lower number of steps (two compared to three). Direct hydroxylation of benzene remains a potentially attractive alternative but up till now yields and/or productivities are much too low to be competitive.

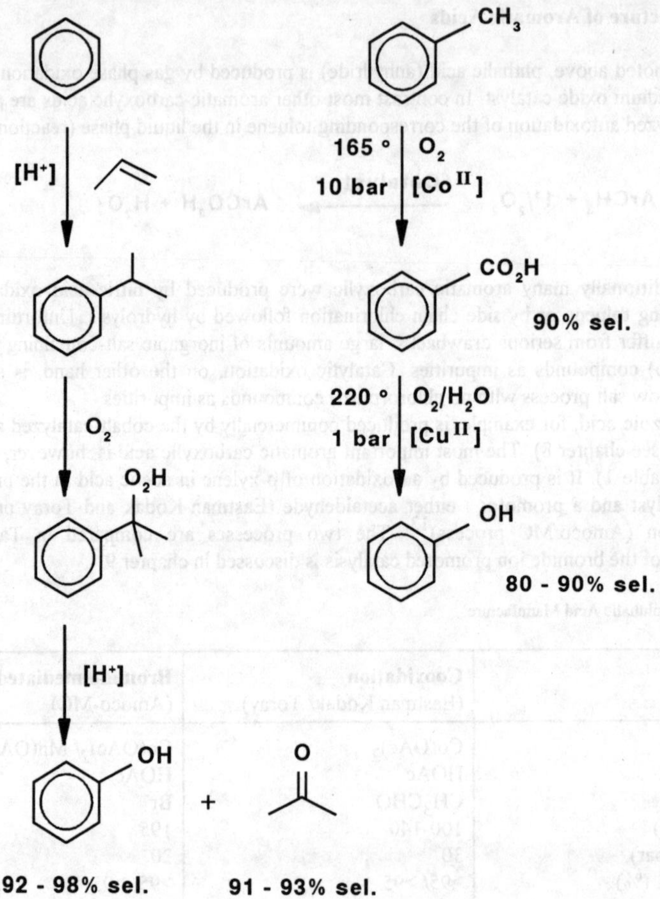


Figure 6. Two Routes to Phenol.

## 7. Caprolactam Manufacture: the Salt Issue

Caprolactam is an excellent example to illustrate the role of catalytic oxidations and reductions in chemicals manufacture. A key intermediate in most processes is cyclohexanone which is produced by the autoxidation of cyclohexane or hydrogenation of phenol (figure 7).

Asahi Chemical[14] has recently commercialized yet another variant involving the selective hydrogenation of benzene to cyclohexene over a ruthenium catalyst, followed by zeolite-catalyzed hydration to cyclohexanol and subsequent dehydrogenation to cyclohexanone. This process has the advantages of higher conversions and selectivities than cyclohexane autoxidation. All of these processes are ultimately derived from either benzene or toluene as the basic building block.

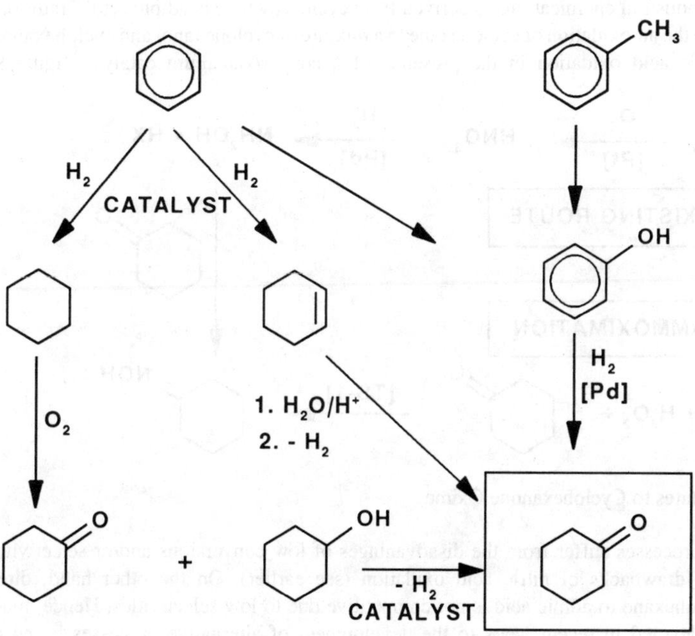


Figure 7. Cyclohexanone Manufacturing Routes.

Cyclohexanone is converted to caprolactam via Beckmann rearrangement of the oxime in the presence of sulfuric acid. The cyclohexanone oxime is made by reaction of cyclohexanone with hydroxylamine sulfate. The overall process leads to the formation of about 4.5 kg of ammonium sulfate per kg of caprolactam. There is currently much interest, therefore, in the development of alternative, low-salt routes to caprolactam. The Enichem company has developed a process involving ammoximation of cyclohexanone with a mixture of ammonia and hydrogen peroxide (figure 8) in the presence of a heterogeneous titanium silicalite (TS-1) catalyst[15,16]. This route employs a more expensive oxidant ($H_2O_2$) but is shorter and circumvents the salt production. In this context it is also worth mentioning that Sumitomo has recently developed a solid catalyst (a high-silica ZSM-5) for the Beckmann rearrangement step[17]. Hence, a salt-free route to caprolactam is now, in principle, available.

## 8. Adipic Acid Production

Another industrial chemical that is derived from cyclohexanone is adipic acid. Traditional processes involve the autoxidation of cyclohexane to a mixture of cyclohexanol and cyclohexanone followed by nitric acid oxidation in the presence of a copper/vanadium catalyst (figure 9).

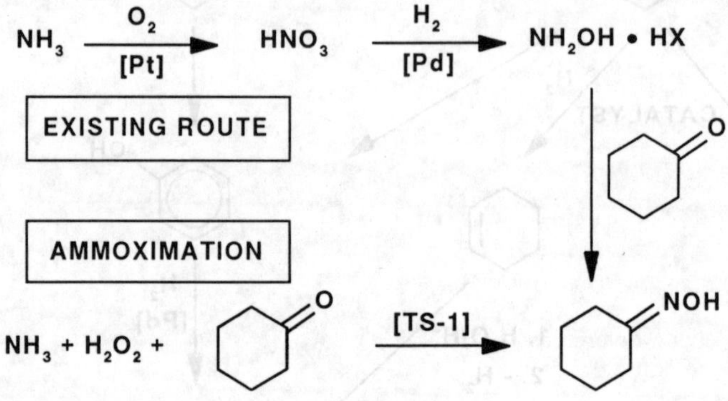


Figure 8. Two routes to Cyclohexanone Oxime

However, these processes suffer from the disadvantages of low conversions and/or selectivities and the inherent drawbacks of nitric acid oxidation (see earlier). On the other hand, direct oxidation of cyclohexane to adipic acid is not competitive due to low selectivities. Hence, much effort has been devoted in recent years to the development of alternative processes based on the catalytic carbonylation of butadiene[10]. In this context it should also be mentioned that DuPont produces adiponitrile via nickel-catalyzed addition of two equivalents of HCN to butadiene.

## 9. Major Trends in Chemicals Manufacture

The current trend in chemicals manufacturing is towards processes that are more efficient in their use of raw materials and energy and more environmentally benign. Catch phrases that are widely heard in the industry are integrated waste management, zero emission plants and benign-by-design. There is a marked trend away from waste remediation, i.e. end-of-pipe solutions, towards the development of processes that do not generate waste in the first place, i.e. primary

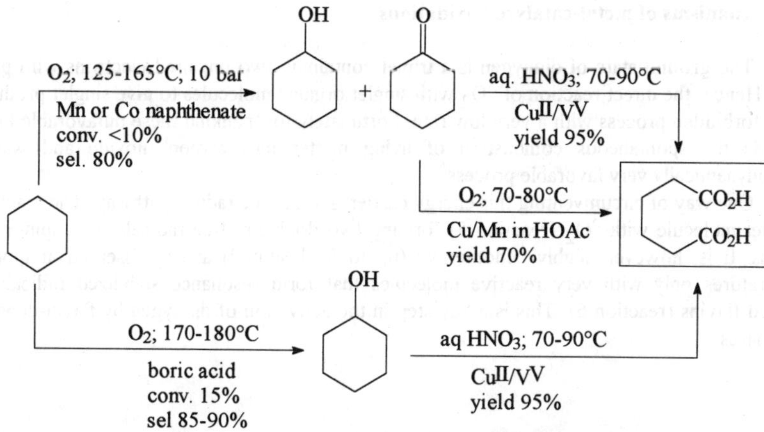


**Problem:** Decarboxation to lower mono- and di-carboxylic acids

30 kg $C_6$ recycle + 0.25 byproducts per kg product

Figure 9. Adipic Acid Production

pollution control. The emphasis is clearly on higher selectivities rather than yields. Furthermore, catalytic oxidations are being widely applied in the manufacture of fine chemicals[5,18]. Another identifiable trend is towards the use of cheaper feedstocks which is exemplified by the current drive to replace alkenes by alkanes, e.g. acrylonitrile by ammoxidation of propane rather than propylene. In this context it is worth mentioning that there are several `dream' reactions (eqns 2-4) for which there is enormous commmercial potential if an efficient catalytic system could be found.

$$CH_3CH{=}CH_2 + {}^1/_2O_2 \longrightarrow CH_3\overset{\frown O \frown}{CH-CH_2} \quad (2)$$

$$CH_4 + {}^1/_2O_2 \longrightarrow CH_3OH \quad (3)$$

$$PhH + {}^1/_2O_2 \longrightarrow PhOH \quad (4)$$

In order to understand the reactivity and selectivity problems inherent in such transformations it is necessary to have insights into the reactivity of dioxygen and the mechanisms of metal-catalyzed oxidations.

**10. Mechanisms of metal-catalyzed oxidations**

The ground state of dioxygen is a triplet containing two unpaired electrons with parallel spins. Hence, the direct reaction of $^3O_2$ with singlet organic molecules to give singlet products is a spin forbidden process with a very low rate. Fortunately for mankind these unfavorable kinetics preclude the spontaneous combustion of living matter into carbon dioxide and water, a thermodynamically very favorable process.

One way of circumventing this energy barrier is via a free radical pathway. The reaction of a singlet molecule with $^3O_2$ (reaction 5) forming two doublets (free radicals) is a spin-allowed process. It is, however, highly endothermic (up to 50 kcal/mol) and is observed at moderate temperatures only with very reactive molecules that form resonance stabilized radicals, e.g. reduced flavins (reaction 6). This is a key step in the activation of dioxygen by flavin-dependent oxygenases.

$$RH + {}^3O_2 \longrightarrow R\bullet + HO_2\bullet \qquad (5)$$

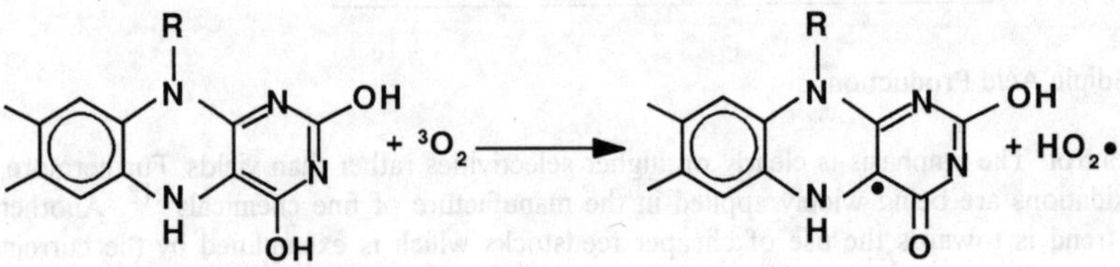

(6)

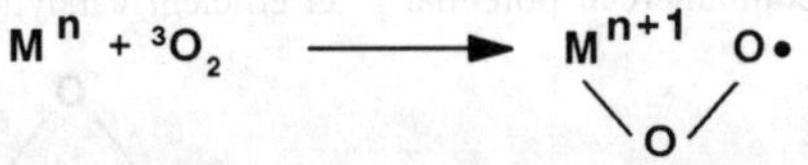

(7)

A second way to overcome the spin conservation obstacle is via complexation of $^3O_2$ with a paramagnetic (transition) metal ion (reaction 7). The expectation that the resulting metal-dioxygen complex could react selectively with hydrocarbons at moderate temperatures provided the stimulus for extensive studies of dioxygen activation by metal complexes during the last three decades[19]. The various oxygenated species that can result from the interaction of metalcomplexes with dioxygen, and may play a role in subsequent reactions with organic substrates, are depicted in figure 10.

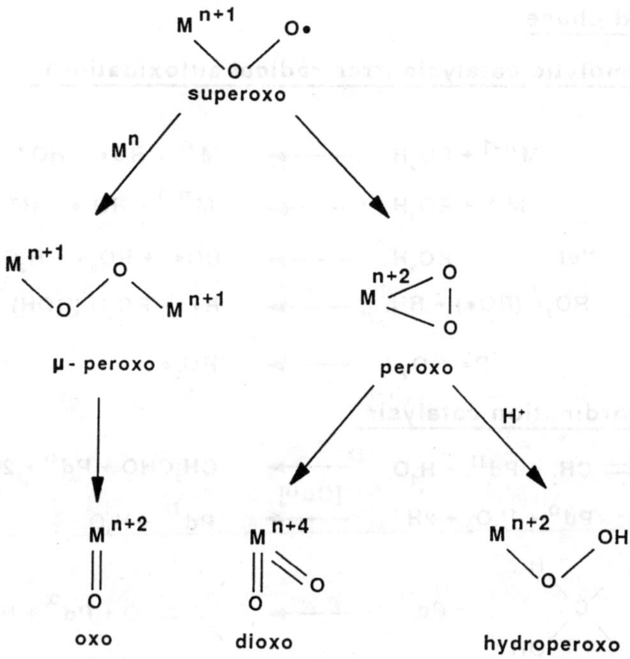


Figure 10. Metal-Oxygen Species.

Basically, three different types of oxidation with dioxygen can be delineated with respect to the mechanism involved (figure 11). The first type involves the generation of chain-initiating radicals via the metal-catalyzed decompostion of alkyl hydroperoxides. The latter are omnipresent in hydrocarbons that have not been rigorously purified by, for example, column chromatography over basic alumina. Since this always constitutes a background reaction in hydrocarbon oxidations it has led to many erroneous interprations of results in studies of `oxygen activation' by metal complexes. This is discussed in more detail in chapter 8.

The second type of mechanism involves the oxidation of a coordinated substrate by a metal ion. The oxidized form of the metal is subsequently regenerated by reaction of the reduced form with dioxygen. Examples include the palladium(II)-catalyzed oxidation of alkenes (Wacker process) and alcohols.

**Liquid phase**

**1. Homolytic catalysis (free radical autoxidation)**

$$M^{n-1} + RO_2H \longrightarrow M^{n} + RO\bullet + HO^-$$

$$M^{n} + RO_2H \longrightarrow M^{n-1} + RO_2\bullet + H^+$$

$$\text{Net}: \quad RO_2H \longrightarrow RO\bullet + RO_2\bullet + H_2O$$

$$RO_2\bullet\ (RO\bullet) + RH \longrightarrow R\bullet + RO_2H\ (ROH)$$

$$R\bullet + O_2 \longrightarrow RO_2\bullet$$

**2. Coordination catalysis**

a) $$H_2C{=}CH_2 + Pd^{II} + H_2O \longrightarrow CH_3CHO + Pd^{0} + 2H^+$$

$$Pd^{0} + {}^1/_2O_2 + 2H^+ \xrightarrow{[Cu^{II}]} Pd^{II} + H_2O$$

b) $$>C(H)(OH) + Pd^{II} \longrightarrow >C{=}O + Pd^{0} + H_2O$$

**Gas phase (Mars - van Krevelen mechanism)**

$$M^{n}{=}O + S \longrightarrow M^{n-2} + SO$$

$$M^{n-2} + O_2 \longrightarrow 2M^{n}{=}O$$

$$M^{n} = V^{V}, Mo^{VI}, \text{etc.} \quad (\text{exception}: Ag)$$

Figure 11. Mechanisms of Metal-Catalyzed Oxidations.

In the third type, which pertains mainly to gas phase oxidations, an oxometal species oxidizes the substrate and the reduced form is subsequently reoxidized by dioxygen. This is generally referred to as the Mars van Krevelen mechanism. With regard to the sequence and type of redox events it closily resembles the type 2 mechanism.

A fourth type of mechanism comprises catalytic oxygen transfer processses whereby a substrate reacts with an oxygen donor, such as $H_2O_2$ or ROOH, in the presence of a metal catalyst (reaction 8).

$$S + XOY \xrightarrow{\text{catalyst}} SO + XY \qquad (8)$$

$$XOY = H_2O_2\ ,\ RO_2H\ ,\ NaOCl\ ,\ etc.$$

The metal-catalyzed epoxidation of propylene with alkyl hydroperoxides referred to earlier constitutes the most important industrial example of a catalytic oxygen transfer. As we shall see in chapter 10 this type of process is also eminently suited to the production of a wide variety of fine chemicals.

**References**


1. R.A.Sheldon and J.K.Kochi, *Metal-Catalyzed Oxidations of Organic Compounds* (Academic Press, New York, 1981).
2. G.Franz and R.A.Sheldon, *Ullmann's Encyclopedia of Industrial Chemistry*, Vol. A18 (VCH, Weinheim, 1991) pp. 261-311
3. R.A.Sheldon, in *Dioxygen Activation and Homogeneous Catalytic Oxidation*, (Elsevier, Amsterdam, 1991) pp. 573-594
4. R.A.Sheldon, *CHEMTECH*, 1994, 38
5. R.A.Sheldon, *CHEMTECH*, 1991, 566
6. R.A.Sheldon, in Precision Process Technology, eds. M.P.C.Weijnen and A.A.H.Drinkenburg (Kluwer, Amsterdam, 1993) pp. 125-138
7. R.A.Sheldon, *Chem.Ind.* (London), 1992, 903
8. R.A.Sheldon, in Industrial Environmental Chemistry, eds. D.T.Sawyer and A.E.Martell, (Plenum Press, New York, 1992) 903
9. J.F.Roth, J.H.Craddock, A.Hershman and F.E.Paulik, *CHEMTECH*, 1971, 600
10. A.Chauvel, B.Delmon and W.F.Hölderich, *Appl.Catal.*,A, **115**(1994) 173-217 and references cited therein.
11. E.Drent, P.H.M.Budzelaar and W.W.Jager, *Eur.Pat.Appl.*, O386833 (1990) to Shell.
12. R.A.Sheldon, in *Aspects of Homogeneous Catalysis*, Vol.4, ed. R.Ugo (Reidel, Dordrecht, 1981) 1 and references cited therein.
13. W.Partenheimer and R.K.Gipe, in *Catalytic Selective Oxidation*, eds.S.T.Oyama and J.W.Highower, ACS Symp.Ser., **523** (1993)81
14. M. Kohono, *Chem. Ind.*,(Japan), *41*(1988) 936; see also M. Misono and N. Nojiri, *Appl. Catal.*, **64**(1990) 1-30
15. A.Zecchina, G.Spoto, S.Bordiga, F.Geobaldo, G.Petrini, G.Leofanti, M.Padovan, M.Montegazza and P.Roffia, in *New Frontiers in Catalysis*, eds. L.Guczi, F.Solymosi and P.Tetenyi (Elsevier, Amsterdam, 1993) 719
16. U.Romano, A.Esposito, F.Maspero, C.Neri and M.G.Clerici, *Chim.Ind.*(Milan) **72** (1990)610
17. H.Sato, K.Hirose, M.Kitamura and Y.Nakamura, in *Zeolites : Facts, Figures, Future*, eds. P.A.Jacobs and R.A. van Santen (Elsevier, Amsterdam, 1989)1213
18. R.A.Sheldon and J.Dakka, *Catal. Today,* **19** (1994)215-245
19. R.A.Sheldon, in *The Activation of Dioxygen and Homogeneous Catalytic Oxidation*, eds. D.H.R.Barton, A.E.Martell and D.T.Sawyer, (Plenum Press, New York, 1993) pp. 9-30

# MECHANISM OF HETEROGENEOUS CATALYTIC OXIDATION

**Jerzy Haber**
Institute of Catalysis and Surface Chemistry,
Polish Academy of Sciences, Krakow, Poland

## ABSTRACT

**The role of electrophilic and nucleophilic oxygen in selective oxidation of hydrocarbons is discussed and their participation in different steps of the reaction is illustrated. Many opposing factors influencing the selectivity of these reactions are then described. Mechanisms of elementary steps involved in electrophilic and nucleophilic oxidation of hydrocarbons are presented and such phenomena as structure sensitivity, synergistic effects in multicomponent oxide systems, and oxygen spill-over are described. Quantum-chemical description of the interaction of a hydrocarbon molecule with electrophilic molecular oxygen and nucleophilic lattice oxide ions are illustrated and the general conclusion is emphasized that oxidation is a concerted reaction, in which the interactions developed on approach of the organic molecule to the oxide surface cause the redistribution of electrons followed by reconstruction of the surface and rearrangement of nuclei, resulting in desorption of the product. The dynamic state of oxide surfaces manifested by their reconstruction on interaction with the gas phase is stressed.**

## 1. ELECTROPHILIC AND NUCLEOPHILIC OXIDATION

Molecular oxygen contains in its ground state two unpaired electrons, which are localized on the degenerate antibonding $\pi 2p$ orbitals, the ground state is thus a triplet. Because of the rule of spin conservation, reactions between this triplet oxygen and organic molecules which are in the singlet state experience high activation energies. This symmetry barrier may be overcome either by activating oxygen to the singlet state or by activating the organic molecule to make it susceptible to the reaction with molecular or atomic ions. Of particular importance is the nucleophilic attack by the oxide ion $O^{2-}$.

Dioxygen, adsorbed at oxide surfaces, is present mainly in form of superoxide ions. The existence of peroxide ions has not been proved directly, however it cannot be excluded a priori. At higher temperatures both adsorbed superoxide and peroxide dioxygen species are unstable and presumably decompose with the formation of the ion radical species $O^-$. All three activated oxygen forms - the neutral singlet $O_2$ and the ionic $O_2^-$ and $O^-$ species- are strongly electrophilic reactants which attack the organic molecule in the region of its highest electron density, i.e. the $\pi$ bonds (Fig.1). At variance with their behaviour in the liquid phase, the peroxy and epoxy complexes formed as the result of an electrophilic attack of $O_2^-$ or $O^-$ species on the $\pi$ bonds of hydrocarbon molecule at the surface of an oxide are intermediates which lead under heterogeneous catalytic reaction conditions to the degradation of the carbon skeleton. In the case of olefins saturated aldehydes are formed in the first stage which are usually much more reactive than unsaturated aldehydes or anhydrides and at higher temperatures undergo a rapid total oxidation.

The presence of $O_2^-$ or $O^-$ at the surface of an oxide may be detected by different techniques such as EPR or IR spectroscopy. The method which permits the quantitative determination of the number of electrons transferred between the solid and the adsorbed layer is the

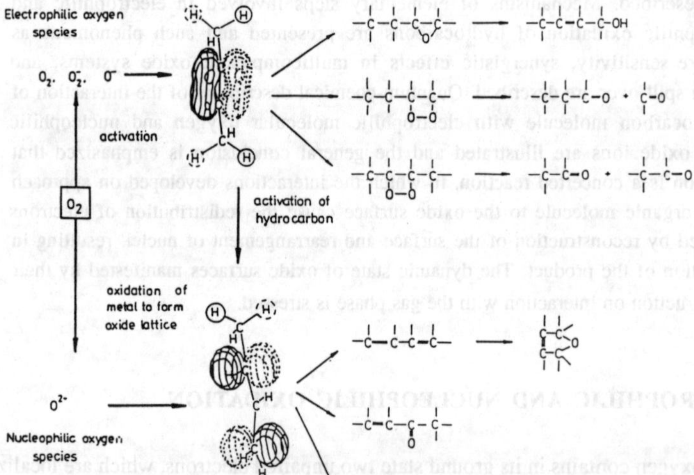


Fig.1. Mechanism of the catalytic oxidation of hydrocarbons

measurement of the changes of work function in the course of adsorption. Thus, when changes of work fucntion due to exposure to oxygen are followed upon temperature variation and the amount of oxygen adsorbed is simultaneously measured, the number of elecrons localized per oxygen atom adsorbed may be determined and hence the type of oxygen species residing at the surface may be found. Results of such experiments carried out with different oxides are summarized in table 1 and compared with catalytic properties of these oxides. They indicate that whenever electrophilic oxygen species $O_2^-$ or $O^-$ are present at the surface, total oxidation is observed in the course of the catalytic oxidation of hydrocarbons.

The second route of heterogeneous oxidation is the reaction with lattice oxide ion

$O^{2-}$. These ions have no oxidizing properties but are nucleophilic reactants which can be inserted into the activated hydrocarbon molecule by a nucleophilic addition, to form an oxygenated product. This reaction path starts with the activation of organic molecules, which thus become prone to undergo an attack by nucleophiles, and consists of a series of consecutive steps of hydrogen abstraction and nucleophilic oxygen addition, with each of these steps requiring different active centres to be present at the catalyst surface. It should be emphasized that it is the cations of the catalyst which act as oxidizing agents in some of

Table 1.
Oxygen species at surfaces of various oxides

| Catalyst | Temperature range (K) | Oxygen species | Catalytic behaviour |
|---|---|---|---|
| $Co_3O_4$ | 293 - 423 | $O_2^-$ | Total oxidation |
| | 573 - 673 | $O^-$ | Total oxidation |
| $V_2O_5$ and | 293 - 393 | $O_2^-$ | Total oxidation |
| $V_2O_5/TiO_2$ | 533 - 653 | $O^-$ | Total oxidation |
| | > 653 | $O^{2-}$ | Selective oxidation of alkylaromatics |
| $Bi_2Mo_3O_{12}$ | 538 - 673 | $O^{2-}$ | Selective oxidation of olefins |

the consecutive steps of the reaction sequence, forming the activated hydrocarbon species. In subsequent steps these undergo a nucleophilic attack by lattice oxygen ions and the oxygenated product is desorbed, leaving oxygen vacancies at the surface of the catalyts. Such vacancies are then filled with oxygen from the gas phase, simultaneously reoxidizing the reduced cations. It should be noted that incorporation of oxygen from the gas phase into the oxide surface does not necessarily take place at the same site from which surface oxygen is inserted into the hydrocarbon molecule, but may occur at a different site, oxygen ions being then transported through the lattice. This mechanism may be represented by the cycles shown in Fig. 2. In the case of complex hydrocarbon molecules, the nucleophilic addition of oxygen may take place at different sites of the molecule. Specifically, it will take place at that site which is most electropositive by appropriate bonding of the molecule at the active centre of the catalyst.

Reactions of catalytic oxidation may be thus devided into two categories: (1) electrophilic oxidation, proceeding through the activation of oxygen, and (2) nucleophilic oxidation, in which activation of the hydrocarbon molecule is the first step, followed by consecutive steps of nucleophilic oxygen addition and hydrogen abstraction. They may be conveniently systematized according to the number of elementary structural transformations introduced

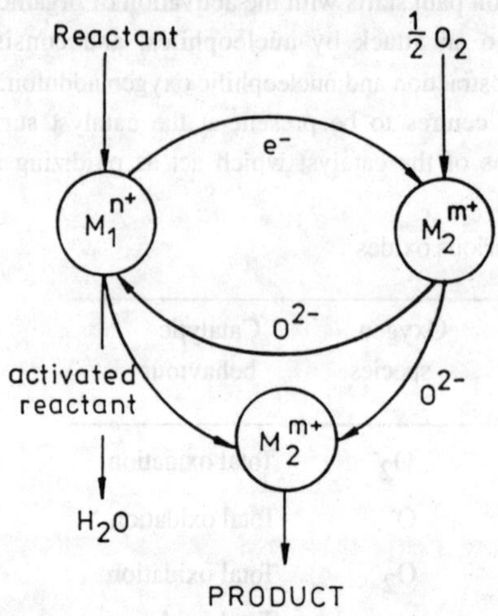


Fig.2. Oxidation -reduction cycles in selective catalytic oxidation

into the reacting molecule (Table 2). The mildest electrophilic oxidation is the addition of oxygen to the double bond resulting in the formation of epoxides, or the oxyhydration of the double bond to form respective saturated ketones (e.g.propene ---- acetone). A more pronounced structural change is the fission of the double C-C bond, saturated aldehydes being formed from olefins (propene ---- acetaldehyde + formaldehyde), or the fission of the aromatic ring resulting in the appearance of anhydrides. The final stage is total oxidation to $CO_2$+$H_2O$. Along the nucleophilic route the first, smallest structural change is the abstraction of hydrogen in the process of oxidative dehydrogenation of alkanes and alkenes to dienes, or the dehydrodimerization and dehydrocyclization. Deeper structural changes are involved in reactions in which a heteroatom is introduced into the hydrocarbon molecule by a nucleophilic addition. This may be oxygen, sulphur, nitrogen, etc. Introduction of the first e.g. oxygen results in the formation of aldehydes; the introduction of the second one - in the formation of acids or anhydrides. In all these processes the carbon skeleton and the $\pi$ electron system remain unchanged.

As mentioned above the oxygen species $O_2$, $O_2^-$, and $O^-$ generated on adsorption at the surface of an oxide catalyst are responsible for electrophilic oxidation. Both reactants of this reaction are located at the gas phase side of the gas/solid interface. Such catalytic reactions are called extrafacial, or reactions without transfer (Fig.3a). On the other hand, the nucleophilic oxidation is a reaction between the adsorbed reactant and the oxide ion of the catalyst lattice,

Table 2
Heterogeneous oxidation of hydrocarbons

| **Electrophilic oxidation** | | **Nucleophilic oxidation** | |
|---|---|---|---|
| Reaction type | Catalyst | Reaction type | Catalyst |
| **With double bond fission** | | **Without introduction of** | |
| - oxidation of olefins | | **heteroatom** | |
| to oxides | $Ag_2O$ | - oxidative dehydrogenation | |
| - oxyhydration of olefins | | of alkanes and alkenes | |
| to saturated ketones | $SnO_2$-$MoO_3$ | to dienes | $Bi_2O_3$-$MoO_3$-$P_2O_5$ |
| | | - oxidative | |
| **With C-C bond fission** | | dehydrodimerization | |
| - oxidation of olefins to | | and dehydrocyclization | |
| saturated aldehydes | $V_2O_5$ | of alkenes | $BiPO_4$ |
| - oxidation of aromatics | $V_2O_5$-$MoO_3$ | | |
| to anhydrides and acids | | **With introduction of** | |
| with ring rupture | | **heteroatom** | |
| - total oxidation to $CO_2$ | $Co_3O_4$ | - introduction of heteroatom | |
| and $H_2O$ | $CuCo_2O_4$ | into hydrocarbon chain | |
| | $CuCr_2O_4$ | a/ introduction of oxygen | |
| | | - oxidation of olefins | |
| | | to unsaturated aldehydes | |
| | | and ketones | $Bi_2O_3$-$MoO_3$ |
| | | - oxidation of alkylaromatics | |
| | | to aldehydes | $Bi_2O_3$-$MoO_3$ |
| | | b/ introduction of nitrogen | |
| | | - ammoxidation of olefins to | |
| | | to nitriles | $UO_3$-$Sb_2O_4$ |
| | | - introduction of heteroatom | |
| | | into acyl group | |
| | | - oxidation of aldehydes to | |
| | | acids | NiO-$MoO_3$ |
| | | - oxidation of alkylaromatics | |
| | | to anhydrides | $V_2O_5/TiO_2$ |

which is transferred across the gas/solid interface. Therefore this type of catalytic reaction is called interfacial (Fig.3b). Two kinds of interfacial reactions may be distinguished. One kind is reactions which are considered to proceed in two steps: (1) reaction between the reactant and the oxide to give oxygenated product and partially reduced catalyst and (2) reoxidation of the reduced catalyst with gaseous oxygen to restore the catalyst to its initial state. Because in this kind of reaction the reduction part and the oxidation part are considered to be separate steps, it is called the redox mechanism. A mechanism of this type was postulated by Mars and van Krevelen to explain the kinetics of the oxidation of aromatics on $V_2O_5$ catalysts, therefore it is also known as the Mars-van Krevelen mechanism. The second kind of interfacial reactions comprises those in which both reduction and oxidation steps are performed in one transformation. They are thus called concerted or push-pull reactions.

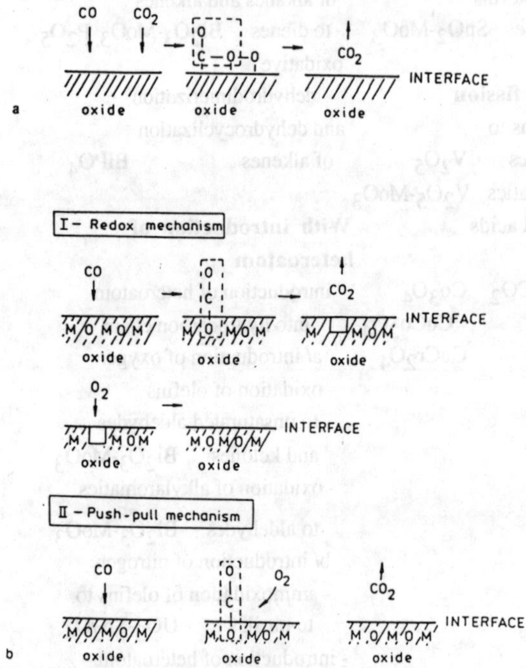


Fig.3. Extrafacial (a) and interfacial (b) reactions

## 2. SELECTIVITY IN OXIDATION REACTIONS

Oxidation of a hydrocarbon molecule with high regio- and/or stereoselectivity requires finding a compromise between many opposing factors, among which the following are the most important:

- in processes of hydrocarbon oxidation thermodynamics favours the ultimate formation of carbon dioxide and water; therefore all products of partial oxidation are intermediates derived by kinetic control of the reaction;
- the hydrocarbon+oxygen mixture can usually react along many different pathways in the network of competing parallel and consecutive reactions and therefore the catalyst must strictly control their relative rates, accelerating the sequence of steps leading to the desired products and hindering those in which unwanted byproducts could be formed;
- the C-H bonds in the initial reactant are usually stronger than those in the intermediate products, which makes these intermediates prone to rapid further oxidation;
- all oxidation processes are strongly exothermal and efficient heat removal must be secured to control the temperature and prevent overoxidation as well as catalyst damage.

The complexity of possible interactions of hydrocarbon molecules may be illustrated by the reaction network of an olefin (Fig.4). When an oxide catalyst is used, the olefin molecule begins to interact with its surface by forming weak hydrogen bonds with surface OH groups. If the surface OH groups show Bronsted acid properties, their protons may form hydrogen bonds with the $\pi$ bonds of the olefin, and when the acid properties are strong enough, the transfer of a proton from the surface to the olefin may take place resulting in the formation of a carbocation. This may start the whole network of reactions proceeding by carbocation mechanism, like isomerization, transalkylation, cracking, etc. Moreover, carbocations may be attacked by surface $OH^-$ groups or adsorbed water molecules to form secondary alcohols, which dehydrogenate to give saturated ketones. As an example propene is transformed into

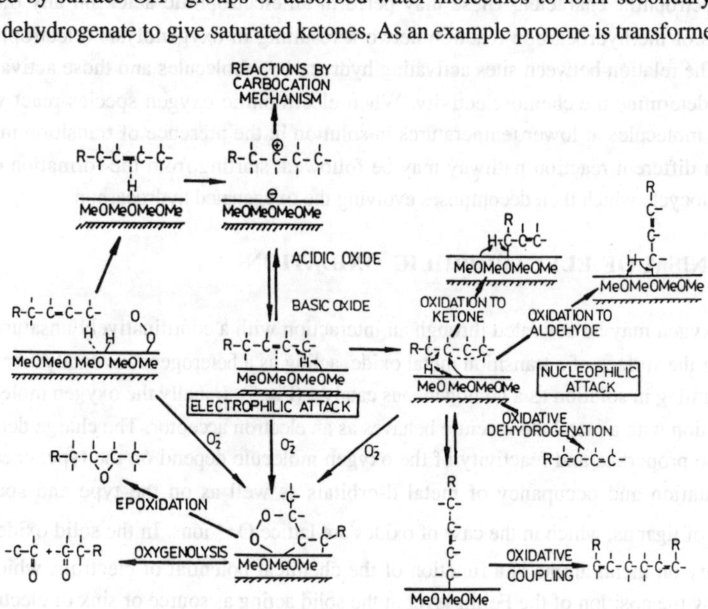


Fig.4. Reaction network in oxidation of an olefin molecule

acetone in a process called oxyhydration. The π bond of the olefin instead of interacting with a surface proton, may react with a transition metal cation showing properties of a Lewis acid site and may form surfce π-complex.

When the basicity of surface oxide ions is high enough, they may perform a nucleophilic attack on the hydrogen atom in the a position which results in the formation of an allylic species.This may be bonded to the transition metal ion either side-on as a π-allyl, or end-on as σ-allyl. They show high mobility at the oxide surface and an equilibrium between these two forms exists. The s-allyl may be an intermediate in oxidative coupling to form dienes and the π-allyl may undergo a nucleophilic attack of surface oxide ion resulting in a surface alkoxide intermediate, which desorbs as an aldehyde in the case of addition to the primary carbon atom or a ketone when secondary carbon atom is involved (nucleophilic oxidation). Thus, the type of adsorbed intermediate complex predetermines the regio selectivity. One could envisage a situation in which active centres abstracting hydrogen from hydrocarbon molecules will be dispersed far apart and the mobility of alkyl and allyl moieties will be restricted. In such circumstances, instead of a nucleophilic addition of one of the surface oxide ions the nucleophilic attack by an adjacent $OH^-$ group may be more probable, resulting in the formation of an alcohol. This is the case in homogeneous oxidation, in which both alkyl or allyl moieties and OH groups are bonded to the same metal atom acting as the active centre. Many oxides contain surface vacancies generating F-centres which may play the role of sites activating oxygen molecules to species of electrophilic character. These may perform an electrophilic atack on any of the intermediates of the hydrocarbon reaction network resulting in oxygenolysis (electrophilic oxidation). The relation between sites activating hydrocarbon molecules and those activating oxygen will determine the chemoselectivity. When electrophilic oxygen species react with hydrocarbon molecules at lower temperatures in solution in the presence of transition metal complexes, a different reaction pathway may be followed, starting from the formation of a peroxo metallocycle which then decomposes evolving the oxygenated hydrocarbon.

## 3. MECHANISM OF ELECTROPHILIC OXIDATION

Molecular oxygen may be activated through an interaction with a coordinatively unsaturated metal atom at the surface of a transition metal oxide, acting as a heterogeneous catalyst or in a complex operating in solution as a homogeneous catalyst (Fig.5). Usually the oxygen molecule in the interaction with a metal atom centre behaves as an electron acceptor. The charge density and hence the properties and reactivity of the oxygen molecule depend on the type, energy, spatial orientation and occupancy of metal d-orbitals as well as on the type and spatial arrangement of ligands, which in the case of oxides are lattice $O^{2-}$ ions. In the solid oxide the charge density on metal atoms is a function of the chemical potential of electrons which is determined by the position of the Fermi level in the solid acting as source or sink of electrons and determining thus the occupancy of orbitals in the surface complex. When the active site at the oxide surface or in the coordination compound has the redox potential sufficient to effect

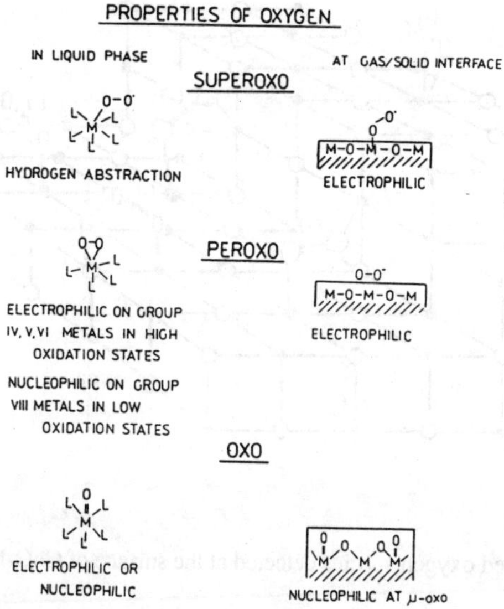


Fig.5. Properties of activated oxygen.

charge transfer to the oxygen molecule, but not strong enough to cause the cleavage of the O-O bond, electrophilic $O_2^-$ species are generated which may be bound to the active site end-on behaving as superoxo species, or side-on to form moieties of the peroxo type. EPR and IR spectroscopies provide information about the existence, location and spatial orientation of these species (Fig.6). When the redox potential of the active site has an appropriate value, cleavage of the O-O bond may take place and highly reactive $O^-$ may also be formed. All these oxygen species interact with the $\pi$-electron system of organic molecules and start the electrophilic oxidation. It should be born in mind that in the case of transition metal oxides electrophilic oxygen species may appear at the surfacee also in the absence of oxygen in the gas phase, as the intermediates in the transfer of oxygen from the lattice into the gas phase in the course of the dissociation of the solid (Fig.7) due to equilibriation of the nonstoichiometric oxide with the gas phase (Fig.8), or in the process of its reduction by e.g., hydrocarbon molecules. Thus, not only the gas phase but also the lattice of the oxide may serve as the source of electrophilic oxygen.

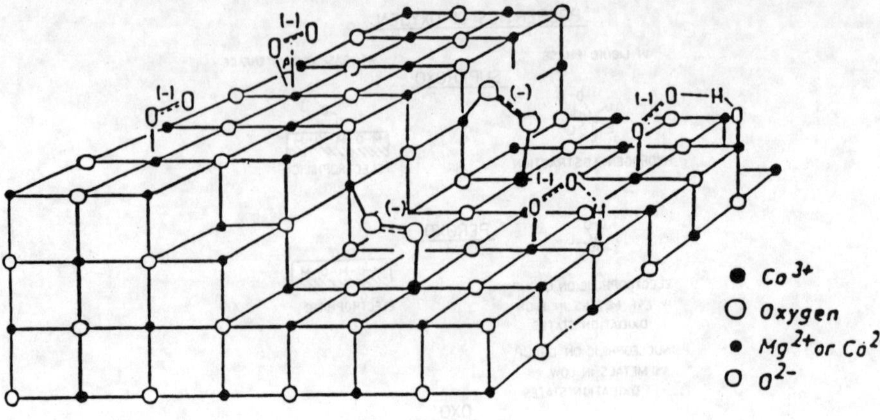


Fig.6. Different adsorbed oxygen species detected at the surface of CoO-MgO solid solution

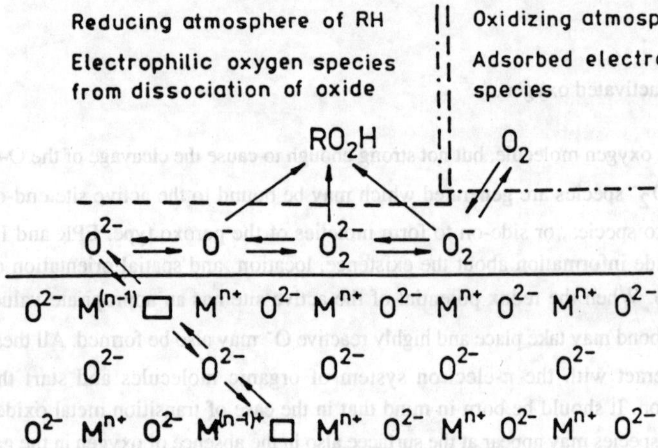


Fig.7. Formation of electrophilic oxygen sepcies at the oxide surface

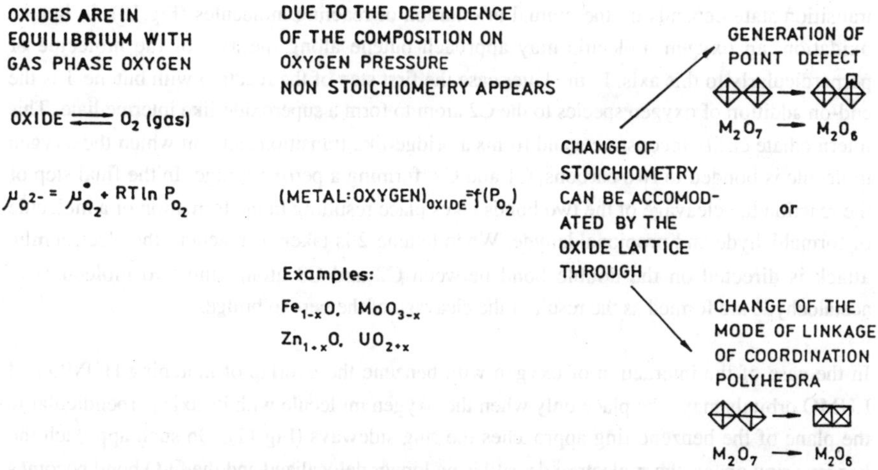


Fig.8. Generation of defects in oxides

Quantum chemical calculations show that when an oxygen molecule approaches a hydrocarbon molecule containing π-bonds, the attack is directed into this region of high electron density (Fig.9). Oxygen in the ground state encounters a very high potential barrier, but if it is activated by stretching the O-O bond, e.g., through a transfer of an electron from the catalyst onto the antibonding orbital and formation of $O_2^-$ species the reaction becomes facile. The character of the

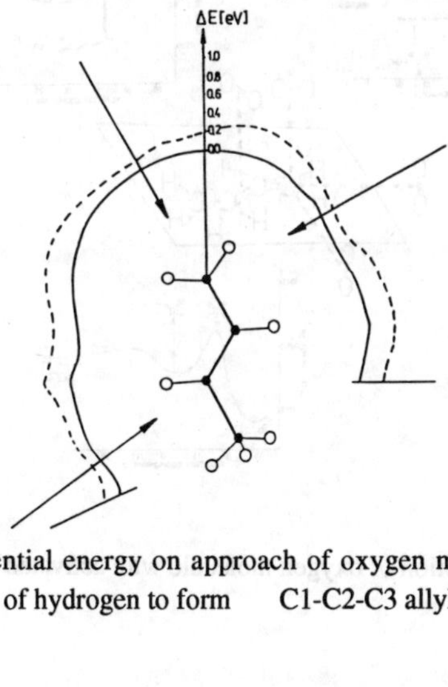


Fig.9. Gradient of potential energy on approach of oxygen molecule to butene-1 molecule activated by abstraction of hydrogen to form C1-C2-C3 allylic species.

transition state depends on the mutual orientation of reacting molecules (Fig.10). In butene oxidation, an oxygen molecule may approach butene along the axis of the molecule or perperdicularly to this axis. In the latter case the first step of the reaction with butene-1 is the end-on addition of oxygen species to the C2 atom to form a superoxide-like intermediate. This intermediate easily reconstructs and forms a bridge-like transition state, in which the oxygen molecule is bonded to two carbons, C1 and C2, forming a peroxo bridge. In the final step of the reaction the cleavage of the two bonds takes place resulting in the formation of a molecule of formaldehyde and propionaldehyde. When butene-2 is taken as reactant, the electrophilic attack is directed on the double bond between C2 and C3 atoms and two molecules of acetaldehyde are formed as the result of the cleavage of the peroxo bridge.

In the case of the interaction of oxygen with benzene the overlap of matching HOMO and LUMO orbitals may take place only when the oxygen molecule with its axis perpendicular to the plane of the benzene ring approaches the ring sideways (Fig.11). On such approach the benzene ring opens, the π-electron density is no longer delocalized and the O-O bond becomes more and more elongated (Fig.12). Further evolution of the system results in the extraction of two carbon atoms of the benzene ring, together with their hydrogens, towards the oxygen molecule so that the system splits into two parts: a four carbon residual fragment of the benzene ring and a nearly cyclic $C_2H_2O_2$ moiety of a composition equivalent to glyoxal. Both are highly reactive

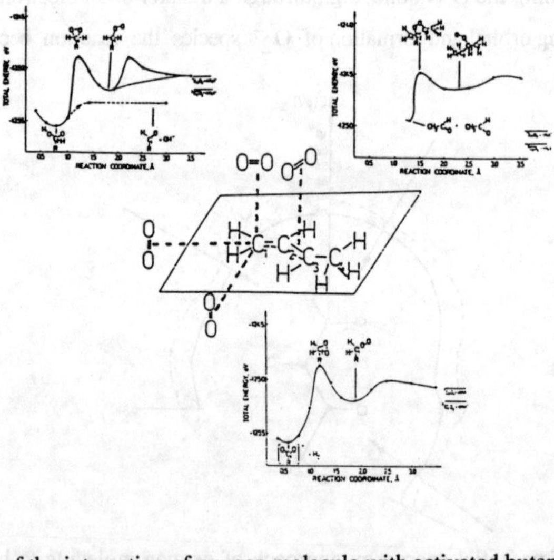

Fig.10. Pathways of the interaction of oxygen molecule with activated butene molecule.

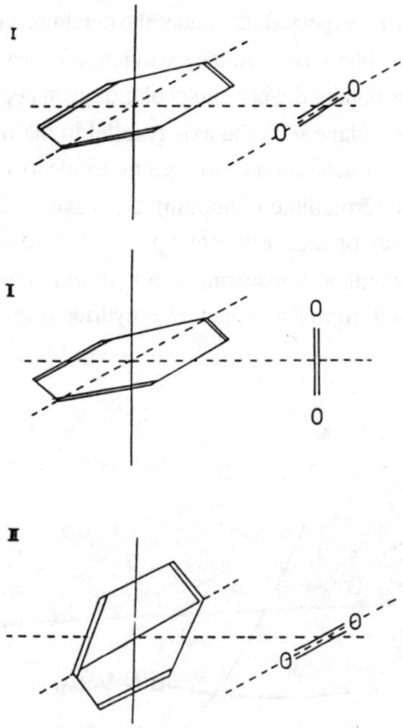


Fig.11. Pathways for the reaction of benzene and oxygen molecules.

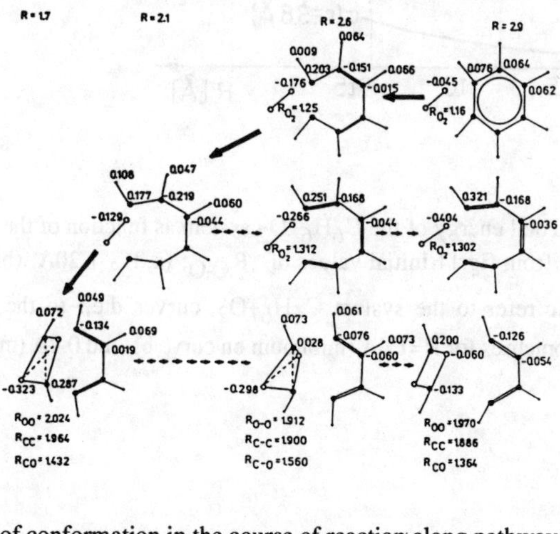


Fig.12. Changes of conformation in the course of reaction along pathway II from fig.11.

species and it may be expected that under the conditions of a heterogeneous catalytic reaction in the presence of gas phase oxygen they will react further to form products of total oxidation. A different picture is obtained when molecular oxygen approaches benzene ring along a pathway perpendicular to its plane with the axis parallel to the ring diagonal. In such case the reaction takes place only after activation of oxygen molecule to stretch the O-O distance and leads to the formation of an intermediate containing a peroxo bridge over the benzene ring (Fig.13) in analogy to the attack on olefins where a peroxo bridge is formed between the π-bonded carbon atoms. This intermediate transforms into hydroquinone, which then may react with the next oxygen molecules to form finaly maleic anhydride (Fig.14).

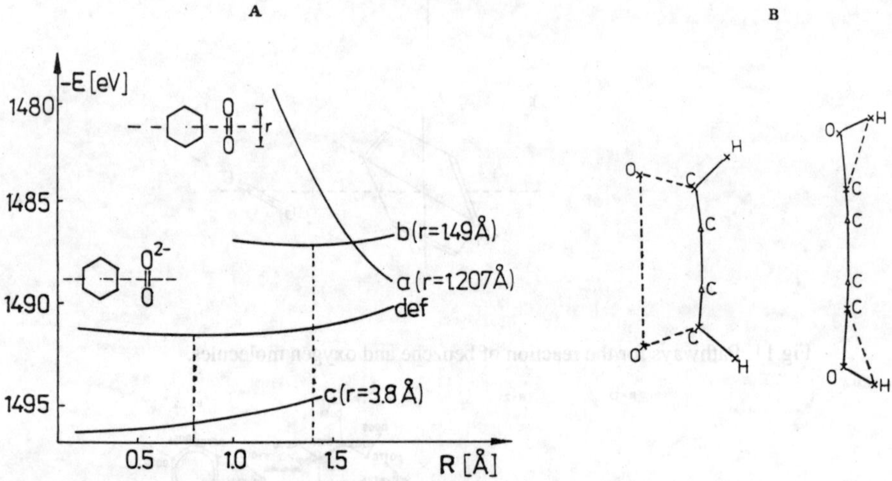


Fig.13. A. Total energy of the $C_6H_6+O_2$ system as function of the reaction coordinate along pathway III from fig.11.Initial values of $R_{O\text{-}O}$: (a,d) - 1.20A, (b,e) - 1.49A, (C,f) - 3.9A. Curves a,b,c refer to the system $C_6H_6+O_2$, curves d,e,f to the system $C_6H_6+O_2^{2-}$. B. Optimal geometries for R=1.4A (minimum on curve b), and 0.2A (minimum on curve c)

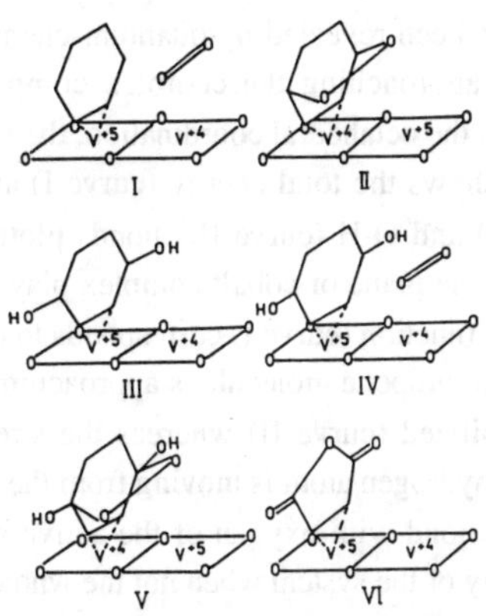


Fig.14. Mechanism of elementary steps in the oxidation of benzene to maleic anhydride

## 4. MECHANISM OF NUCLEOPHILIC OXIDATION

The second route of oxidation reactions starts with the activation of the organic molecule, followed by the nucleophilic addition of oxide ion. The classical studies by Adams using deuterated propene and olefins $C_4$ to $C_8$, and of Sachtler and de Boer with $C^{14}$-labelled propenes showed unequivocally that activation of the olefin molecule consists in abstraction of α-hydrogen and formation of a symmetric allyl intermediate, activation being the rate determining step.

$$E = \sum_{A} E_A + \sum_{A<B} \sum E_{AB}$$

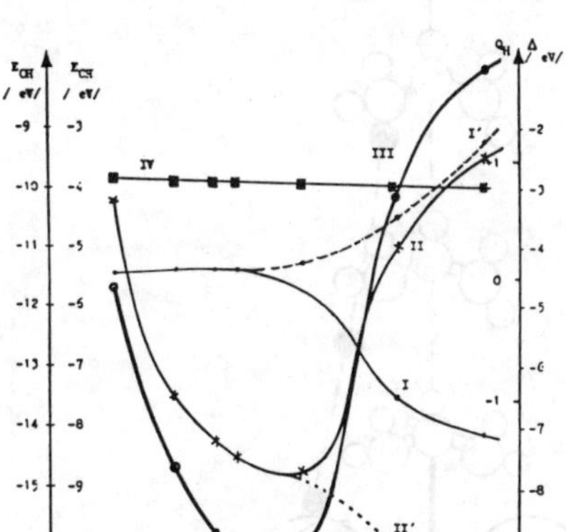

Fig.15. Total energy (curve i) and diatomic contributions of C-H (curve II) and O-H (curve III) interactions as well as charge on hydrogen atom (curve IV) as a function of the distance of propene molecule from the plane of $CoO_5$ complex. Dotted curves refer to allyl species.

The mechanism of this process has been revealed by quantum-chemical calculations of the interaction of propene molecule, approaching the complex composed of e.g. cobalt ion surrounded by five oxygen atoms in the octahedral coordination, the sixth site being occupied by the incoming propene. Fig.15 shows the total energy (curve I) and the diatomic energy contributions of the C-H (curve II) and O-H (curve III) bonds plotted as a function of the distance of propene molecule from the plane of cobalt complex playing the role of an active site. A minimum of the total energy function (curve I) corresponds to the formation of a stable intermediate surface complex. When propene molecule is approaching the complex, the C-H bond is being continuously destabilized (curve II) whereas the strength of the O-H bond increases (curve III) indicating that hydrogen atom is moving from the C-H bond in the methyl group of propene to form the O-H bond with oxygen of the active site. The dotted curve I represents the changes of total energy of the system when not the whole propene molecule, but only the allylic species formed after abstraction of hydrogen is being removed away from the surface, one hydrogen atom remaining at the surface as the OH group. It may be seen that this process is energetically much more favourable than removal of propene. It may be thus concluded that on contacting propene with the surface of e.g. cobalt oxide or oxysalt, its reactive chemisorption takes place, consisting in the formation of allyl species. It is noteworthy that in the course of the approach of propene molecule to the active site the charge on hydrogen atom remains practically constant and amounts to about 0.7 (curve IV in Fig.15). This indicates that the abstraction of hydrogen is accompanied by a very rapid redistribution of electrons. It should be also emphasized that the movement of hydrogen begins already when propene molecule is still quite far from the active site, the elementary catalytic transformation being thus a concerted redistribution of electrons and rearrangment of nuclei (Fig.16). We shall underline this point on many occasions.

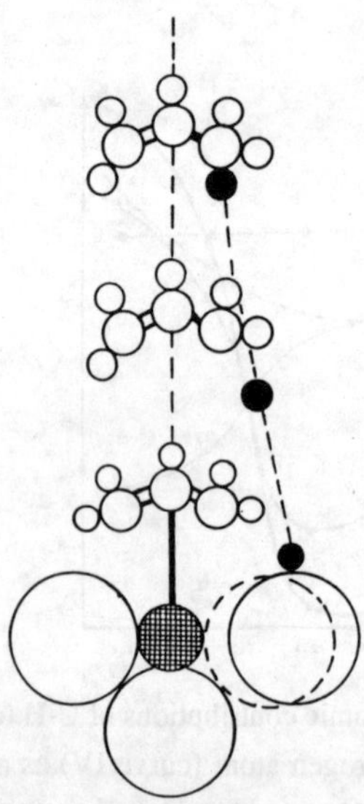

Fig.16. Reactive chemisorption of propene.

The question may be now raised as to where the activation of hydrocarbon and the subsequent nucleophilic addition of oxygen takes place at the surface of an oxide catalyst. Because in partial oxidations of propene and majority of other hydrocarbons it is the first step which is rate determining, studies of these reactions cannot yield any information on the next step in which addition of oxygen takes place. One of the ways by which insertion of oxygen could be investigated in more details is to by-pass the first step generating the allyl radicals by some other more efficient route. This was accomplished by using allyl iodide which readily decomposes into allyl radicals. The first step of the selective oxidation of propene is thus facilitated making possible the examination of conditions which are necessary for the insertion of oxygen in the next step of the reaction.

Important conclusions could be drawn from the comparison of the behaviour of $Bi_2O_3$ and $MoO_3$ as components of the classical catalyst for oxidation and ammoxidation of propene, in the reactions with propene and allyl iodide (Table 3).When allyl iodide was passed over $MoO_3$ practically total conversion is observed with 98% selectivity to acrolein already at 310°C. At the same conditions $MoO_3$ is completely inactive with respect to propene. On contacting allyl iodide with $Bi_2O_3$ total conversion at 310°C was also observed, in this case however 70% of the product formed was 1,5-hexadiene, practically no acrolein being detected. 1,5-hexadiene as the product was also obtained in the reaction of propene on $Bi_2O_3$ indicating that activation of hydrocarbon takes place resulting in the formation of allyl species, but there are no sites to insert oxygen and in the absence of the next step they simply dimerize to give 1,5-hexadiene. These results clearly demonstrate that in the molybdate catalysts it is the Mo-O sublattice which performs the insertion of oxygen into the organic molecule. $MoO_3$ itself is inactive in propene oxidation because no centres are available for efficient generation of allylic species. When however such species are formed by some other route, their total conversion to acrolein at the surface of $MoO_3$ takes place. In bismuth molybdate catalysts $Bi^{3+}$ions play the role of active site where allyl species are generated by activation of propene.

Table 3.
Interaction of propene and allyl with $Bi_2O_3$, $MoO_3$ and $Bi_2MoO_6$

| | Propene | | | | Allyl iodide | | | |
|---|---|---|---|---|---|---|---|---|
| Catalyst | Temp. (°C) | Yield,% diene | benz | acr | Temp. (°C) | Yield,% diene | benz | acr |
| $Bi_2O_3$ | 480 | 8.6 | - | - | 310 | 70.0 | - | 5.0 |
| $MoO_3$ | 480 | - | - | - | 310 | - | - | 98.0 |
| $Bi_2MoO_6$ | 480 | - | - | 13.0 | 310 | 12.0 | 1.0 | 15.0 |

One of the problems in studies of oxide catalysts is the difficulty in counting the number of active sites at the surface of oxide crystallites, whereat the elementary catalytic transformations take place. In particular no such estimate is available concerning the activation of hydrocarbon molecules. In order to obtain some relevant information following experiments were performed. As already mentioned it is now well established that in the case of molybdate catalysts the activation of hydrocarbon molecules takes place at the cationic sites such as $Bi^{3+}$, $Co^{2+}$, $Ni^{2+}$ etc of the $Bi_2(MoO_4)_3$, $CoMoO_4$, $NiMoO_4$ etc. catalysts respectively. Let us now support the defined number of isolated cations at the surface of a carrier. We could determine the turn-over frequency of the activation of hydrocarbon molecules if we had a method to count the number of activated molecules, i.e. gernerated allyl species. At this point it should be reminded that experiments with allyl compounds have shown that allyl radicals when contacted with the surface of $MoO_3$ in appropriate conditions (c.f.Table 4)- pick up oxygen and are totaly converted to acrolein. Oxygen from $MoO_3$ may thus serve as a probe to detect activated propene molecules, their number being measured by determining the number of acrolein molecules formed. Following these ideas isolated bismuth ions were supported at the surface of $MoO_3$ and their activity in the oxidation of propene was measured by pulse technique as a function of their surface concentration, expressed as number of bismuth atoms per surface molybdenum atom. Results are shown in Fig. 17. Yield of acrolein observed when allyl iodide was introduced was constant and independent of bismuth

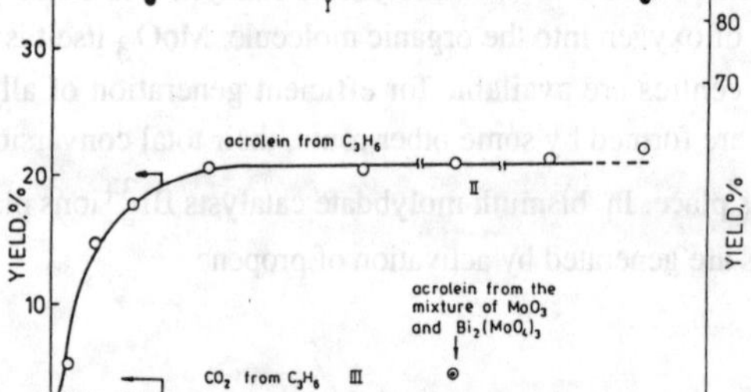

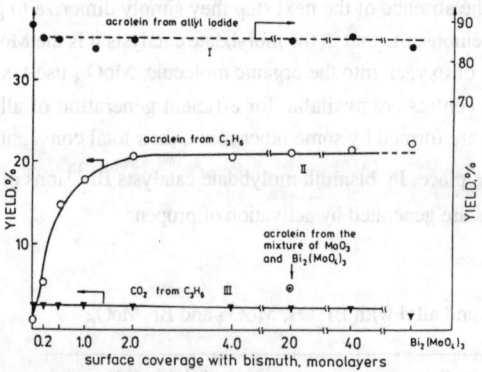


Fig.17. Yield of acrolein and $CO_2$ in oxidation of propene and allyl iodide as function of coverage of $MoO_3$ surface with Bi ions.

coverage confirming the assumption that once allyl radicals have been generated they rapidly undergo a nucleophilic attack by oxide ions from $MoO_3$ lattice. On introducing the mixture of propene and oxygen the activity at low surface coverage with bismuth increased proportionally to this coverage, the turn-over frequency per bismuth ion being thus constant. In the conditions of experiment it amounted to

0.5 propene molecule per bismuth ion per pulse. At higher bismuth coverages the activity levels off because once $Bi^{3+}$ ions formed a monolayer no more change of the number of active sites could take place. The yield of acrolein observed at the plateau is similar to that observed with $Bi_2(MoO_4)_3$ phase. It is noteworthy that the amount of $CO_2$ formed remains constant which indicates that the stray reaction of total combustion is not due to consecutive oxidation of acrolein, but proceeds at some other sites, resulting from the properties of $MoO_3$ itself. Results of these experiments clearly demonstrated that activation of hydrocarbon molecules takes place at the active sites composed of coordinatively unsaturated transition metal cations coordinated by oxide ions.

## 5. MECHANISM OF THE OXIDATION OF ALKYLAROMATICS

Oxidation of benzene and toluene on vanadium oxide monolayer catalysts have been subjects of many studies, in which attempts to identify the reaction intermediates by IR spectroscopy and to elucidate the mechanism of the reaction have been undertaken. These studies led to the conclusion that in the case of toluene the reaction starts with the formation of the benzyl intermediate, which interacts with surface lattice oxygen to form, consecutively, adsorbed benzaldehyde and benzoic acid precursors. These may either desorb or may be further oxidized to carbon oxides, or undergo degradation of the aromatic ring with the formation of maleic anhydride and carbon oxides. In order to find the mechanism of the initial activation of the molecule and specify the factors determining the choice of the pathway by the reacting system quantum chemical calculations were carried out of the interactions developing on approach of benzene and toluene molecule to a cluster composed of six vanadium-oxygen square pyramids assumed to be a model of supported vanadium oxide monolayer catalyst.

SINDO method was used for calculations. Toluene was approached side-on with the ring plane parallel to the plane of the cluster or end-on with the molecular axis perpendicular along a trajectory perpendicular to the plane of six edge- and corner-linked vanadium-oxygen square pyramids, which represent an element of the (010) plane of $V_2O_5$. The trajectory was chosen to point either at vanadium ion or bridging oxygen ions. In all cases the plot of total energy vs reaction coordinate showed a minimum, corresponding to the formation of an adsorbed complex (Fig.18). Side-on adsorption was much stronger, with strong interactions of all carbon atoms of the ring and methyl group with oxygen atoms of the cluster, accompanied by simultaneous weekening of C-C bonds (Fig.19). This indicates that side-on adsorption leads to complete destruction of the aromatic molecule and formation of coke or carbon oxides. End-on adsorption is most facile on bridging oxygen, and on approaching the toluene molecule two hydrogen atoms of the methyl group move simultaneously away to form finally OH groups with oxygens of the cluster and the methine group becomes linked to the bridging oxygen, forming the precursor of Ar-CHO (Fig.20). The bonds between vanadium and bridging oxygen are considerably weekened,resulting in reconstruction of the catalyst surface which enables a facile desorption of benzaldehyde. It may be thus concluded that selective oxidation of toluene is a concerted reaction, in which the interactions developed on approach of toluene to vanadium oxide cluster cause the rearrangement of electrons and nuclei resulting in desorption of benzaldehyde (Fig.21). Total oxidation starts from a different surface complex.

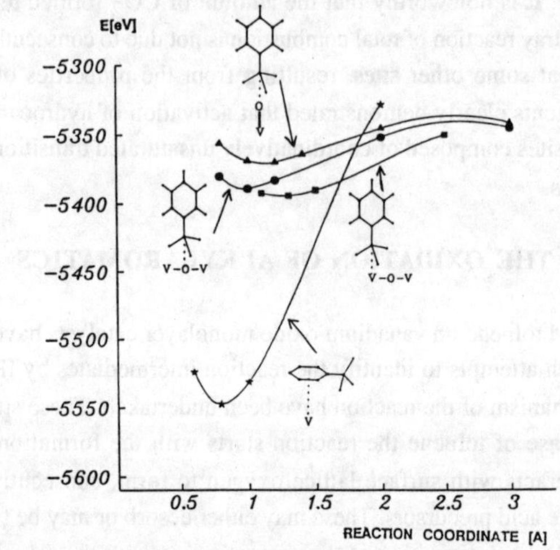


Fig.18. Total energy as function of the distance of toluene molecule from the plane of $V_6O_{20}$ cluster for different orientations and adsorption sites.

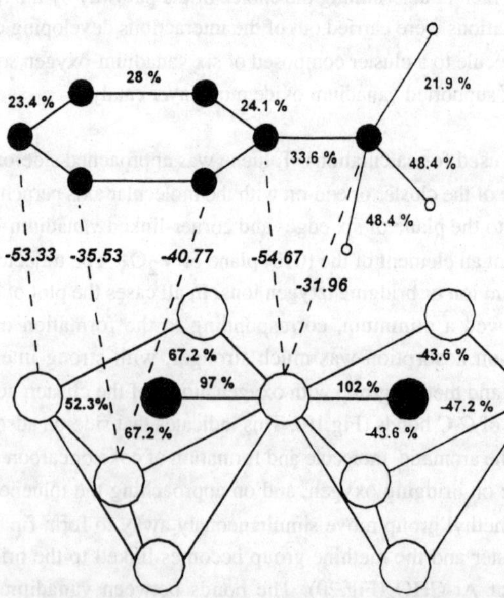


Fig.19. Changes (in %) of the diatomic energy contributions in side-on adsorbate complex $C_6H_5CH_3$--bridging oxygen site in respect to the isolated cluster and toluene molecule

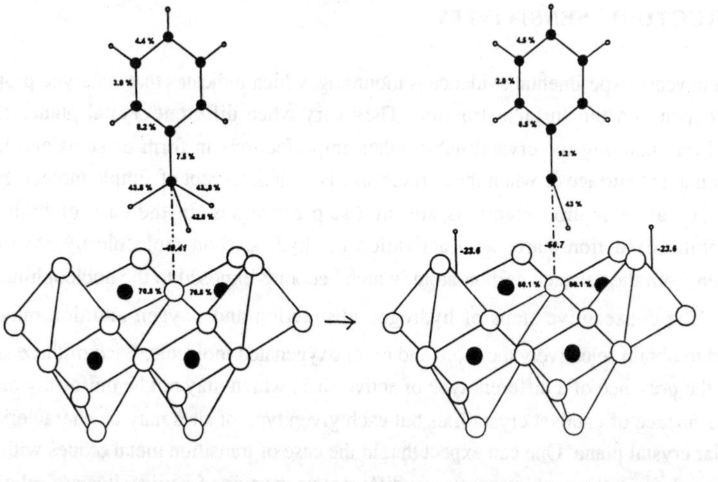

Fig.20. Changes (in %) of the diatomic energy contributions in end-on adsorbate complex $C_6H_5CH_3$--bridging oxygen site in respect to the isolated cluster and toluene molecule.

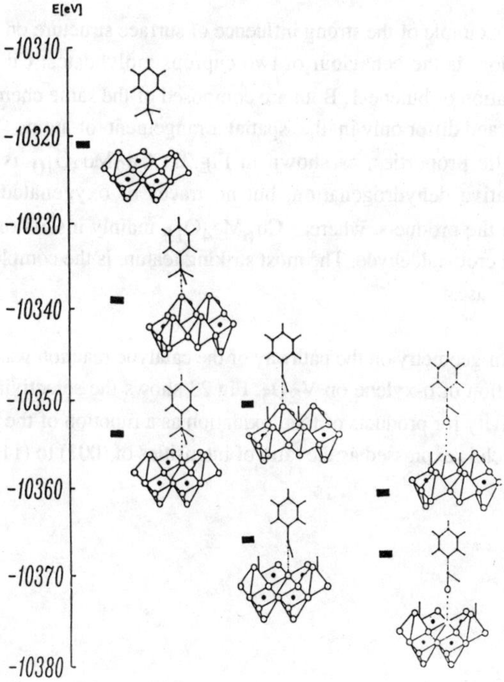


Fig.21. Mechanism of the oxidation of toluene

## 6. STRUCTURE SENSITIVITY

In recent years experimental evidence is mounting, which indicates that catalytic properties of oxides depend on the surface structure. They vary when different crystal planes are being exposed on changing the crystal habit, when imperfections in form of steps and kinks are generated at the surface or when the particle size is varied. Except of simple molecules such as CO or $H_2$ all oxidation reactions are multistep processes. In the case of hydrocarbons nucleophilic oxidation starts with activation of hydrocarbon molecule by abstraction of hydrogen from the selected carbon atom, which becomes exposed to the nucleophilic addition of $O^{2-}$. The consecutive steps of hydrogen abstraction and oxygen addition may be then repeated to obtain selectively the more and more oxygenated molecule. Each of these steps may require the presence of a different type of active sites, which may not be uniformly distributed over the surface of catalyst crystallites but each given type of sites may be characteristic for a particular crystal plane. One can expect that in the case of transition metal oxides with strongly pronounced crystallographic anisotropy different properties of active sites are related to the differences of the surface structure of various crystal faces which results in structure sensitivity of oxidation reactions.

The most spectacular example of the strong influence of surface structure on the direction of the oxidation reaction is the behaviour of two cuprous molybdates: $Cu_2Mo_3O_{10}$ and $Cu_6Mo_4O_{15}$ in the oxidation of butene-1. Both are composed of the same chemical elements in the same valence state and differ only in the spatial arrangement of atoms. Yet they show entirely different catalytic properties, as shown in Fig.22: $Cu_2Mo_3O_{10}$ is active in the isomerization and oxidative dehydrogenation, but no traces of oxygenated hydrocarbon molecules are present in the products, whereas $Cu_6Mo_4O_{15}$ mainly inserts oxygen into the organic molecule to form crotonaldehyde. The most striking feature is the complete absence of isomerization in the latter case.

A pronounced influence of geometry on the pathway of the catalytic reaction was revealed also for the case of the oxidation of o-xylene on $V_2O_5$. Fig.23 shows the selectivity for phthalic anhydride and the selectivity for products of total oxidation as a function of the textural factor of $V_2O_5$ crystallites, which is expressed as the ratio of intensities of (001) to (110) reflections.

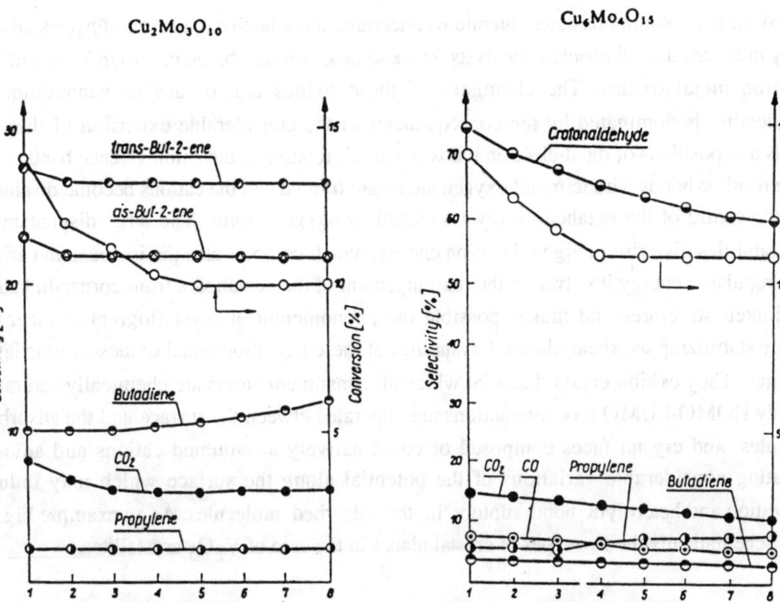


Fig.22. Conversion and selectivities to different products as a function of the number of pulses of butene-1 introduced on $Cu_2Mo_3O_{10}$ and $Cu_6Mo_4O_{15}$ catalysts at 643K.

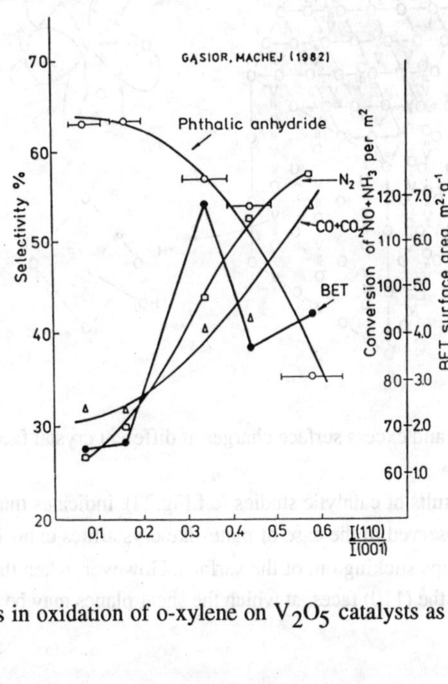


Fig.23. Selectivities in oxidation of o-xylene on $V_2O_5$ catalysts as function of the morphological factor.

Analysis of the voluminous patent literature concerning the selective oxidation of hydrocarbons clearly indicates that all efficient catalysts for these processes are based on group V,VI and VII transition metal oxides. The chemistry of these oxides e.g. oxides of vaanadium or molibdenum, is dominated by the consequencies of the considerable extension of their d-orbitals and positions of the d-electron redox potentials relative to the anion valence band edge. As the result π-bonds with terminal oxygen atoms are formed and the cations become displaced from the centre of the octahedron towards terminal oxygen atoms. The large displacement polarizabilities give rise to high relaxation energy, which compensates the increase in cation-cation repulsion energy involved in the rearrangement of the ocatahedra from corner-linked to edge linked structures and makes possible the phenomenon of crystallographic shear by strongly stabilizing the shear planes. Crystallites of these transition metal oxides assume layer structures. They exhibit crystal faces, at which all constituent atoms are chemically saturated and only HOMO-LUMO type interactions may operate between the surface and the adsorbed molecules, and crystal faces composed of coordinatively unsaturated cations and anions, generating considerable variations of the potential along the surface which may induce polarization and heterolytic bond rupture in the adsorbed molecules. As an example Fig.24 shows schematically the two types of crystal planes in the case of $V_2O_5$ crystallites.

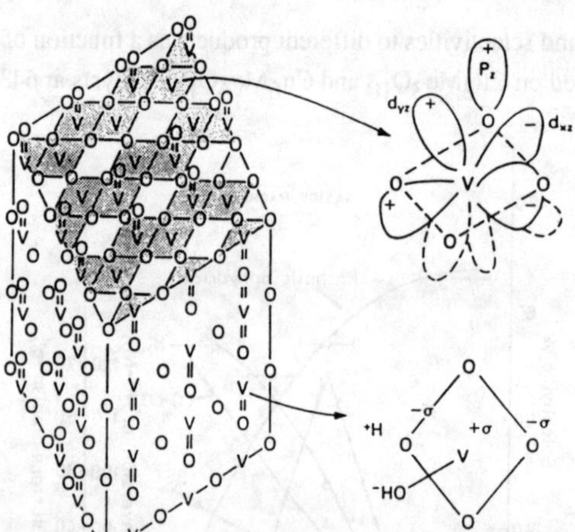

Fig.24. Orbital structure and excess surface charges at different crystal faces of $V_2O_5$.

Comparison with the results of catalytic studies (c.f.Fig.23) indicates that high selectivity for phthalic anhydride is observed in the case of plate-like crystallites exposing mainly the (001) faces with the V=O groups sticking out of the surface. However, when the crystallites expose to a considerable degree the (110) faces, at which the shear planes may be nucleated and whole perpendicular layers of oxygen may be extracted, total oxidation becomes predominant reaction pathway. A general conclusion may thus be formulated that in compounds of those transition elements, in which the phenomenon of the displacement stabilization results in the strong anisotropy of properties, differences in surface and catalytic properties of different crystal faces may be encountered.

Foreign ions present at the surface as impurities or additives constitute point defects which may play the role of new active sites or modify the properties of the existing ones by shifting the defect equilibria of the oxide. Moreover, these ions may preferentially accumulate only at certain crystal faces, e.g. charged ions will segregate to polar crystal surfaces, a phenomenon which may be called structure sensitivity of deposition. Unravelling of the role of these parameters in determining the rate of elementary steps of catalytic oxidation reactions is a great chalenge for the science of catalysis in the future.

After nucleophilic addition of the surface oxide ion to the carbon atom of the hydrocarbon molecule, resulting in the formation of a precursor of the oxygenated species, the latter is desorbed generating a surface oxygen vacancy. As mentioned above, oxides of group IV-VII transition metals show a strong tendency to annihilate the vacancies by the formation of shear planes, which are nucleated at the surface with the simultaneous release of oxygen by the crystal (Fig.25). As one of the possible explanations of the fact that selective catalysts for partial oxidation are always based on group V-VII transition metal oxides a hypothesis was advanced that this tendency is the driving force facilitating the desorption of the oxygenated product (Fig.26). A facile and efficient route is thus provided for the addition of a nucleophilic lattice oxygen into the hydrocarbon molecule. Little is however known about the mechanism of the release of an oxide ion from the surface layer of the oxide into the gas phase and about the parameters which determine the rate of this process. It may be hoped that with the further development of surface science techniques it will be possible to answer many of these intriguing questions.

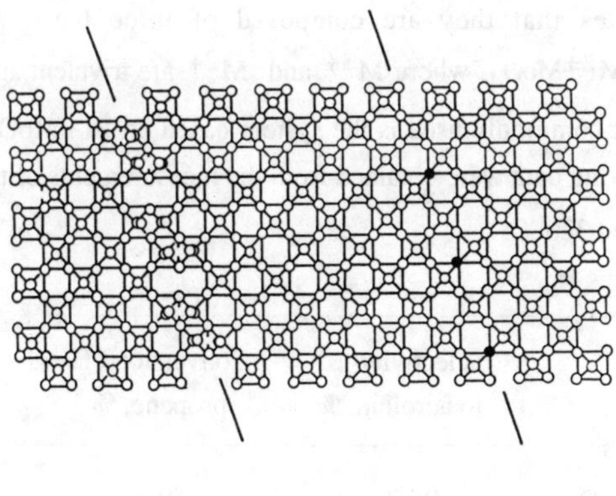

Fig.25. Formation of a shear plane

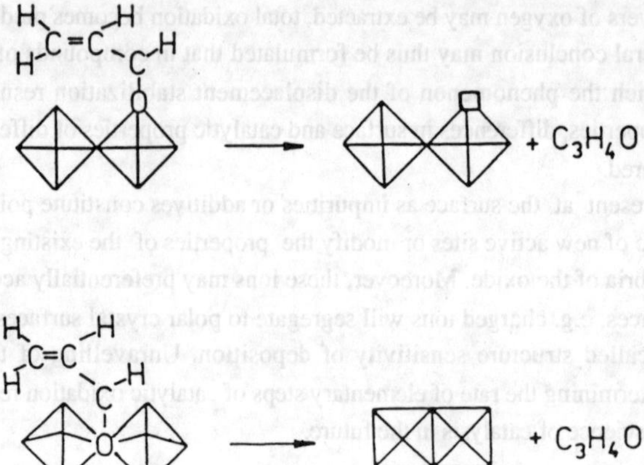


Fig.26. Mechanism of the nucleophilic addition of oxygen

## 7. SYNERGY OF CATALYTIC PROPERTIES IN OXIDE SYSTEMS

As discussed above heterogeneous oxidation reactions are multistep processes which require multifunctional catalysts. Therefore multicomponent oxide systems are usually used as catalysts in form of heterogeneous mixtures or supported oxide monolayers. In both cases strong synergistic effects are often observed, non-existent in solid solutions which indicates that they may be related to the presence of interfaces. The origin of these effects is one of the most fascinating questions of catalysis to be answered in future studies.

One of the spectacular examples are multicomponent molybdate catalysts for oxidation and ammoxidation of propene, some of them containing 10 or more components. X-ray examination indicates that they are composed of three basic phases: $Bi_xMo_yO_z$, $M^{3+}{}_2(MoO_4)_3$ and $M^{2+}MoO_4$, where $M^{3+}$ and $M^{2+}$ are trivalent and divalent transition metal ions. The most commonly used is the system based on $Bi_2(MoO_4)_3$, $Fe^{3+}{}_2(MoO_4)_3$ and $CoMoO_4$. Following data were obtained from the measurements of their behaviour in the oxidation of propene at 320$^o$C:

| Catalyst | Selectivity to acrolein, % | Conversion of propene, % |
|---|---|---|
| $M^{2+}{}_aM^{3+}{}_bBi_xMo_yO_z$ | 95.7 | 69.7 |
| $Bi_2(MoO_4)_3$ | 90.3 | 7.4 |

It may be seen that addition of trivalent and divalent cations to the bismuth molybdate system, which is a selective catalyst, increases the catalytic activity by an order of magnitude without much influencing the selectivity. Apparently the mechanism of the reaction remains the same, but the rate determining step is accelerated.

As already discussed, in bismuth molybdate catalysts for oxidation of olefins the bismuth ions play the role of sites activating the olefin molecule by abstraction of hydrogen and formation of the allyl species, whereas the molybdate sublattice is responsible for the nucleophilic addition of oxygen. It has been argued that introduction of the redox pair $Fe^{2+}/Fe^{3+}$ promotes oxygen and electron transfer. However, it is well established that activation of the hydrocarbon molecule is the rate determining step of the reaction and it is not clear how the redox pair interferes in this step.

Basing on the in situ studies of XRD and Mossbauer spectra it was shown (Fig.27) that in the conditions of the catalytic reaction $Fe^{3+}$ ions in $Fe_2(MoO_4)_3$ are partially reduced to $Fe^{2+}$ ions and nuclei of $Fe^{2+}MoO_4$ are formed. The latter serve as active sites for binding $O_2$ molecules and reducing them to $O^{2-}$ ions which are then transported through the defected $Fe_2(MoO_4)_3$ phase to replenish the active sites at the surface of $Bi_2(MoO_4)_3$, reduced during the abstraction of hydrogen from the olefin and subsequent addition of oxygen. The role of $CoMoO_4$ consists in stabilizing the isomorphous $FeMoO_4$ nuclei.

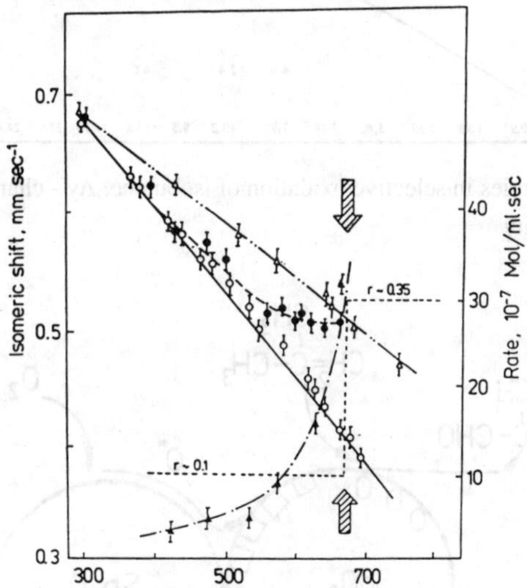


Fig.27. Dependence of the isomeric shift on the Fe(III)molybdate in the Mossbauer resonance spectrum (curves 1-3) and dependence of the rate of propene oxidation on temperature (curve IV). o-1, o-2, Δ-3, Δ-4.

Further studies are needed to confirm this hypothesis and to unravel the mechanism of elementary steps involved in such complex performance of the multicomponent oxide catalysts and the role of individual components.

Synergy of catalytic properties in mechanical oxide mixtures seems to be a general phenomenon, its existence having been established in the case of a number of catalytic reactions for mixtures of many different oxides. A hypothesis was advanced that the synergistic effects are due to the spillover of oxygen. All oxides were devided into two groups: oxygen donors and oxygen acceptors (Fig.28). Oxygen becomes activated at the surface of the donor type oxide and is supplied through a spillover to the surface of the acceptor-type oxide, where it generates new active centres accelerating thus the catalytic reaction (Fig.29). The molecular mechanism of such phenomena remains as yet to be explained.

| (Δy) | $MoO_3$ | CoMo | NiMo | CuMo | MgMo | FeMo | FeSb | BiP | ZnFe | $SnO_2$ | $Bi_2O_3$ | $TeO_2$ | (Cu)Sb | $Sb_2O_4$ (ΔS) |
|---|---|---|---|---|---|---|---|---|---|---|---|---|---|---|
| $Sb_2O_4$ | 2.62 | | 0.11 | | | 1.19 | 0.28 | 0.93 | | 0.55 | | 2.51 | | |
| (Cu)Sb | 2.43 | | | | | | | | | | | | | |
| $TeO_2$ | 5.28 | | | | | | | | | | | | | 7.30 |
| $Bi_2O_3$ | 6.60 | | | | | | | | | | | | | |
| $SnO_2$ | 1.11 | | 0.09 | | | | 0.04 | -0.16 | | | | | | 17.0 |
| ZnFe | 0.56 | | | | | | | | | | | | | |
| BiP | 1.87 | | 0.04 | | | | -0.05 | | | -3.7 | | | | 24.1 |
| FeSb | 0.23 | | 0.07 | | | | | -3.6 | | 2.0 | | | | 6.70 |
| FeMo | 1.39 | | | | | | | | | | | | | 39.4 |
| MgMo | 0.12 | | | | | | | | | | | | | |
| CuMo | 1.12 | | | | | | | | | | | | | |
| NiMo | 0.21 | | | | | | 4.6 | 2.4 | | 4.6 | | | | 3.9 |
| CoMo | 0.26 | | | | | | | | | | | | | |
| $MoO_3$ | | -0.91 | 1.93 | 3.59 | 5.40 | 7.18 | 7.93 | 10.2 | 8.5 | 10.8 | 14.74 | 25.4 | 24.4 | 13.09 |

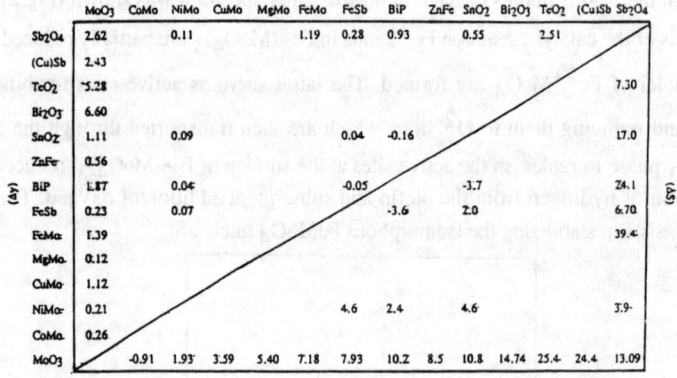

Fig.28. Catalytic synergies in selective oxidation of isobutene. Δy - change of intrinsic yield, ΔS - change in selectivity.

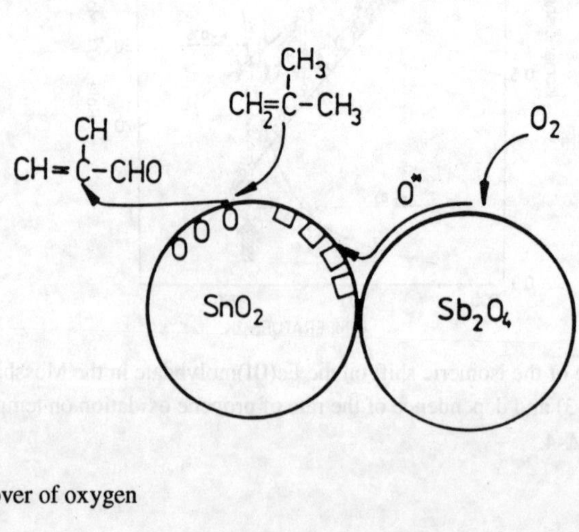


Fig.29. Spill-over of oxygen

It should be however borne in mind that on heating the oxide mixtures spreading of one of the oxides over surfaces of the others may take place. Simple considerations of equilibrium conditions at the interface between two solid phases and the gas phase show that when the energy of cohesion of the clusters of supported oxide is smaller than the energy of adhesion of this oxide to the support spontaneous spreading of the former over the surface of the latter will take place as manifested in the phenomenon of wetting of one solid by another one. As a result, there is always a tendency of oxide surfaces to become covered by a thin layer of other components present in the mixture. The formation of such overlayer, too thin to be detected by many of the standard experimental techniques, may profoundly modify the catalytic properties. Therefore the phenomenon of wetting in oxide systems is of paramount importance for preparation of catalysts. Little data concerning the surface free energy of oxides are available at present and practically no information exists relating to the mechanism of surface migration.

## 8. THE DYNAMIC STATE OF OXIDE SURFACES

The behaviour of oxide monolayers and three dimensional clusters, which may be considered as colloidal particles, deposited at the surface of an oxide support, will be controlled to a large extent by the surface free energy relations at the interfaces between the support, the clusters of the monolayer and the gas phase. As the energy of cohesion of e.g. $V_2O_5$ is smaller than its energy of adhesion to anatase or alumina, it is wetting these supports and spreads over their surface. This is illustrated in Fig.30, in which the degree of coverage of an anatase support with vanadium oxide monolayer is plotted as function of time of calcination of a mechanical mixture of $V_2O_5$ and anatase at 450$^{o}$C. In the same figure the results are also plotted in the coordinates q vs. t. The linear dependence of parabolic coordinates indicates that the kinetics of thermal spreading is diffusion controlled.The model of this phenomenon is shown in Fig.31. When $V^{5+}$-O clusters are reduced to $V^{3+}$-O clusters, the energy of the cohesion increases to such an extent that it becomes greater than that of its adhesion to the support and the monolayer of vanadium oxides shrinks and coalesces into three dimensional particles. Thus, exposure of the vanadium oxide monolayer catalysts to alternating oxidation and reduction cycles will entail dispersion and shrinking of the monolayer. These processes may be followed by measuring the ir-spectra of appropriate probe molecules.

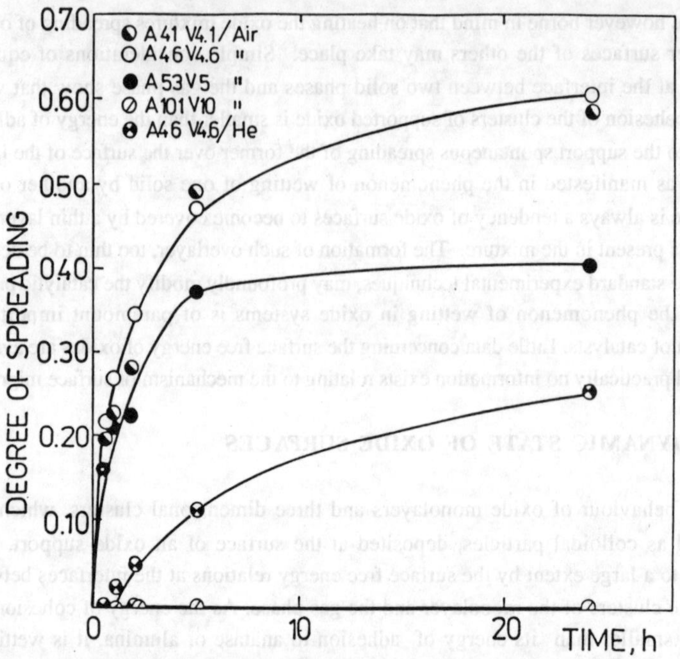


Fig.30. Changes of surface coverage of $TiO_2$ with $VO_x$ monolayer as function of the time of heating of mechanical mixture of $V_2O_5$+$TiO_2$.

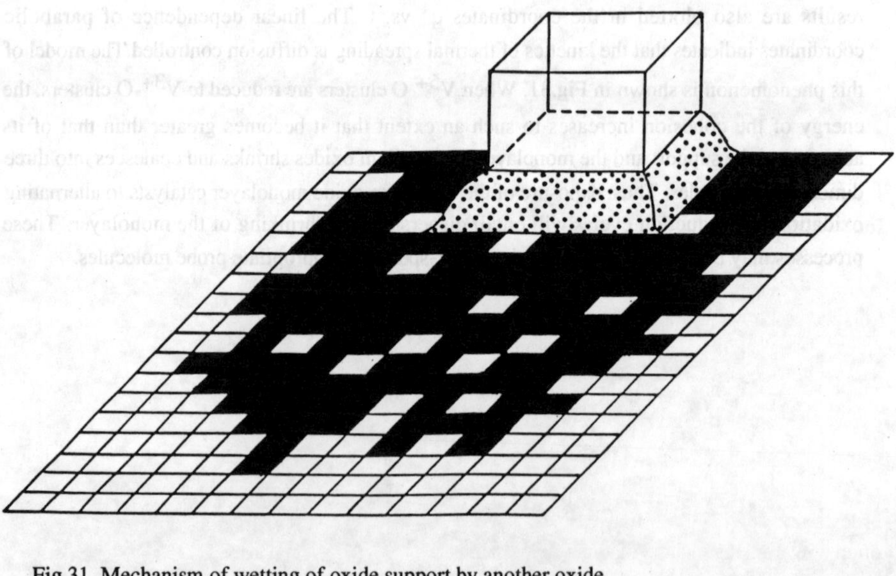

Fig.31. Mechanism of wetting of oxide support by another oxide

It has been shown that the basic OH groups of the g -alumina surface interact selectively with $CO_2$, which leads to the formation of surface bicarbonate species. These species give rise to IR bands at about 1235, 1480 and 1640 $cm^{-1}$, which can be used to monitor changes of the number of hydroxyl groups of the alumina support when active phase of a transition metal oxide is deposited and then treated in different atmospheres. On covering the support with e.g. vanadia the amount of basic hydroxyls on the alumina surface is rapidly decreasing, as revealed by the diminishing intensity of the 1235 $cm^{-1}$ band (Fig.32) so that no more groups are visible at vanadia coverage of 6.6 V $atoms.nm^{-2}$. When however the sample is reduced, the surface hydroxyl groups are restored and the 1235 $cm^{-1}$ reappears indicating that coalescence of vanadia monolayer into clusters took place and free alumina surface was uncovered. When the sample was exposed to oxygen, redispersion of vanadia took place. Use of ammonia as probe molecules revealed also the existence of two types of Lewis acid centers at the surface of alumina, their acid strength depending on the coverage with vanadia and degree of reduction of the latter. The mechanism of transformations of the vanadia monolayer are summarized in Fig.33.

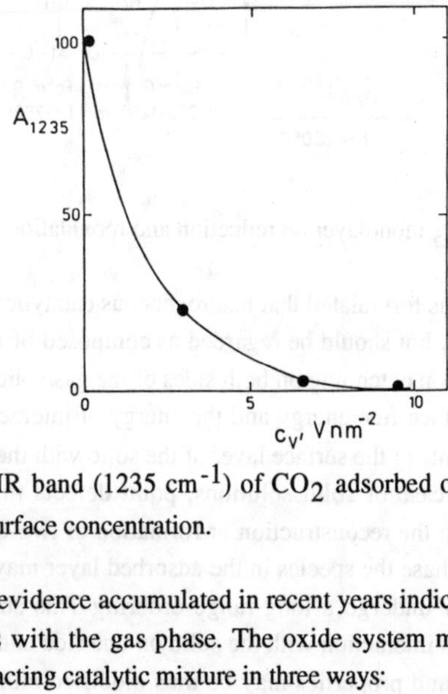


Fig.32. Intensity of IR band (1235 $cm^{-1}$) of $CO_2$ adsorbed on $V_2O_5/Al_2O_3$ catalysts as a function of vanadia surface concentration.

Ample experimental evidence accumulated in recent years indicates that oxide surfaces are in dynamic interactions with the gas phase. The oxide system may respond to the change of composition of the reacting catalytic mixture in three ways:

- defect equillibria at the oxide surface or in the whole bulk may be shifted and the change of concentration of the given type of active sites involved in the catalytic transformation may cause the change of catalytic properties;
- when the concentration of defects at the oxide surface surpasses certain critical value, ordering of defects or formation of a new bidimentsional surface phase may occur resulting often in a dramatic change of catalytic properties;

- when redox mechanism operates in the catalytic reaction, the ratio of the rates of catalyst reduction and its reoxidation may be different for various oxide phases and hysteresis of the dependence of catalytic properties on the composition of the gas phase may appear, these properties being then strongly influenced by the type of pretreatment.

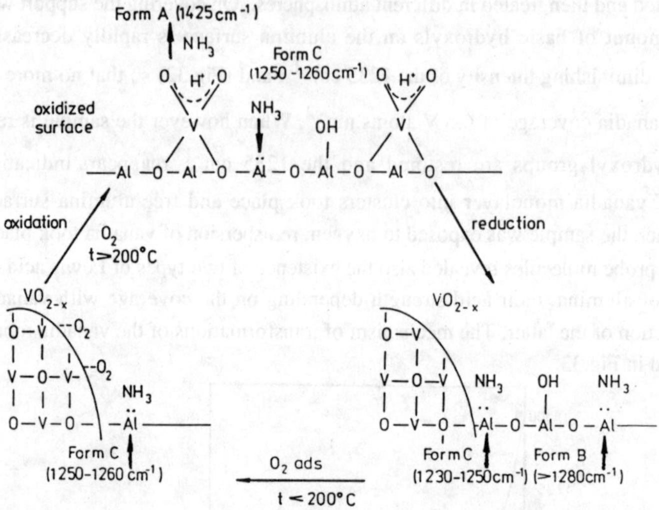


Fig.33. Transformations of $V_2O_5$ monolayer on reduction and reoxidation.

A general conclusion may be thus formulated that heterogeneous catalytic systems should not be treated as two phase systems, but should be regarded as composed of three parts: gas and solid phases and the surface region extending on both sides of the gas/solid interface (Fig.34). On the side of the solid the surface free energy and the energy of interaction with adsorbed species may cause the enrichments of the surface layer of the solid with the constituents of the lattice (atoms of the solute in case of solid solutions, point defects in nonstoichiometric compounds etc) or may result in the reconstruction or formation of two-dimensional surface phases. On the side of the gas phase the species in the adsorbed layer may aggregate to form two-dimensional liquid or may undergo a long-range ordering This surface region is not autonomous but is in continuous interaction with the solid on one side and with the gas phase on the other side. Its structure and properties may be thus modified either by changing the composition of the gas phase or by altering the properties of the solid.

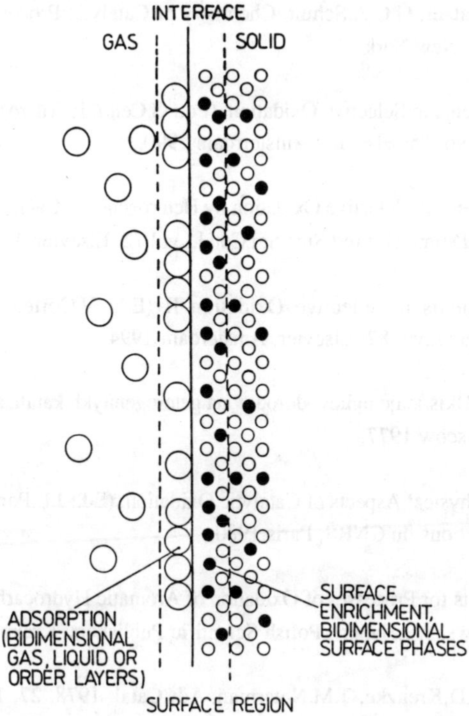


Fig.34. Heterogeneous catalytic system.

**Further reading**

A.Bielanski, J.Haber, Oxygen in Catalysis, Marcel Dekker Inc., New York 1991.

D.J.Hucknall, Selective Oxidation of Hydrocarbons, Academic Press, New York 1974.

H.H.Kung, Transition Metal Oxides: Surface Chemistry and Catalysis, Elsevier, Amsterdam 1989.

Adsorption and Catalysis on Oxide Surfaces, (Ed.: M.Che, G.C.Bond) Elsevier, Amsterdam 1985.

Surface Properties and Catalysis by Non-metals, (Ed.: J.P.Bonnelle, B.Delmon, E.Derouane), Reidel Publ.Co., Dordrecht 1983.

B.C.Gates, R.Katzer, G.C.A.Schuit, Chemistry of Catalytic Processes, Academic Press, New York

New Developments in Selective Oxidation, (Ed.: G.Centi, F.Trifiro), Stud. Surf.Sci.Catal., vol.55, Elsevier, Amsterrdam 1990.

New Developments in Selective Oxidation by Heterogeneous Catalysis, (Ed.: P.Ruiz, B.Delmon), Stud.Surf.Sci.Catal., vol.72, Elsevier,Amsterdam 1992.

New Developments in Selective Oxidation II (Ed.: V.Cortez Corberan, S.Vic.Bellon), Stud.Surf.Sci.Catal.,vol.82, Elsevier, Amsterdam 1994.

L.Ya.Margolis, Okisljenje uglievodorodov na geterogennykh katalizatorakh, Izd.Khimja, Moscow 1977.

Chemical and Physical Aspects of Catalytic Oxidation, (Ed.: J.L.Portefaix, F.Figueras), Editions du CNRS, Paris 1980.

Vanadia Catalysts for Processes of Oxidation of Aromatic Hydrocarbons, (Ed.: B.Grzybowska, J.Haber), Polish Scientific Publishers, Krakow 1984.

G.W.Keulks, L.D.Krenzke, T.M.Noterman, Adv.Catal. 1978, 27, 183.

R.K.Grasselli, J.D.Burrington, Adv.Catal. 1981, 30, 133

D.B.Dadyburjor, S.S.Jewur, E.Ruckenstein, Catal.Rev.Sci.Eng. 1979, 19, 293

M.S.Wainwright, N.R.Foster, Catal.Rev.Sci.Eng., 1979, 19, 211.

R.Higgins, P.Hayder, in Specialist Periodical Report-Catalysis vol.1, The Royal Society of Chemistry, London 1977, p.168.

P.J.Gellings, in Specialist Periodical Report-Catalysis vol.7, The Royal Society of Chemistry, London 1983.

C.F.Cullis, D.J.Hucknall, in Specialist Periodical Report-Catalysis vol.5, The Royal Society of Chemistry, London 1981, p.273.

J.Haber, in Proc.8th Intern.Congr.Catalysis, Berlin 1984, Verlag Chemie-Dechema, Frankfurt 1984, Plenary Lectures vol.1, p.85

J.Haber, in Solid State Chemistry in Catalysis, (Ed.:R.K.Grasselli,
J.F.Brazdil), ACS Symposia Series No 279, Washington D.C., 1985, p.3.

O.V.Krylov, L.I.Margolis, in Problems of Kinetics and Catalysis (in russian), Izd.Nauka, Moscow 1985, vol.19, p.5

# *HETEROGENEOUS OXIDATION CATALYSIS ON METALLIC OXIDES*

JACQUES C. VEDRINE
*Institut de Recherches sur la Catalyse, CNRS*
*2, avenue Albert Einstein F-69626 VILLEURBANNE Cédex, France*

**ABSTRACT**

**As a complement of the general presentation by J. Haber in this book, this article describes general features of oxidation reactions in heterogeneous catalysis on metallic oxides. Emphasis is placed on the Mars and van Krevelen catalytic mechanism and on its implication in oxidation catalysis. The concept of active sites as ensembles of $MO_x$ species of limited size is developped and examplified. The dynamical aspect of an oxide surface during catalysis including the wetting phenomenon of an oxide spreading over another oxide (multi component-type catalysts) or on a support is described. Such properties are shown to greatly depend on the preparation procedure and on the activation and catalytic reaction conditions and emphasis is placed on the importance of such parameters for the catalyst performances in oxidation reactions.**

## 1. Introduction

Oxidation catalytic reactions are of prime importance at an industrial level since they correspond to a huge market. For instance in the US in 1991, 31.2% of the catalytic production of major organic chemicals corresponded to oxidation catalytic processes (18.1% heterogeneous, 13.1% homogeneous) and 17.8% to oxychlorination[1]. The market corresponds to 20 billions US $ in the USA and world wide such numbers have roughly to be multiplied by a factor of 2.5.

The concepts of oxidation began with Lavoisier's disproving of the phlogeston theory in 1773. One usually defines two groups of reactions namely homolytic and heterolytic. The first type involves radicals formed by homolytic cleavage of interatomic bonds. The second type involves an active oxygen compound or a metal ion which oxidizes the starting material in a two electrons transfer reaction. The reduced oxidizing agent must be reoxidized in a second step. One distinguishes several types of oxygen species as described in the previous paper of this course by J. Haber.

Some important industrial processes and some great intermediates in industrial chemistry are given in tables 1 and 2, respectively.

Majority of the catalysts correspond to metallic oxides with V or Mo as one of the key elements. Some metals (mainly Ag for ethylene epoxidation), noble metals (as Pt, Pd for total oxidation, etc) zeolites (Titano silicalite TS-1 from ENI for phenol oxidation) and heteropolyoxometallates (e.g. $H_4PMo_{11}VO_{40}$ for isobutene oxidation to methacrolein) may also be used.

In majority of cases catalytic properties in oxidation reactions involve a redox mechanism between reactant molecules and surface active sites as represented in the scheme in fig. 1 as suggested by Mars and van Krevelen in 1953. Such a scheme necessitates a catalyst which contains a redox couple as for instance transition metal ions and which exhibits high electrical conductivity to favour electron transfer and at last which has a high lattice oxygen anion mobility within the material to insure the reoxidation of the reduced catalyst.

From this feature arises the idea that the active sites are not isolated ions as in the Taylor's model but rather an ensemble of ions in a kind of "inorganic oxide cluster". A molecular concept of the active sites has then to be defined. Several examples are chosen below to show how such a concept may be valid in oxidation reaction.

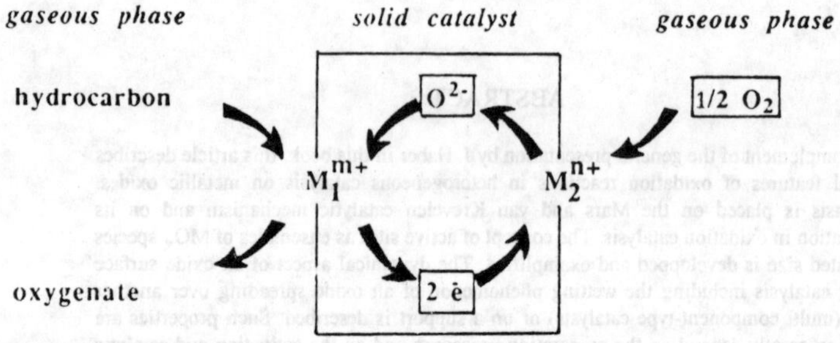


**Figure 1** : Scheme of the Mars and van Krevelen mechanism

Table 1: **Some industrial processes for the formation of oxygenates by gas-solid heterogeneous reactions**

| Reactant | Product | Catalyst | Conditions (°C) | Yield % |
|---|---|---|---|---|
| methanol | formaldehyde | iron molybdate | 250-300 | 90-95 |
| | | Ag or Ag/support | | |
| propylene | acrolein | Bi Mo Co Fe K oxide (multi component-type) | 350-400 | 90-95 |
| acrolein | acrylic acid | $V_2O_5/MoO_3$ | 400-450 | 90-95 |
| propylene | acrylonitrile | $V_2O_5/Sb_2O_5/Al_2O_3$ | | |
| benzene | maleic anhydride | VPO | 400-450 | 70-75 |
| butane | maleic anhydride | VPO | 400-450 | 60-65 |
| cyclohexane | caprolactane | $Pd/Al_2O_3$ | | |
| O-xylene | phthalic anhydride | $V_2O_5/TiO_2$ | 350-550 | 75-80 |
| $SO_2$ | $SO_3$ | $V_2O_5/SiO_2$ | 420 | 95 |
| ethylene | ethylene oxide | $Ag/Al_2O_3$ | 200-300 (10-30 atm) | 70-75 |
| ethylene | vinyl chloride | $CuCl_2, MgCl_2, C$ | 230 | |
| ethylene | vinyl acetate | Pd-Cu | 175-200 (2-10 atm) | 91-95 |

Table 2: Some great intermediates produced by heterogeneous oxidation catalysis and their more important final products

| | | |
|---|---|---|
| formaldehyde | → | glues, thermo hard resins |
| acetaldehyde | →. | acetic acid and anhydride |
| ethylene oxide | → | ethylene glycol |
| acrolein | → | methionine (amino acid based food for animals) |
| acrylic acid | → | paints, adhesives |
| methacrylic acid | → | altuglass, plexiglass |
| maleic anhydride | → | thermo hard resins |
| phthalic anhydride | → | plastifiers |
| acrylonitrile,methacrylonitrile | → | rubber, fibers |
| benzoic acid | → | phenol, dyes, fragances |

## 2. General features of oxidation catalysts and oxidation reaction[2]

### *2.1 Oxidation catalysts*

The oxidation catalysts are schematically mixed oxides which operate according to the redox process suggested by Mars and van Krevelen (fig. 1). According to this mechanism the substrate is oxidized by the solid and not directly by molecular oxygen of the gaseous phase. The rôle of such dioxygen is to regenerate or maintain the oxidized state of the catalyst. The oxygen introduced in the substrate (or giving $H_2O$ for oxidative dehydrogenation reactions) stems from the lattice. The mechanism involves the presence of two types of distinct active sites: an active site which oxidises the substrate and another site active for oxygen reduction. An adequate structure of the material should also facilitate both electrons and oxygen species transfer.

### *2.2 Oxygen species*

The oxygen atom incorporated into the substrate stems from the lattice and is at -2 oxidation state. Its replacement by molecular oxygen necessitates electrons according to: $O_2 + 4e^- \rightarrow 2O^{2-}$. This process has its own kinetics related to the reactivity of the sites with oxygen, their concentration, the efficiency of electron transfer, the partial pressure of oxygen, etc.. Usually, it is much faster that the oxidation of the substrate i.e. it is generally admitted that the rate determining state is the substrate activation.

Let's take some examples which will be considered in more details later on:

$$CH_3 - CH = CH_2 + 2(O^{2-}) \rightarrow CHO - CH = CH_2 + H_2O + 4e^- \quad \text{(acrolein)}$$

$$CH_3 - CH = CH_2 + NH_3 + 3(O^{2-}) \rightarrow CN - CH = CH_2 + 3H_2O + 6e^- \quad \text{(acrylonitrile)}$$

$CH_3 - CH_2 - CH_2 - CH_3 + 7(O^{2-}) \rightarrow$ 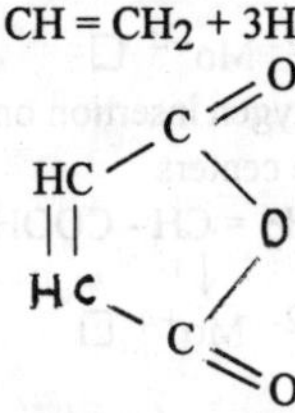 $+ 4\,H_2O + 14e^-$ (maleic anhydride)

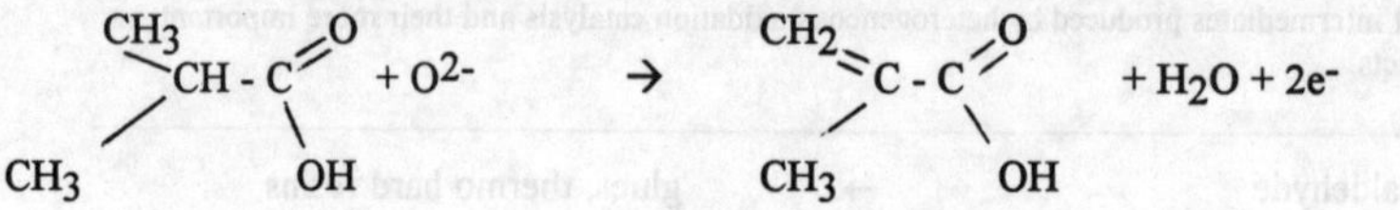


isobutyric acid methacrylic acid

It clearly appears that a single and isolated metallic ion site cannot take into account all the necessary transformations involved in the reactions since a fast replenishing of oxygenated species H atom extraction and a fast electron transfer are concerned. For instance n butane oxidation reaction to maleic anhydride necessitates 7 lattice oxygen ions, 8 hydrogen abstraction from the substrate 3 oxygen atoms insertion and 14 electrons transfer!

The homolytic fragmentation of a C-H bond in the coordination sphere of the acceptor metal ion may occur via a transfer of the hydrogen to the oxygen ion at -2 oxidation state. This is a concerted action with homolytic breaking of metal - oxygen bond which transfers one electron to the metal. Without any hypothesis about the nature of the metal oxygen bond one can write with formation of a II alkyl complex (as usually admitted) or a δ-alkyl complex.

Depending on the nature, oxidation state of the metal ion and its environment (coordination structure), the metal oxygen bonds may be more or less polarised and therefore the oxygen ion may exhibit electrophilic or nucleophilic properties. One may distinguished three extreme cases:

a: $\overset{\delta+}{M} = \overset{\delta-}{O}$ (nucleophilic) b: M = O and c: $\overset{\delta-}{M} = \overset{\delta+}{O}$ (electrophilic).

In the first case reaction of protonation or deprotonation will be favoured, according to $M = 0 + H^+ \rightarrow M^+ - OH$.

In the second case with non or weak polarisation, homolytic concerted reactions will be favoured as allylic dehydrogenation of olefins, according to the following scheme 1.

$CH_2 = CH - CH_3$

↓ First hydrogen abstraction on transition metal cation

$CH_2 - CH - CH_2 + CH_2 - CH - CH_2$ <u>dimerization</u> $CH_2 = CH - CH_2 - CH_2 - CH = CH_2$

↓ ↓

$O^{2-}\ Me^{+n}\ O^{2-}$ $O^{2-}\ Me^{+n}\ O^{2-}$

↓ First oxygen insertion on Mo O polyhedra and second hydrogen abstraction

$CH_2 = CH - CH_2$ ⟶ $CH_2 = CH - \overset{H}{CO}$

$O^{2-}\ Mo^{+6}\ O^{2-}$ $O^{2-}\ Mo^{+4}$ □

↓ Second oxygen insertion on acid - base centers

$CH_2 = CH - C - O$ ⟶ $CH_2 = CH - COOH$ and

↓ ↓

$O^{2-}\ Mo^{+6}\ O^{2-}$ $O^{2-}\ Mo^{+4}$ □

<u>Scheme 1</u>: Mechanism of propene oxidation into hexadine or acrolein or acrylic acid according to Mars and van Krevelen scheme

In the third case the electrophilic character of oxygen ion allows it to proceed to a direct attack of a double bond or of an aromatic ring according to:

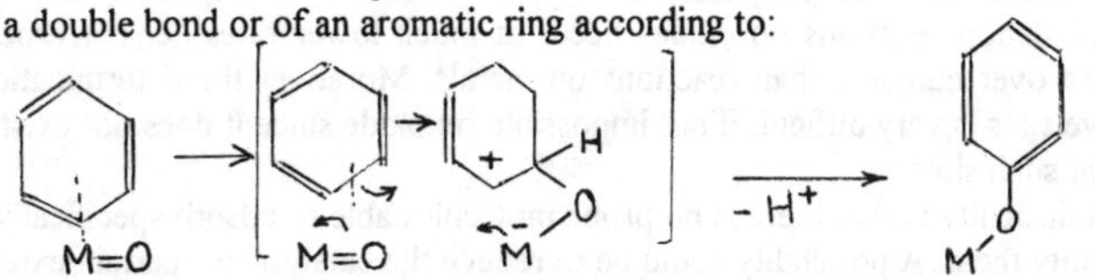


One may then obtain acetone from propene or even a double bond rupture giving acetaldehyde and formaldehyde from propene or the anhydride from an aromatic ring. Final stage could be total oxidation.

Obviously other oxygen species may exist as electrophilic surface species $O^-$, superoxo (electrophilic) M - O - O°, peroxo (electrophilic) O - $O^-$ ($O_2^-$) and oxo (nucleophilic) as

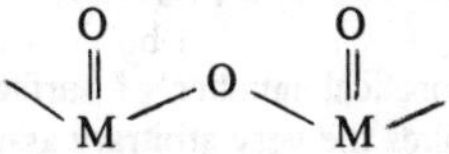


*2.3 Selectivity and bond strength*

The above statements let us imagine that the strength of the metal oxygen bond will play a determining role in the selectivity of oxidation reactions. The Russian school from the Boreskov Institute of Catalysis in Novosibirsk has made strong effort in the 60 s to correlate selectivity with the heat of metal oxygen bond formation in the oxide itself. The correlation was not really clear and not general but was valid in certain cases. More recently H. Kung et al[3] have tried to correlate the heat of reoxidation of more or less reduced catalysts with their catalytic selectivity. The reaction studied was the oxidative dehydrogenation of n-butane at 500°C with $C_4$:$O_2$:He = 4:8:88 ratios and the catalyst was $V_2O_5$ / γ-$Al_2O_3$ at different V coverages. The authors have reduced the catalysts under $H_2$ at 400 up to 480°C (depending on V coverage) at different extents (0 to 0.4 (O) atom per V atom) and measured by microcalorimetry the heat of reoxidation. They have observed that the selectivity was better when reoxidation heat was higher. Moreover such reoxidation heat was higher at low V coverage, i.e. for well dispersed vanadium and thus for species more isolated vanadium-oxygen species.

*2.4 Reducibility of the cations*

This parameter is obviously important too and should be related to the reoxidability of the catalysts as described above. It is then interesting to relate the selectivity with the redox potential of the metal cation involved in the reaction. The previous authors[3] have compared several orthovanadate compounds as $M_3(VO_4)_2$ with M = Mg, Zn, Ni or Cu and $M(VO_4)$ with M = Fe, Sm, Nd or Eu and studied the oxidative dehydrogenation of butane at 500°C. They have observed that the selectivity increases when the reduction potential decreases from + 0.77 up to - 2.4 volts (values taken in aqueous medium). These data clearly show that redox ability of a catalyst is an important parameter for oxidation reaction.

In the reaction of oxidative dehydrogenation of propane over magnesium vanadate sample J.C. Volta et al[4] have used temperature programmed reduction and temperature programmed oxidation techniques. They have shown that when the reducibility of the catalyst is easier (and subsequently its reoxidability) the selectivity was enhanced which supports the above statements. In such a case the $\alpha Mg_2V_2O_7$ phase which exhibits more reductive bonds was more selective for propane oxidative dehydrogenation at 500°C than the other two phases as $\beta MgV_2O_6$ and $Mg_3V_2O_8$.

*2.5 Turn over number, turn over frequency*

Usually oxidation reactions on oxides occur at much lower rates i.e. corresponding to much lower turn over numbers than reactions on metals. Moreover the determination of the number of active sites is very difficult if not impossible on oxide since it does not exist a simple method to count such sites.

The main difficulty is that there is no probe molecules able to adsorb specifically on such sites i.e. to identify them. A possibility could be to reduce the catalyst to a certain extent and to irreversibly adsorb a probe molecule as di oxygen at - 78°C. Assuming an atomic ratio O/M equal to one one could evaluate the number of accessible metallic ion. Such a method has been used for instance by T. Oyama et al[5] on unsupported and silica supported $V_2O_5$ catalysts as a function of V loading. The values obtained appeared to be reasonable but they obviously depend on the reduction extent (at 368°C in the previous work) which in the same reducing conditions will obviously depend on the catalyst (vide supra) and on the adsorption stoichiometry (O/V = 1!).

Majority of the time one considers the theoretical number of surface metallic atoms which is known for each cristalline face and one makes the very arbitrary assumption that each metallic surface ion is a potential active site. This may hold also true for supported oxides assuming in addition that the free support remains inactive or keeps its low starting activity. We will see in the following examples that the TON values may vary by several orders of magnitude depending on the reaction, on the temperature and more importantly on the support itself for supported catalysts.

**3 Structure sensitivity of oxidation reactions on oxides**

Such a concept has been introduced by M. Boudart on metals. It has been introduced for oxides in the late seventies by J.C. Volta et al[6-8], or early 80's by J.E. Germain[9, 10], J. Haber et al[11] and it is widely accepted at present. For instance in the work by J.C. Volta et al [6-8] it was shown that single crystal type samples of $MoO_3$ exhibiting different relative amounts of the different faces (010) basal, (100) side and (101) and ($\bar{1}$01) apical exhibited different activity and selectivity in the oxidation of propene to acrolein and $CO_x$. The originality of the work was to synthesize crystals of various shapes by epitaxial growth via oxyhydrolysis of $MoCl_5$ inserted between the layers of graphite. Table 3 summarizes the main results obtained for propene, but 1 ene and isobutene oxidation on $MoO_3$ crystals . It clearly appears that for propene oxidation the (100) face is selective for acrolein formation and the (010) for total oxidation. Such specificity depends on the hydrocarbon molecule. It may thus be proposed that stereo chemistry of the hydrocarbon molecule and that of the oxide face play a determining rôle.

A more precise analysis and characterization of the $MoO_3$ crystallites shape has shown that in fact the better plane for propene oxidation to acrolein corresponds to the (1kO) plane as shown in fig. 2[12]. It is then suggested that the propene activation (H abstraction) into the II-allyl intermediate occurs on the side (100) plane while the O atom insertion occurs on the (010) basal plane. The layered structure of $MoO_3$ makes that lattice oxygen atoms are much more labile in the (010) plane than perpendicularly to it.

Such a structure sensitivity is also described by J. Haber in his paper (preceding article) and more recently on $Cu_2O$ surface for propene oxidation[13]. This aspect is particularly important since it shows that by modifying preparation conditions of the metallic oxides one may develop some faces rather than others and then one may obtain some modifications in selectivities. A concept of epitaxial fitting of an oxide on a support developping specific cristalline faces was developped by P. Courtine on $V_2O_5$ / $TiO_2$ catalyst[14]. This concept is interesting but was not confirmed later one.

Table 3: **Structure sensitivity of the different faces of $MoO_3$ crystals in olefine oxidation at 380°C (from ref. 8).**

| | | Relative selectivity | | |
|---|---|---|---|---|
| Reactant Olefine | Products | Basal (010) | Side (100) | Apical (101), ($\bar{1}$01) |
| Propene | Acrolein | 0.06 | 2.3 | 0.7 |
| | CO, $CO_2$ | 1 | 0 | 0 |
| But - 1 - ene | Butadiene | 3 | 9.3 | 2 |
| | CO, $CO_2$ | 1 | 0 | 0 |
| Iso butene | Methacrolein | 0 | 0.6 | 0.1 |
| | acetone | 0.06 | 0 | 0.06 |
| | CO, $CO_2$ | 0 | 1 | 0 |

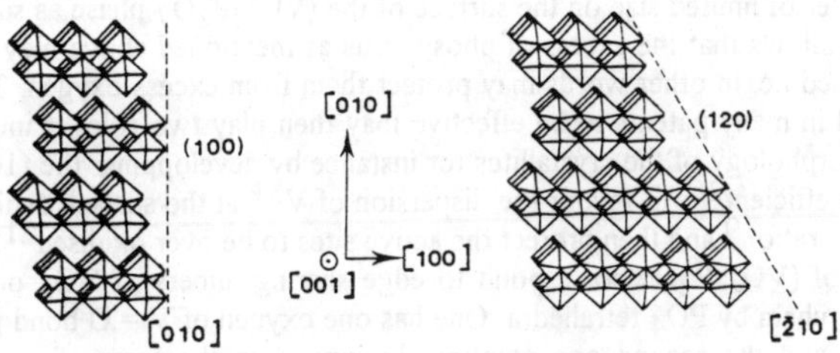


Figure 2: **Cross section view of αMoO3 (100) and (120) planes (projection of the lattice on the (001) plane) (from ref. 12).**

## 4 Vanadyl pyrophosphate[15]

Such a catalyst is well known for the oxidation of n-butane into maleic anhydride[16]. The preparation necessitates the formation of $VOPO_4$, $0.5H_2O$ as a precursor synthesized in an aqueous or better in an organic medium and its activation in a flow of 1 to 2% butane in air at the reaction temperature (ca 380°C).

Here too the preparation and the activation of the samples appeared to be particularly important to obtain a good catalyst. In all cases whatever the catalysts being good or exceptionally good the $(VO)_2P_2O_7$ phase ($V^{4+}$ cations) as a main constituent was detected by X Ray Diffraction and by in situ Laser Raman spectroscopy in addition to small amounts of some $VOPO_4$ phases ($V^{5+}$ cations) as $\alpha_{II}$, β, γ or δ[17-19]. Moreover $VOPO_4$ pure phases were observed to be more or less active and selective[20]. It turned out that the presence of some $V^{5+}$ cations on the $V^{4+}$ catalyst surface of $(VO)_2P_2O_7$ was necessary although an excess (particularly when some of $VOPO_4$ phases were present) was detrimental. Moreover the catalyst surface is richer in P than the bulk by a ratio of about 2, presumably to protect the active sites to be too much oxidized[21].

Many studies have been performed showing several features:

- the precursor has a low activity and no selectivity into maleic anhydride. Maleic anhydride formation only started when $(VO)_2P_2O_7$ was formed as shown by in situ Laser Raman[22] and Xray diffraction[23] in situ studies;

- well dispersed $V^{5+}$ ions on $(VO)_2P_2O_7$ phase were identified by NMR studies of spin echo mapping[24] and were shown to arise from $VOPO_4$ phase spreading over the $(VO)_2P_2O_7$ phase playing the rôle of a support[20];

- the well cristallized precursor as $VOHPO_4\ 0.5H_2O$ when activated was shown to dehydrate, to become amorphous and then to transform into the active phases mainly $(VO)_2P_2O_7$)[22, 23, 25];
- the (100) plane was shown to exhibit the best catalytic properties[26].

From the large amount of works[15] devoted to such a system a concerted mechanism as proposed by F. Trifiro[16] involving alkoxy intermediate was suggested as schematised in figure 3. A rattle type mechanism may occur as schematized in fig. 4 a and b. In fact we have known recently[20] that a direct process was occuring as represented in fig. 5 corresponding to scheme b in fig. 4 from a detailed kinetics study. The direct route as schematized in part a fig. 4 is occuring also but minoritarily. The mechanism 4 b does not imply that some of adsorbed intermediates do not desorb under specific conditions for instance at short contact time or in vacuum as in a TAP (temporary analysis product) reactor[27]. In such a model the 8 hydrogen atoms abstraction, 14 electrons transfer and 3 oxygen atom insertion in the butane molecule should occur at the same location i.e. without desorption and readsorption of the intermediate species as in a true rattle-type mechanism. Such a site should necessitates at least 4 vanadium atoms i.e. an oxide cluster of limited size on the surface of the $(VO)_2P_2O_7$ phase as suggested by Grasselli et al[28]. It follows that the excess of phosphorus as mentioned above may prevent such sites to be reoxidised i.e. in other words may protect them from excess oxygen. The rôle of additives as reported in many patents to be effective may then play two roles, namely they may (i) influence the morphology of the crystallites for instance by developping the (100) face assumed to be the most efficient[26] (ii) favour the dispersion of $V^{5+}$ at the surface or(iii) adapt a right surface $V^{5+}/V^{4+}$ ratio[19] and then protect the active sites to be over oxidised[21].

The (100) face of $(VO)_2P_2O_7$ correspond to edge sharing dimers of $VO_6$ octahedra bonded to the following chain by $PO_4$ tetrahedra. One has one oxygen of V = O bond pointing away from the surface and the second one pointing downward in the form of a dimer as schematized in fig. 3.

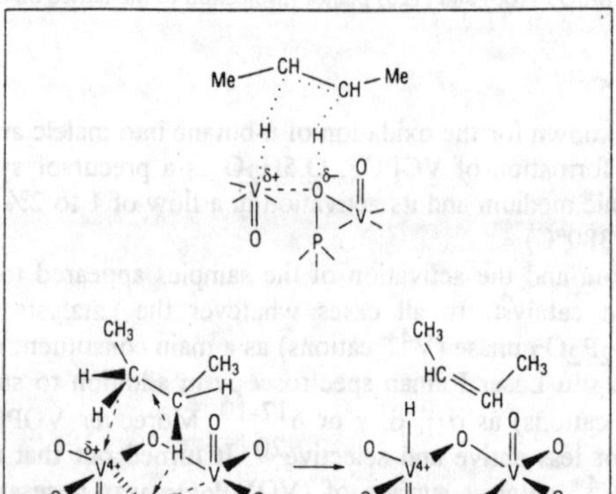


Figure 3: Proposed mechanism for the first step of the n-butane oxidation over $(VO)_2P_2O_7$ (from ref. 16) n-butane adsorbs on the free Lewis site and reacts with lattice oxygen.

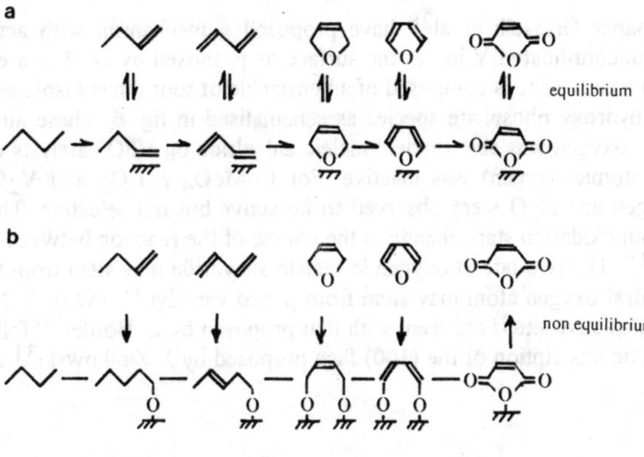


**Figure 4**: The two routes for butane oxidation in a ratle-type mechanism: (a) olefinic and (b) alkoxide.

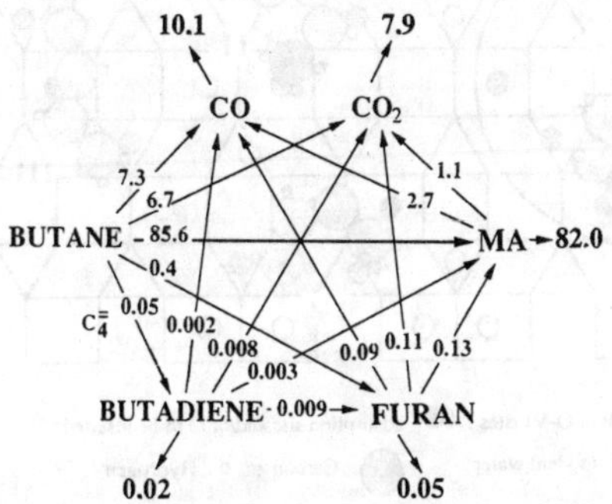


**Figure 5**: Reaction scheme for butane oxidation on VPO catalyst at 350°C deduced from kinetics study given in ref. 20.

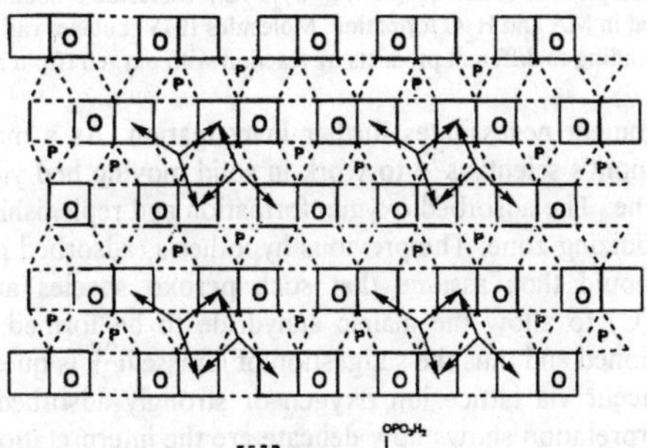

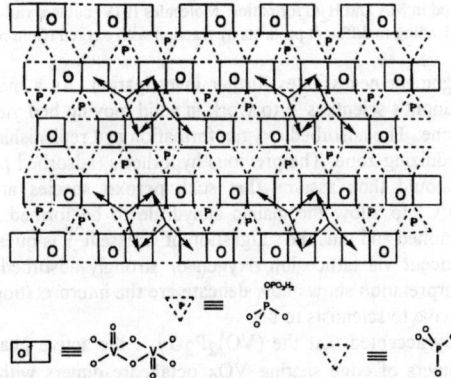

**Figure 6**: Schematic representation of the surface structure of one polytype of $(VO)_2P_2O_7$. The arrows represent the possible pathways for facile exchange of surface bound oxygen, either monoatomic or diatomic, between the active sites. The "site-isolation" due to the diffusion barrier posed by the pyrophosphate groups is clearly shown by these arrows (from ref. 28).

In a recent paper Grasselli et al[28] have proposed a mechanism with activated $O_2$ (peroxo) species on uncoordinated V ion at the surface as proposed by G. Busca et al[29] and have proposed that the active site is composed of an ensemble of four dimers isolated one from the other by excess hydroxy phosphate species as schematised in fig. 6. These authors have shown that molecular oxygen was able to yield maleic anhydride on VPO catalysts while $N_2O$ (which gives mono atomic oxygen) was inactive. For $CoMoO_4$ / $TiO_2$ and $V_2O_5$ / $SiO_2$ catalysts both dioxygen and $N_2O$ were observed to be active but not selective. The reaction mechanism involves an oxidation state change in the course of the reaction between $V^{5+}$, $V^{4+}$ or $V^{5+}$, $V^{4+}$ and $V^{3+}$. The two lateral oxygen in maleic anhydride may stem from the peroxo species while the central oxygen atom may stem from μ oxo vanadyl V = O or VOV oxygen. Such description of the active site is coherent with that proposed by E. Bordes[30] following the geometric and energetic description of the (100) face proposed by J. Ziolkowski[31] and shown in fig. 7.

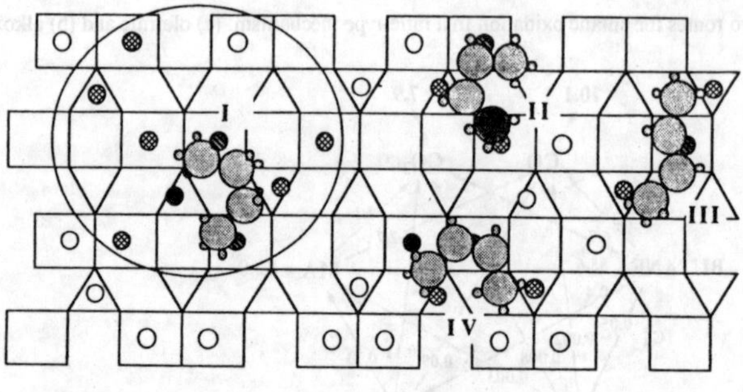


Figure 7: Model of adsorption of butane on (100) $(VO)_2P_2O_7$. Molecule I=butane; encircled area: cluster of sites involved in MA and $H_2O$ formation. Molecules II-IV=butene; various configurations of adsorption leading to different products by reaction with oxygen (from ref. 30).

Such assignment necessitates further investigation. As a matter of fact such catalyst was shown by Dupont's scientists[32] to work in solid moving bed yielding maleic anhydride in the oxygen free zone. The adsorbed oxygen formation and replenishing oxygen vacancies were occuring in the oxidizing zone. The previous hypothesis (adsorbed peroxo species rather than lattice oxygen) should then assume that such peroxo species are stable at the reaction temperature (380°C) to allow the maleic anhydride to be formed. Such a stability may be nevertheless questioned and thus the suggestion of Grasselli[28] is questionable since the oxygen insertion should occur via lattice ion oxygen or strongly adsorbed oxygen species. Such a controversary interpretation shows how delicate are the interpretations of experimental results and how modest have to scientists to be

It is usually accepted that the $(VO)_2P_2O_7$ is the active phase, particularly the (100) face exhibiting dimers of edge sharing $VO_6$ octahedre dimers with one vanadyl μ oxo site pointing outwards the other downwards the (100) surface. However it one synthesizes $(VO)_2P_2O_7$ sample at high temperature e.g. 850°C in order to get better cristallized sample the catalytic properties are poorer[33] than if the sample are prepared at lower temperature (750°C or below). It may be quite possible that the actual and optimum catalyst presents some local defects which may then isolate better the domains I was describing above. Such a conclusion is coherent with a work by Overbeek et al[23] in which VPO catalyst was deposited in a support. It was then shown that the catalytic properties were very satisfactory even if $(VO)_2P_2O_7$ could not be detected by X Ray diffraction, presumably because of its too high dispersion, i.e. its too small domains. Recall that XRD pattern can be obtained only if the lattice contains at least above 10 unit cells in continuous without defect in between. For lower ensemble sizes broadening of XRD peaks is known to occur even beyond detection for very small ensembles. The presence of defects along chains obviously gives rise to the same phenomenon. This is also a phenomenon frequently met in heterogeneous catalysis. Too well organized i.e. well cristallized samples are not as good catalysts as defectuous catalysts. Note that the characterization of such defects is particularly difficult.Such a characterization may be possible by studying the magnetic susceptibility of the $(VO)_2P_2O_7$ catalyst versus the temperature. Such a susceptibility is very sensitive to the spin pairing of the $V^{4+}$ ions in the double chains of the structure and changes with the presence of $V^{4+}$ defect densities[34].

**5.Iron phosphates and hydrophosphates[35-38]**

Such catalysts appeared to be potentially important catalysts for the oxidative dehydrogenation of isobutyric acid (IBA) to methacrylic acid. The industrial type catalyst contains iron hydroxyphosphate of uncertain nature and Cs as an additive and unfortunately necessitates a large amount of water in the feed (namely 10 to 12 mol. $H_2O$ per mol. of IBA) to remain stable with time on stream. Taking into account the phase diagramme FeO, $Fe_2O_3$, $P_2O_5$ it could be possible to select several phases which contain both $Fe^{2+}$ and $Fe^{3+}$ cations able to insure the redox mechanism necessary for the reaction to take place (see fig. 8).

| | | |
|---|---|---|
| 1 : FeO | 6 : $Fe(PO_3)_2$ | 11 : $Fe_2O_3$ |
| 2 : $Fe_3(PO_4)_2$ | 7 : $Fe_4(P_2O_7)_3$ | 12 : $Fe_3O_4$ |
| **3 : $Fe_2P_2O_7$** | **8 : $FePO_4$** | 13 : $Fe_9(PO_4)O_8$ |
| 4 : $Fe_2P_4O_{12}$ | **9 : $Fe_5P_3O_{15}$** | 14 : $Fe_5(PO_4)_3O$ |
| 5 : $P_2O_5$ | 10 : $Fe_3(PO_4)O_3$ | **15 : $Fe_2(PO_4)O$ α et β** |
| | | **16 : $Fe_7(PO_4)_6$** |

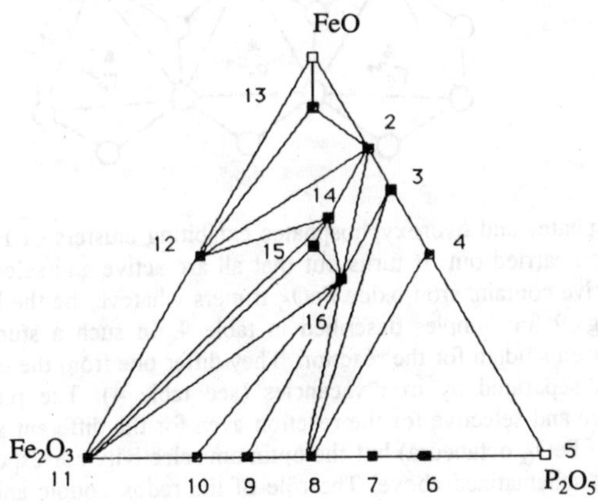


Figure 8: Phase diagramme FeO, $Fe_2O_3$, $P_2O_5$ ternary oxides

Such a reaction is important as a first step to form methyl methacrylate monomer used for altuglass or plexiglass formation by polymerisation.

The present process consists in preparing acetone cyanhydride by reacting HCN with acetone. By acting $H_2SO_4$ ($CH_3COHCN\ CH_3$) one gets $CH_2 = C(CH_3) - COONH_2, H_2SO_4$ which by action of $CH_3OH$ gives $CH_2 = C(CH_3)COOCH_3 + NH_4HSO_4$. Such a process not only uses environment unfriendly reactants as HCN and $H_2SO_4$ but also results in large amount of ammonium sulfate (2 kg per kg of acid formed)*.

Other ways to prepare methacrylic acid consits on oxidising methacrolein (synthesized by partial oxidation of isobutene for instance by hetero polyoxometallates) ($C_4$ process) or by propene carbonylation on HF medium ($C_3$ process) or by hydroformylation of ethylene giving propionaldehyde followed by action of formaldehyde giving methacrolein ($C_2$ process).

The way under study here consists in oxidative dehydrogenation of isobutyric acid into methacrylic acid by either iron phosphate catalysts using large excess of water ($12H_2O$ per IBA molecule) or heteropolyoxometallates such as $H_4PMo_4VO_{40}$ which are less stable but necessitate much less water 1 to 2 $H_2O$ molecules per IBA molecule.

In the case of iron phosphate several phases have been studied[35-38] and it was shown that the active phases consist of hydroxy phosphate with iron ions at two oxidation states +2 and +3 as summarized below.

It also appears that the best catalysts are composed of trimers (schematised below) of edge sharing $FeO_6$ octahedra separated one from the other by a vacancy and bonded to the following chain by $PO_4$ tetrahedra.

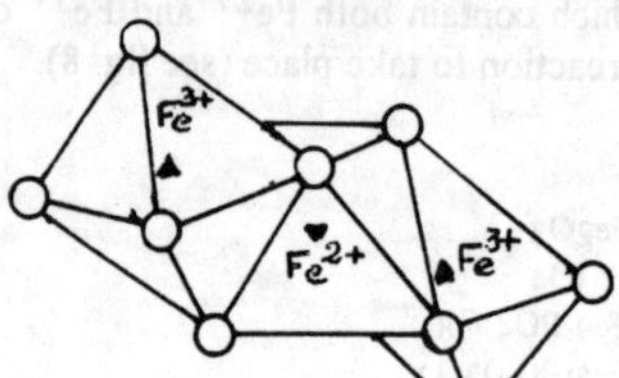


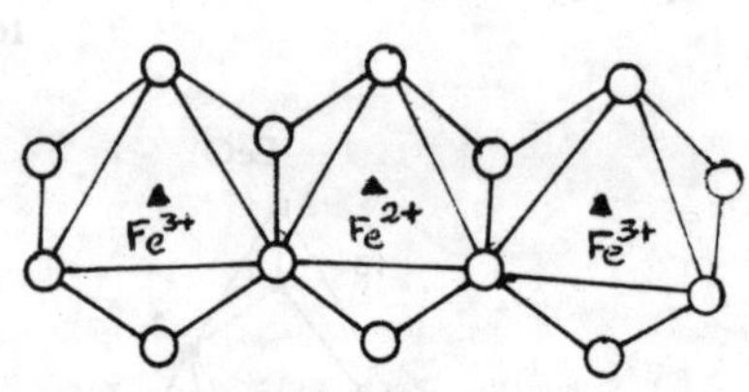


A study of many iron phosphates and hydroxyphosphates exhibiting clusters of $FeO_6$ octahedra of different sizes has been carried out. It turns out that all are active and selective but the most active and most selective contains iron oxide $FeO_6$ trimers whatever be the P/Fe atomic ratio value as shown in fig. 9 for samples described in table 4. In such a study[39] different hydroxyphosphates have been studied for the reaction. They differ one from the other by the size of the $FeO_6$ clusters separated by iron vacancies (see table 4). The results showed[39] that all samples are active and selective for the reaction even for the different sizes of the clusters (continuous chains of $FeO_6$ octahedra) but the optimum selectivity corresponds to a limited size mainly the trimers schematized above. The rôle of the redox couple and of hydroxylation is shown in fig. 10.

* see the notion of atom utilization proposed by R. A. Sheldon one of his papers in this book.

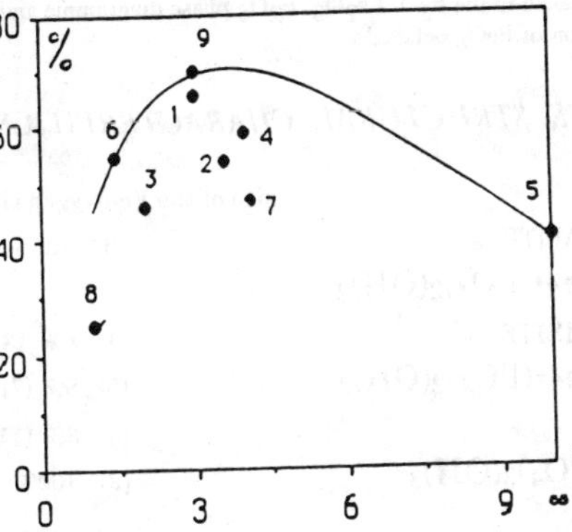


**Figure 9**: Variations of the selectivity in methacrylic acid (%) as a function of the size of the $(FeO_6)n$ clusters. The numbers refer to samples in Table 4 (from ref. 39).

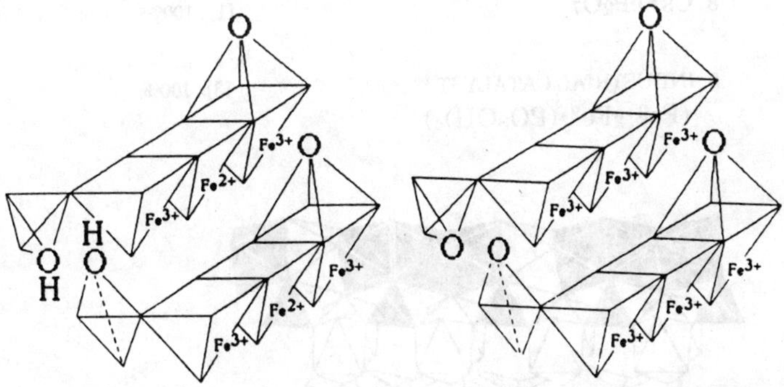


**Figure 10**: Scheme of the redox couple and hydroxylation extent suggested for the isobutyric acid oxidation as:

$$Fe_2^{3+}Fe^{2+}(P_2O_7)_2 \underset{420°C}{\overset{H_2O}{\longleftrightarrow}} Fe_2^{3+}Fe^{2+}(PO_3OH)_4 \xrightarrow{O_2} Fe_{2+x}^{3+}Fe_{1-x}^{2+}(PO_3OH)_{4-x}(PO_4)_x$$

Table 4: Some inorganic phases taken in the FeO, $Fe_2O_3$, $P_2O_5$ phase diagramme and different by the Fe/P ratio and the clusterisation of $FeO_6$ octahedra.

*CATALYSTS STRUCTURAL CHARACTERISTICS*

| | | size of the Fe-oxygen clusters [n] |
|---|---|---|
| 1 | BARBOSALITE $Fe^{3+}{}_2Fe^{2+}(PO_4)_2(OH)_2$ | [3] : 100% |
| 2 | LIPSCOMBITE $Fe^{3+}{}_2Fe^{2+}(PO_4)_2(OH)_2$ | [1] : 8% [3] : 48% [5] : 9% [7] : 24% [9] : 3% [11] : 6% |
| 3 | $Fe^{3+}{}_4(PO_4)_3(OH)_3$ | [2] : 100% |
| 4 | $Fe^{3+}{}_{3.87}Fe^{2+}{}_{0.38}(PO_4)_3(OH)_{2.62}O_{0.38}$ | [2 : 55% [5] : 28% [8] : 11% [11] : 4% |
| 5 | $\beta Fe^{3+}Fe^{2+}(PO_4)O$ | [∞] : 100% |
| 6 | OXIDIZED VIVIANITE $Fe^{3+}{}_{0.87}Fe^{2+}{}_{2.13}(PO_4)_2(OH)_{0.87},7.13H_2O$ | [1] : 33% [2] : 66% |
| 7 | ROCKBRIDGEITE $Fe^{3+}{}_4Fe^{2+}(PO_4)_3(OH)_5$ | [4] : 100% |
| 8 | $CsFeP_2O_7$ | [1] : 100% |
| 9 | INDUSTRIAL CATALYST $(Fe^{3+}{}_2Fe^{2+}(PO_3OH)_4)$ | [3] : 100% |

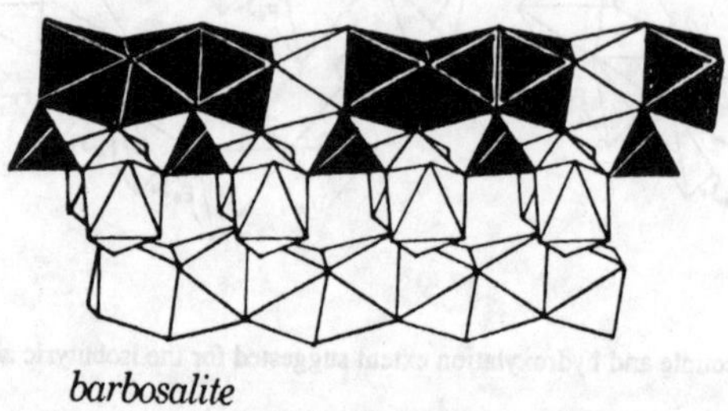

*barbosalite*

Such polymers of $FeO_6$ octahedra also exist in other inorganic compounds. For instance they exist in ilvaite ($CaFe^{3+}Fe^{2+}{}_2Si_2O_7OOH$) where silicate layers replace phosphate anions and $FeO_6$ octahedra form ribbons. It is interesting to note that such material is active and selective for the reaction although less than the previous hydroxyphosphates[40]. This is presumably due to the presence of Ca cations in the structure and of silicate counter anion whose basicity in the sense of Pearson is different from that of phosphate anion.

The main idea we can remember from this study is that inorganic clusters of iron octahedra with iron at two oxidation states are active for the reaction studied and that one has to consider the active sites as these clusters, preferentially as ensemble of two trimers but other sizes (dimers, tetramers, pentamers...) are also active and selective but to a lesser extent. The iron oxidation state is changing during the reaction in a similar way that the $V^{4+}/V^{5+}$ redox couple which was identified by spin echo mapping on VPO catalysts (vide supra §4). This again shows that metallic oxides have to be considered with a dynamical view during the oxidation reaction.

### 6 Heteropolyoxometallates[41-42]

Such materials are constituted from a central atom X attached tetrahedrally by an isopolyanion $M_xO_y^{n-}$ with:

M = Mo(VI), W(VI), V(V)

X = P(V), Si(IV), Bi(III), As(V) giving an heteropolyanion $HPA^{n-}$.

The negative charges are compensated by cations as protons, alkaline, alkaline earth or transition metal ions. The main structures are those of Keggin $(XM_{12}O_{40})^{n-}$ and Dawson $(X_2M_{18}O_{62})^{p-}$. The M cation can be substituted partly by other cations as V leading to compounds as $H_4PMO_{11}V_1O_{40}$. Moreover the protons can be exchanged by metallic cation as $VO^{2+}$, $Cu^{2+}$, $Mn^{2+}$ etc. It follows that such materials may be used in oxidation reactions as summarized below in table 5 and developed in more details in ref. 41, 42 and in one of R. Sheldon chapters in this book.

Table 5: Some oxidation reactions performed on hetero polyoxometallates

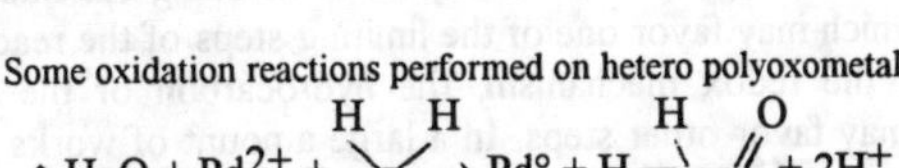

$2HPA^{n-} + Pd° \rightarrow 2HPA^{(n+1)-} + Pd^{2+}$

$2H^+ + 2HPA^{(n+1)-} + 1/2O_2 \rightarrow H_2O + 2HPA^{n-}$

This is a substitute of the Wacker process as proposed by Catalytica. The HPA playing the rôle of $Cu^{2+}$ as a reoxidizing partner. To my knowledge because of secondary transformation of the HPA this process has not yet been commercialized.

- Alkene → Ketone
- Benzene + Ethylene → Styrene
- $1/2O_2 + PhH + CH_3COOH \rightarrow Ph-O-COCH_3 + H_2O$
- Methacrolein + $O_2$ → Methacrylic acid on $P_{1.5}Mo_{12}V_{1.1}K_{1.5}Ce_{0.9}O$ catalyst at 290 - 320°C with ratios values $Ald./O_2/H_2O/N_2 = 3/7.5/31.5/58$
  Space velocity 1 000 - 2 000$h^{-1}$ P = 2 - 3 atm
  Conversion 80% Selectivity 80 - 85%
- Oxidative dehydrogenation of isobutyric acid to methacrylic acid at around 300°C
  $H_3PMo_{12}O_{40}$ Conversion 96%, Selectivity 40%
  $H_5PMo_{10}V_2O_{40}$ Conversion 94%, Selectivity 73%

### 7 Zeolitic materials

Such materials are mainly known for acid type reactions such as fluidised bed cracking catalysis (FCC), alkylation of aromatics, methanol conversion to hydrocarbon mixture in the gasoline range (MTG process) or to low molecular weight olefins as $C_2$ or $C_3$ (MTO process), etc. The size and shape of the cavities and channels influence the acidity strength, the acid sites density and thus the catalytic properties.

In the case of oxidation reactions in gaseous phase zeolitic matrices usually are not useful because the residence time of reactant molecules within the cavities is too long and favors total oxidation.

At variance ENI company has developed a titanosilicate zeolite which is non acidic, has a ZSM-5 structure (tridimensional framework with channels 0.52 x 0.56 nm in size) and is working in liquid phase with $H_2O_2$ as the oxidant reactant. It is used for phenol oxidation to anthraquinone and catechol and for many other reactions as alkanes oxidation or hydroxylation. Many other reactions were studied on such catalysts designated TS-1 (Titano silicate) and on other systems as VS-1, CrS-1 (with ZSM-5 structure or other structures and and V or Cr elements substituting Ti), CrAPO-5 (El APO or MeAPO molecular sieves developped by Union Carbide in the 1980's (E. Flanigen) with different known or unknown structure and different element or metal at framework lattice positions), etc. More details are given in the lecture #3 by R.A. Sheldon in this book.

**8 Synergy effect in multi component catalysts[45-50]**

It is known that the industrial catalyst formulations contain usually a large number of elements and are composed of mixtures of several phases: they are designated as multi component catalysts. For instance, for propene oxidation to acrolein the catalysts are based on mixed phases of Fe(Co, Ni)$MoO_4$ and $Bi_2(MoO_4)_3$ and several additives as alkalines as K and elements as P. The chemical formulation is then very complex and the rôle of each constituent is not well established.

It is known also that for mixed oxides the mixture of them, even mechanical mixtures, is much more active and selective than each component itself[46]. This phenomenon is designated by the term "synergy" which usually is hiddening our ignorance. Such a synergy effect could be related to the intimate contact between two phases including epitaxial fitting between two crystalline structures which may favor one of the limiting steps of the reaction[45]. One phase may favor for instance the redox mechanism, the hydrocarbon or the oxygen activations and the other phase (s) may favor other steps. In a large amount of works dealing with mixtures of oxides B. Delmon et al[45] have introduced the concept of remote control effect of oxygen. This corresponds to a spill over phenomenon of oxygen similar to that known for metals (H spill over). The oxygen is activated on one oxide and by remote effect affects the insertion of oxygen which occurs on the activated hydrocarbon adsorbed on the other phase. B. Delmon et al could even establish a relative scale of oxygen acceptor or donor properties for many metallic oxide and then predict the catalytic properties of any mixed oxide[45].

Another possibility was proposed long time ago by Matsuura, Schuit and other scientists considering the cherry-like structure of a mixture of oxides. In such a model the active phase as bismuth molybdate for propene oxidation to acrolein lies on the surface of the other oxide as in a cherry. Such a model was reconsidered recently by O. Legendre et al[46], J.C. Vedrine et al[47, 48]. It was shown for instance that $\gamma Bi_2MoO_6$ phase catalyst which is not as active and selective for the propene partial oxidation reaction than $\alpha Bi_2Mo_3O_{12}$ phase catalyse contains in fact excess of bismuth oxide which is not selective on the surface (XPS study)[50]. By mixing with $MoO_3$ the catalyst performances were very much exhanced. This was assigned to the reaction of $MoO_3$ with excess surface bismuth leading to active bismuth molybdate.

A more striking example corresponds to a mechanical mixture of a solid solution Fe(Co)$MoO_4$ and $Bi_2(MoO_4)_3$[47, 48]. For Fe(Co)$MoO_4$ solid solution Mössbauer, Xray diffraction,DTA and electrical conductivity studies showed[51-53] that when Fe was introduced under reducing conditions in $Co^{II}MoO_4$, $Fe^{II}$ ions were dissolved but also some $Fe^{III}$ ions. The excess positive charge resulting from $Fe^{III}$ ion dissolution is compensated by free electron resulting in a large increase in electrical conductivity (multiplied by 3 orders of magnitude). Moreover it was shown that the temperature of the $\alpha \rightarrow \beta$ phase transition of $CoMoO_4$ varied with Fe (total) and $Fe^{III}$ contents. This was shown to be important in catalysis, the $\beta$ phase being more efficient and the reaction temperature occuring in the same temperature domain was usually increased with catalyst ageing. One can thus realize how $Fe^{III}$ content could be important for catalytic behaviour and catalyst life time.

When a mechanical mixture of large crystallites of $Bi_2(MoO_4)_3$ phase and small crystallites of Fe(Co)$MoO_4$ was prepared and placed in a reactor for propene oxidation to acrolein the following properties were observed[47, 48]:

- the activity was increased with time on stream and at steady state reached a multiplication factor of 200 to 300 with respect to that of the starting oxides. This is a huge effect.

- The selectivity into acrolein increased also from 30% for $Fe(Co)MoO_4$ to more than 95% as shown in fig. 11. A detailed study using XPS and EDX-STEM analysis clearly showed that:

(i) before catalytic testing one has a mechanical mixture i.e the presence of grains. closeby but with no overlapping of those of the solid solution $Fe(Co)MoO_4$ and of $Bi_2(MoO_4)_3$. The EDX elemental analysis was found to correspond to the chemical analysis;

(ii) after catalytic testing (steady state) one still detects large crystallites of $Bi_2(MoO_4)_3$ (several µm in size) but also $Bi_2(MoO_4)_3$ spread over the small crystallites of $Fe(Co)MoO_4$ (50-150 nm in size). The latter ones are obviously the actors of the catalytic reaction since because of their small size the surface accessible to the reactants is much larger than for large $Bi_2(MoO_4)_3$ crystallites. The active catalyst is then composed of the active phase $Bi_2(MoO_4)_3$ spread on the surface of the high electrical conductor solid solution $Fe(Co)MoO_4$ crystallites.A question still remains. Why does $Bi_2(MoO_4)_3$ supported on $Fe(Co)MoO_4$ so active? Is it a question of better dispersion of the active phase as suggested by our experiments? Is it a change is the active site on $Bi_2(MoO_4)_3$ phase itself due an epitaxial contact with the support[54]?

An important observation is that during the catalytic reaction in the presence of reactants and products the bismuth molybdate has spread over the other oxide in a process of "wetting".

Such a wetting phenomenon was proposed by J. Haber et al[55] and was shown previously to also depend on the more or less reducing atmosphere over the catalyst, on the presence of water in the feed and to be more or less reversible.

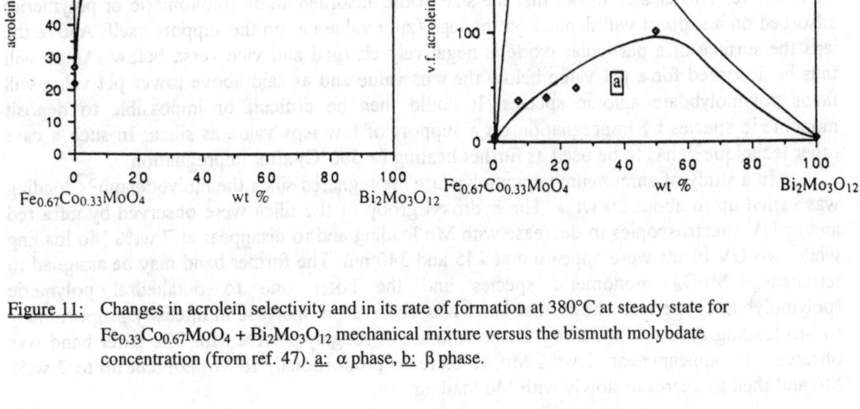

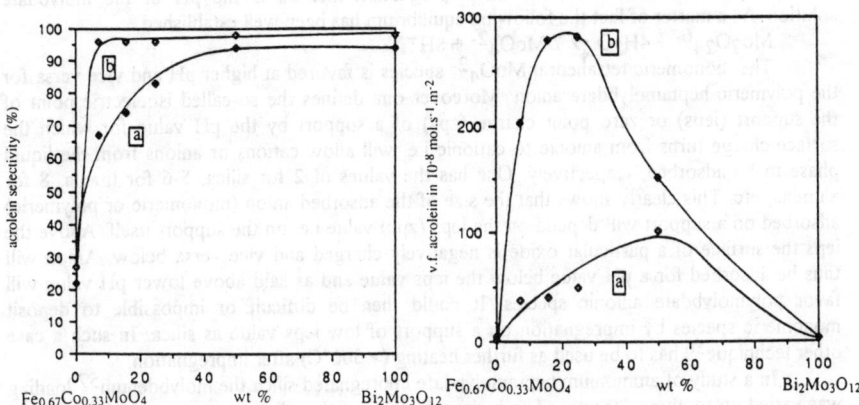


Figure 11: Changes in acrolein selectivity and in its rate of formation at 380°C at steady state for $Fe_{0.33}Co_{0.67}MoO_4 + Bi_2Mo_3O_{12}$ mechanical mixture versus the bismuth molybdate concentration (from ref. 47). a: α phase, b: β phase.

## 9 Supported oxide catalysts

Supports are very often used in catalysis, particularly for noble metals to spend less metal per g. catalyst due to their elevated price. However several properties of a support may have also to be account for, namely:

(i) Dispersion of the active phase in order to increase the surface to volume ratio since heterogeneous catalysis is occuring at the solid surface.

(ii) Heat transfer: this is a particularly important aspect in oxidation reactions because of the high exothermicity involved. This holds particularly true at industrial scale since the hot spot problem is very crucial and should be monitored with precision within a few degrees to avoid local overactivity and subsequently overheating and presumably irreversible phase transformation of the catalyst.

(iii) Attrition: this aspect is of main importance for fluidized or solid transported beds. One very usually uses silica as a binder or as a coating to limitate attrition.

(iv) Formation of new catalytic sites: this point will be emphasised below in some examples. One will see that well dispersed species of limited size could be formed and exhibit peculiar catalytic properties.

(v) Modification of the active phase properties due to its chemical interaction with the support, including epitaxial induced modifications. In some cases the chemical effect of the support will be determining for catalytic properties. An example of several oxides ($V_2O_5$, $MoO_3$, $Re_2O_7$, $Cr_2O_3$...) deposited on several oxide supports ($SiO_2$, $TiO_2$, $Al_2O_3$) will be developped below.

### *9.1 Molybdenum oxide supported on silica*

Different procedures may be used as impregnation of silica with a molybdate salt or grafting molybdenum chloride or molybdenum based organo metallic compounds as Mo carbonyls on the hydroxyl groups of silica or solid-solid between $MoO_3$ and $SiO_2$ at temperatures near or above 500°C.

A parameter important for the impregnation method is the pH of the molybdate solution. As a matter of fact the following equilibrium has been well established.

$$Mo_7O_{24}^{6-} + 4H_2O \rightleftarrows 7\,MoO_4^{2-} + 8H^+.$$

The monomeric tetrahedral $MoO_4^{2-}$ species is favored at higher pH and vice versa for the polymeric heptamolybdate anion. Moreover one defines the so-called isoelectric point of the support (ieps) or zero point charge (zpc) of a support by the pH value for which the surface charge turns from anionic to cationic i.e. will allow cations or anions from the liquid phase to be adsorbed, respectively. One has the values of 2 for silica, 5-6 for titania, 8 for alumina, etc. This clearly shows that the size of the adsorbed anion (monomeric or polymeric) adsorbed on a support will depend on the ieps (zpc) value i.e. on the support itself. Above the ieps the surface of a particular oxide is negatively charged and vice versa below. Anion will thus be adsorbed for a pH value below the ieps value and as said above lower pH value will favor polymolybdate anionic species. It could then be difficult or impossible to deposit monomeric species by impregnation on a support of low ieps value as silica. In such a case other technique[56] has to be used as further heating (> 500°C) after impregnation.

In a study of ammonium heptamolybdate impregnated silica the molybdenum[57] loading was varied up to about 20 wt%. The hydroxyl group of the silica were observed by infra red and/or UV spectroscopies to decrease with Mo loading and to disappear at 7 wt% Mo loading while two UV bands were appearing at 245 and 340 nm. The former band may be assigned to tetrahedral $MoO_4$ monomeric species and the latter one to octahedral polymeric (polymolybdate) species. The former band was observed to increase in intensity proportionaly to Mo loading at low Mo loading and to saturate at roughly 3 wt% Mo. The latter band was observed to appear near 2 wt% Mo, to increase proportionaly to Mo content up to 7 wt% Mo and then to decrease slowly with Mo loading.

The former species was assigned to the following equation:

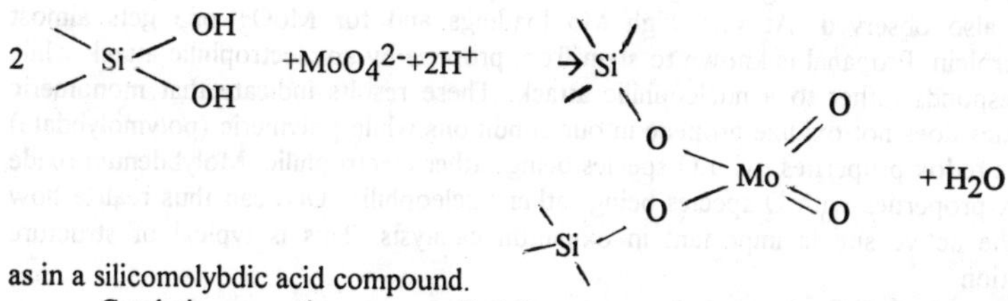

as in a silicomolybdic acid compound.

Catalytic properties were studied for two reactions namely isopropanol conversion and propene partial oxidation. The first reaction is a test reaction which allows to characterize acidic, basic or redox properties of a catalyst. One gets dehydration to propene or di-isopropylether for acid catalyst, acetone for basic catalyst in absence of air and acetone and water for redox type catalyst in presence of air. The experimental results at 100°C clearly show that at low Mo loadings acidic features are favored while redox features are favored at higher loadings. This indicates that monomeric $MoO_4$ species are acidic (presumably as in silicomolybdic acid) while polymeric species exhibit redox properties (see fig. 12).

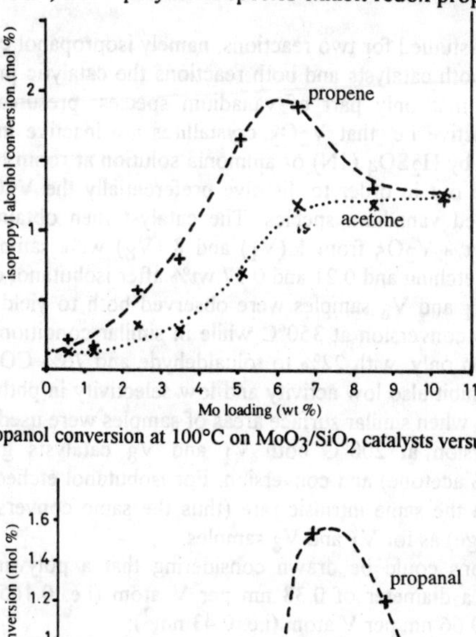


Figure 12: Isopropanol conversion at 100°C on $MoO_3/SiO_2$ catalysts versus Mo loading (from ref. 57) in air.

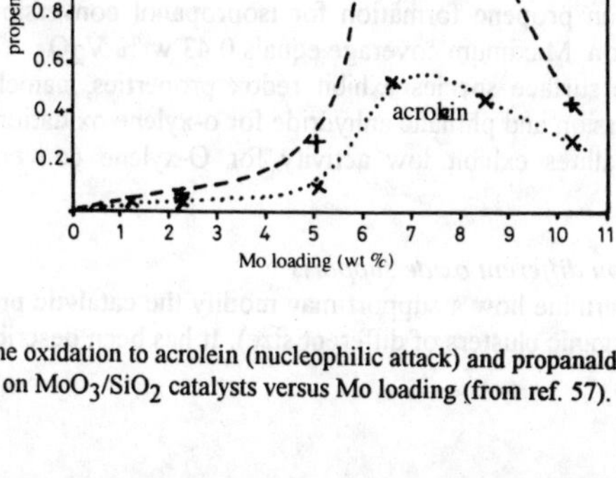


Figure 13: Propene oxidation to acrolein (nucleophilic attack) and propanaldehyde (electrophilic attack) at 400°C on $MoO_3/SiO_2$ catalysts versus Mo loading (from ref. 57).

The second reaction studied is propene oxidation at 380°C (see fig. 13). Weak activity was observed for low Mo loading. At higher Mo loading propanal was the major product while acrolein was also observed. At very high Mo loadings and for $MoO_3$ one gets almost exclusively acrolein. Propanal is known to stem from propene by an electrophilic attack while acrolein corresponds rather to a nucleophilic attack. These results indicate that monomeric $MoO_4^{2-}$ species does not oxidize propene in our conditions while polymeric (polymolybdate) species exhibit redox properties, with O species being rather electrophilic. Molybdenum oxide exhibits redox properties with O species being rather nucleophilic. One can thus realise how the size of the active site is important in oxidation catalysis. This is typical of structure sensitive reaction.

*9.2 Vanadium oxide on $TiO_2$ support*

Such a catalyst is well known for several reactions, such as O-xylene oxidation to phthalic anhydride and selective catalytic reduction (SCR) of NO by ammonia. The anatase form appears to be better than the rutile form of $TiO_2$. Such a catalyst with 1 and 8 wt% $V_2O_5$ / anatase was prepared by Rhône-Poulenc ($S \approx 10\ m^2\ g^{-1}$) for an exercise of characterization by 25 different european laboratories. All results are assembled in a special issue of Catalysis today published in May 1994 vol. 20 n°1[58]. Surface vanadium species were observed to exist in three different forms: monomeric $VO_4^{3-}$ species, polymeric vanadate species and $V_2O_5$ crystallites.

Catalytic properties were studied for two reactions, namely isopropanol conversion and O-xylene oxidation[59, 60]. For both catalysts and both reactions the catalytic properties were found to be similar indicating that only part of vanadium species, presumably dispersed vanadium oxide species were active i.e. that $V_2O_5$ crystallites are inactive in our reaction conditions. Chemical treatments by $H_2SO_4$ (1N) or ammonia solution at room temperature or isobutanol at 80°C were carried out in order to dissolve preferentially the $V_2O_5$ crystallites without dissolving well dispersed vanadium species. The catalyst then obtained had much lower V content namely: 0.12wt% $V_2O_5$ from 1 ($V_1$) and 8 ($V_8$) wt% samples after $NH_3$ etching, 0.24 wt% after $H_2SO_4$ etching and 0.21 and 0.57 wt% after isobutanol etching for $V_1$ and $V_8$ samples respectively. $V_1$ and $V_8$ samples were observed both to yield 74% phthalic anhydride, 23.5% COx, at 96% conversion at 350°C while in similar conditions $NH_3$ etched samples exhibited 6% conversion only, with 22% to tolualdehyde and 76% COx selectivities. Bulk $V_2O_5$ was observed to exhibit also low activity and low selectivity in phthalic anhydride and high selectivity in COx even when similar surface areas of samples were used.

For isopropanol conversion at 200°C both $V_1$ and $V_8$ catalysts gave the same selectivities (50% propene, 50% acetone) and conversion. For isobutanol etched samples only propene was observed but with the same intrinsic rate (thus the same conversion also since surface area values did not change) as for $V_1$ and $V_8$ samples.

The following conclusions could be drawn considering that a polyvanadate species occupies a ca circular zone of a diameter of 0.38 nm per V atom (i.e. 0.165 $nm^2$) and an isolated monovanadate species 0.66 nm per V atom (i.e. 0.43 $nm^2$):

(i) Monomeric $VO_4^{3+-}$ species exhibit acidic character with OH groups (Brönsted acidity) and result in propene formation for isopropanol conversion and in total oxidation for O-xylene oxidation. Maximum coverage equals 0.43 wt% $V_2O_5$.

(ii) Polyvanadate surface species exhibit redox properties, namely give rise to acetone for isopropanol conversion and phthalic anhydride for o-xylene oxidation.

(iii) $V_2O_5$ crystallites exhibit low activity for O-xylene conversion and high selectivity in total oxidation.

*9.3 Various oxides deposited on different oxide supports*

The idea is here to determine how a support may modify the catalytic properties of the different oxide species (as inorganic clusters of different size). It has been described above how the size of the deposited oxide species (monomeric, polymeric, bulk-type) results in different catalytic properties.

Vanadium oxide has been deposited on several supports as silica, alumina, titania but also zirconia, niobia, zirconium hydroxyphosphate, etc. Its reductibility as described in the general features (§1) was studied by reduction by hydrogen at 400°C by J. Haber et al[61]. It was observed that V on $TiO_2$ and $\gamma.Al_2O_3$ was reduced rather fast and one reached an O/N atomic ratio near to 1 for $TiO_2$ and 0.65 for $Al_2O_3$. At variance for silica the reduction was much slower and an O/V ratio of 0.57 was obtained. Such differences were interpreted by the authors as due to different species on the surface: mainly monomeric $VO_4^{3-}$ species for $TiO_2$, dimeric species $V_2O_7^{4-}$ for $\gamma.Al_2O_3$ and $V_2O_5$ crystallites for silica. Even if such a tendancy is correct, I am of the opinion that the reality is much more complex.

The size of the polyvanadate species is known in solution particularly from UV spectra data. For instance, it corresponds to $V_3O_9^{3-}$, $V_4O_{12}^{4-}$ with tetrahedrally coordinated V at pH = 7 and 4.5 respectively and $V_{10}O_{28}^{6-}$ or $V_{10}O_{28}H^{5-}$ with octahedrally coordinated V at pH = 2.5. Deposition of such polyvanadates of different sizes was performed carefully on $\gamma.Al_2O_3$ support[62] and their initial structure despicted above were shown, to remain stable even after calcination in flowing air at 500°C and to be only partly modified after catalytic reaction of oxidative dehydrogenation of propane in the 350 to 450°C range. Moreover in such cases the selectivity towards propene was observed to be the same at the same conversion level, indicating that in this size domain of polyvanadate species the catalytic selectivity was not changed, only the activity was observed to increase with vanadium loading.

Ethane oxidative dehydrogenation at 500/550°C has been studied for $V_2O_5/SiO_2$[63] and $V_2O_5/Al_2O_3$[64] catalyst as a function of V content. It was observed that the activity was much higher for alumina support with respect to silica support. Moreover it remained weak at low V content and increased sharply and linearly with V content for a loading between 2 and 10 wt% $V_2O_5$ and more slowly above 10%. It is then suggested that polyvanadate species may be as dimeric species as suggested by J. Haber et al[61], are the active sites. Note also that the selectivity into ethylene is higher at lower V content for the same ethane conversion level (fig. 14).

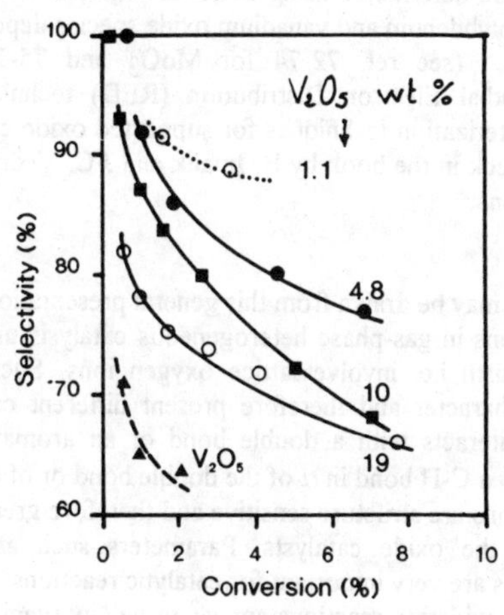


Figure 14: Variation of ethene selectivity in ethane conversion at 550°C versus conversion at different V loading

This indicates that smaller clusters are more selective in ethane oxidative dehydrogenation. This feature has been observed to hold true for other supported oxides as $Cr_2O_3$ on different supports as $SiO_2$, $Al_2O_3$, zirconium hydroxyphosphate[65, 66], etc. It is a new example of structure sensitivity showing how the size of a deposited "inorganic cluster" influences the catalytic selectivity. The reducibility and reoxidability of $V_2O_5/SiO_2$ and $V_2O_5/Al_2O_3$ catalysts are studied in ref. 63 and 64 using microcalorimetry technique.

In a more general analysis, I. Wachs et al[67-69] have studied the conversion of methanol at 230°C. Such a reaction as proposed by J.M. Tatibouet et al[70, 71] from Paris VI university gives formaldehyde on redox sites, dimethyl ether on acid sites and CO, $CO_2$ on basic sites. This is a test reaction to be compared to isopropanol conversion described above. We have reported in table 6 the TON values obtained by the authors[68] for different supported oxides with 1 wt% loading of the active phase corresponding to less than the theoretical monolayer coverage.

It clearly appears that the properties of supported oxides depend on both the nature of the oxide support (a ratio of $10^3$ in TON values were observed) and the nature of the surface metal oxide (a ratio of 10 in TON values was observed) but remains about constant with the metal oxide loading at least at relatively low loading (< 10 wt%)[68].

Table 6: Turn over number values in $(seconde)^{-1}$ for methanol reaction at 230°C on 1wt % metallic oxide deposited on several supports (from ref. 67).

| Oxide support | Supported oxide | | | |
|---|---|---|---|---|
| | $V_2O_5$ | $MoO_3$ | $CrO_3$ | $Re_2O_7$ |
| $SiO_2$ | 2 | 39 | 160 | 20 |
| $Al_2O_3$ | 20 | - | 2 | - |
| $Nb_2O_5$ | 700 | 32 | 58 | 12 |
| $TiO_2$ | 1800 | 310 | 300 | 1200 |
| $ZrO_2$ | 2300 | 92 | 1300 | 170 |

It is worthwhile noting that the environmental symmetry of such inorganic cluster deposited on oxide support may be determined using EXAFS, XANES and RED techniques. This has been done already for molybdenum and vanadium oxide species deposited on supports as $SiO_2$,$TiO_2$ and $Al_2O_3$ (see ref. 72-74 for $MoO_3$ and 75-77 for $V_2O_5$ by XANES/EXAFS and 78 by Radial Electron Distribution (RED) technique). The reader interested in the use these characterization techniques for supported oxide catalysts is advised to read the chapter by B. Moraweck in the book by B. Imalik and J.C. Vedrine in its French[79] (1988) or English[80] (1994) versions.

**10 Conclusions**

Some general conclusions may be drawn from this general presentation:

(i) Oxidation reactions in gas-phase heterogeneous catalysis usually proceed via Mars and van Krevelen mechanism i.e. involve lattice oxygen ions. Such ions exhibit an electrophilic or a nucleophilic character and therefore present different catalytic properties since the electrophilic oxygen interacts with a double bond or an aromatic ring while the nucleophilic oxygen interacts with a C-H bond in $\alpha$ of the double bond or of an aromatic ring.

(ii) Oxidation reactions are structure sensitive and therefore greatly depend on the local and surface structure of the oxide catalysts. Parameters such as reducibility and reoxidability features of the oxides are very important for catalytic reactions.

(iii) Active sites for oxidation reactions appear to be "inorganic ensembles"[81] of metallic oxide atoms whose size greatly influences the catalytic properties. In some examples the number of atoms constituting the active sites could be established. For instance double trimers of edge sharing $FeO_6$ octahedra are particularly active and selective for oxidative dehydrogenation of isobutyric acid to methacrylic acid on iron hydroxyphosphates; ensembles of four dimers of $VO_6$ octahedra are suggested to be the active sites for butane oxidation to maleic anhydride in vanadium pyrophosphate catalysts. Usually monomeric species as $MoO_4^{2-}$ or $VO_4^{3-}$ exhibit acidic feature and no partial oxidation properties. At variance low size polymeric species exhibit better selectivity for many partial oxidation reactions than large size species or bulk-type oxide.

(iv) Oxidation catalysts have to be considered with a dynamical view under reaction conditions. This is related to the Mars and van Krevelen mechanism and also to the mobility of the oxide lattice. This latter phenomenon results from the wetting effect particularly for multicomposent and supported oxide catalysts. It follows that for many catalysts a certain time on stream is necessary before the catalyst reaches its steady state. It is frequent that in an industrial plant a steady state is reached only after one or two hundreds hours, the catalysts lasting several years before being replaced. It is then proposed than the active catalyst for multicomponent materials is composed of the active phase spread over the surface of the other (s) oxide in a sherry-like morphology as suggested long time ago by scientists as Schuit et al from the Netherlands or Matsuura from Japan. For more simple catalysts as doped vanadyl pyrophosphates used for butane oxidation to maleic anhydride the right size of the active sites (e.g. tetramers of vanadyl dimers) is monitored by the reactants in catalytic reaction conditions leading to the right $V^{5+}/V^{4+}$ ratio, by the preparation procedure to change the material morphology (the (100) face of $(VO)_2P_2O_7$ being developped), and by the adequate addition of additive elements which regulate the site size and $V^{5+}/V^{4+}$ ion ratio. The view of an oxidation catalyst as dynamical under catalytic reaction conditions is essential for our understanding of its functioning.

**References**


1. S. T. Oyama, A. N. Desikan and J. W. Hightower, in *Catalytic Selective Oxidation*, ed. S. T. Oyama and J. W. Hightower ACS Sympos. ser. **523**, Washington (1993) 1.
2. See for instance: S. T. Oyama, A. N. Desikan and W. Zhang, Ibid p. 16.
3. H. H. Kung, P. Michalakos, L. Owens, M. Kung, P. Andersen, O. Owen and I. Jahan, Ibid p. 389.
4. A. Guerrero Ruiz, I. Rodriguez Ramos, J. L. G. Fierro, V. Soenen, J. M. Herrmann and J. C. Volta, in *New Developments in Selective Oxidation by Heterogeneous Catalysis*, ed. P. Ruiz and B. Delmon, Stud. in Surf. Sci. and Catal. ser. **72,** Elsevier, Amsterdam (1992) 203.
5. S. T. Oyama, G. T. Went, K. B. Lewis, A. T. Bell and G. A. Somorjai, J. Phys. Chem. **93** (1989) 6786.
6. J. C. Volta, W. Desquesnes, B. Moraweck and G. Coudurier, React. Kinet. Catal. Lett. **12** (1979) 241.
7. J. C. Volta and B. Moraweck, J. Chem. Soc. Chem. Commun. (1980) 338.
8. J. C. Volta and J. L. Portefaix, Appl. Catal. **18** (1985) 1.
9 J. M. Tatibouet and J. E. Germain, J. Catal. **72** (1981) 365.
10. J. M. Tatibouet, J. E. Germain and J. C. Volta, J. Catal. **83** (1983) 24.
11. See paragraph 6 in the chapter of J. Haber in this book.
12. M. Abon, B. Mingot, J. Massardier and J. C. Volta, in *Structure Activity and selectivity. Relationships in Heterogeneous Catalysis*, ed. R. K. Grasselli and A. W. Sleight, Stud. in Surf. Sci. and Catal. ser. **67**, Elsevier, Amsterdam (1991) 67.
13. K. M. Schulz and D. F. Cox, in *Catalytic Selective oxidation*, ed. S. T. Oyama and J. W. Hightower, ACS Sympos. ser. **523,** Washington (1993) 122.
14. P. Courtine in *Solid State Chemistry in Catalysis*, ed. R. K. Grasselli and J. F. Brazdil, ACS Symp. ser. **279,** Washington (1985) 37.

15. G. Centi, Catalysis Today **16** n°1 (1993).
16. F. Trifiro and F. Cavani, Chem. Techn.(april 1994) p. 18.
17. N. Harrouch Batis, N. Batis, A. Ghorbel, J. C. Vedrine and J. C. Volta, J. Catal. **128** (1991) 248.
18. J. C. Volta, K. Bere, Y. Zhang-Lin and R. Olier, in *Catalytic Selective Oxidation*, ed. S. T. Oyama and J. W. Hightower, ACS symp. ser. **523**, Washington (1993) 216.
19. Y. Zhang-Lin, M. Forissier, J. C. Vedrine and J. C. Volta, J. Catal. **145** (1994) 267.
20. Y. Zhang-Lin, M. Forissier, R. P. Sneeden, J. C. Vedrine and J. C. Volta, J. Catal. **145** (1994) 256.
21. J. L. Callahan and R. K. Grasselli, AICh E **9** (1963) 755.
22. G. J. Hutchings, A. Desmartin-Chomel, R. Olier and J. C. Volta, Nature **368** (1994) 41.
23. R. A. Overbeek, M. Vershuijs-Helder, P. A. Warringa, E. J. Bosma and J. W. Geus, in *New Developments in Selective Oxidation II*, ed. V. Cortés Corberan and S. Vic Bellon, Stud. in Surf. Sci. and Catal. ser. **82**, Elsevier, Amsterdam (1994) 183.
24. M. T. Sananes, A. Tuel and J. C. Volta, J. Catal. **145** (1994) 251.
25. M. R. Thompson, A. C. Hess, J. B. Nicholas, J. C. White, J. Anchell and J. R. Ebner, in *New Developments in Selective Oxidation II*, ed. V. Cortés Corberan and S. Vic Bellon, Stud. in Surf. Sci. and Catal. ser. **82**, Elsevier, Amsterdam (1994) 167.
26. E. Bordes, Catalysis Today **16** (1993) 27.
27. J. T. Gleaves, J. R. Ebner and T. C. Knechler, Catal, Rev. -Sci. Eng. **30** (1988) 49.
28. P. A. Agaskar, L. De Caul and R. K. Grasselli, Catal. Lett. **23** (1994) 339.
29. G. Centi, in *Elementary Reaction Steps in Heterogeneous Catalysis*, ed. R. W. Joyner and R. A. van Santen, Kluwer Acad. Publ., Amsterdam (1993) 93.
30. E. Bordes, in *Structure Activity and Selectivity. Relationships in Heterogeneous Catalysis*, ed. R. K. Grasselli and A. W. Sleight, Stud. in Surf. Sci. and Catal. ser. **67**, Elsevier, Amsterdam, (1991) 21.
31. J. Ziolkowski, J. Catal. **100** (1986) 45.
32. R. M. Contractor, D. I. Garnett, H. S. Horowitz, H. E. Bergna, G. S. Patience, J. T. Schwartz and G. M. Sisler, in *New Developments in Selective Oxidation II*, ed. V. Cortés Corberan and S. V. Bellon, Stud. in Surf. Sci. and Catal. ser. **82** (1994) 233.
33. G. Emig, K. Wihlein and C. J. Häcker in *New Developments in Selective Oxidation II*, ed. V. Cortés Corberan and S. V. Bellon, Stud. in Surf. Sci. and Catal. ser. **82**, Elsevier, Amsterdam (1994) 243.
34. D. C. Johnston and J. W. Johnson, J. Chem. Soc., Chem. commun. (1985) 1720.
35. J. M. Millet, J. C. Vedrine and G. Hecquet in *New Developments in Selective Oxidation*, ed. G. Centi and F. Trifiro, Stud. in Surf. Sci. and Catal. ser.**5**, Elsevier, Amsterdam (1990) 833.
36. C. Virely, M. Forissier, J. M. Millet and J. C. Vedrine, J. Molec. Catal. **71** (1992) 199.
37. J. M. Millet and J. C. Vedrine, Appl. Catal. **76** (1991) 209.
38. J. M. Millet, Ph D University of Lyon n° **259-90** (1990).
39. D. Rouzies, Ph D University of Lyon n° **48-92** (1992).
40. P. Bonnet, J. M. Millet, J. C. Vedrine and G. Hecquet, in *New Developments in Selective Oxidation II*, ed. V. Cortés Corberan and S. V. Bellon, Stud. in Surf. Sci. and Catal. ser. **82,** Elsevier, Amsterdam (1994) 829.
41. M. Misono, Catal. Rev. -Sci. Eng. **29** (1987) 269.
42. R. J. Jansen, H. M. van Veldhuizen, M. A. Schwegler and H. van Bekkum, Rec. Trav. Chim. Pays-Bas **113** (1994) 115.
43. C. Marchal, A. Davidson, R. Thouvenot and G. Hervé, J. Chem. Soc., Faraday Trans. **89** (1993) 3301.
44. A. Aboukaïs, D. Ghoussoub, E. Blouet-Crusson, M. Rigole and M. Guelton, Appl. Catal. **A111** (1994) 109.

45. L. T. Weng and B. Delmon, Appl. Catal. A **81** (1992) 141 and references therein.
46. O. Legendre, Ph. Jaeger and J. P. Brunelle, in *New Developments in Selective Oxidation by Heterogeneous Catalysis*, ed. P. Ruiz and B. Delmon, Stud. in Surf. Sci. and Catal. **72**, Elsevier, Amsterdam (1992) 387.
47. J. M. Millet, H. Ponceblanc, G. Coudurier, J. M. Hermann and J. C. Vedrine, J. Catal. **142** (1993) 381.
48. H. Ponceblanc, J. M. Millet, G. Coudurier and J. C. Vedrine, in *Catalytic Selective Oxidation*, ed. S. T. Oyama and J. W. Hightower, ACS Symp. ser. **523**, Washington (1993) 262.
49. D. Carson, G. Forissier, G. Coudurier and J. C. Vedrine, J. Chem. Soc., Faraday Trans. 1 **79** (1983) 1617.
50. M. El Jamal, M. Forissier, G. Coudurier and J. C. Vedrine, in *Prodeed 9th Inter,. Cong. on Catal.*, ed. M. J. Phillipps and M. Ternan, Chem. Institute of Canada, Ottawa (1988) 1617.
51. H. Ponceblanc, J. M. Millet, G. Coudurier, O. Legendre and J. C. Vedrine, J. Phys. Chem. **96** (1992) 9462.
52. H. Ponceblanc, J. M. Millet, G. Thomas, J. M. Herrmann and J. C. Vedrine, J. Phys. Chem. **96** (1992) 9466.
53. B. Benaïchouba, P. Bussiere and J. C. Vedrine, Hyperfine Inter. **69** (1991) 739.
54. P. Courtine and A. Vejux, C. R. Acad. Sci. Paris **286** (1978) 135.
55. J. Haber, E. Mielczarska and W. Turek, Z. Phys. Chem. Neue Folge **144** (1985) 69.
56. M. de Boer, Ph D Thesis, University of Utrecht (1992).
57. T. C. Liu, M. Forissier, G. Coudurier and J. C. Vedrine, J. Chem. Soc., Faraday Trans. I **85** (1989) 1607.
58. *Eurocat Oxide*, ed. J. C. Vedrine, Catal. Today **20** (1) (1994).
59. B. Grzybowska-Swierkosz, G. Coudurier, J. C. Vedrine and I. Gressel, Ibid p. 165.
60. G. Golinelli, F. Trifiro, M. Baerns, F. Majunke, M. Messori, B. Grzybowska, J. Czekaj, B. Majka, C. Dias and M. Portela, Ibid p. 153.
61. J. Haber, A. Kozlowska and R. Kozlowski, J. Catal. **102** (1986) 52.
62. J. G. Eon, R. Olier and J. C. Volta, J. Catal. **145** (1994) 318.
63. J. Le Bars, A. Auroux and J. C. Vedrine in *New Developments in Selective Oxidation by Heterogeneous Catalysis*, ed. P. Ruiz and B. Delmon, Stud. in Surf. Sci. and Catal.. ser. **72**, Elsevier, Amsterdam (1992) 181.
64. J. Le Bars, J. C. Vedrine, A. Auroux, B. Pommier and G. M. Pajonk, J. Phys. Chem. **96** (1992) 2217.
65. M. Loukah, G. Coudurier and J. C. Vedrine in *New Developments in Selective Oxidation by Heterogeneous Catalysis*, ed. P. Ruiz and B. Delmon, Stud. in Surf. Sci. and Catal. ser. **72**, Elsevier, Amsterdam (1992) 111.
66. M. Loukah, Ph D Thesis University of Rabat (1992).
67. I. E. Wachs, G. Deo, M. A. Vuurman, H. Hu, D. S. Kim and J. M. Jehng, J. Molec. Catal. **82** (1993) 443.
68. G. Deo and I. E. Wachs, J. Catal. **146** (1994) 323.
69. D. S. Kim, J. M. Tatibouet and I. E. Wachs, J. Catal. **136** (1992) 209.
70. J. M. Tatibouet and J. E. Germain, J. Catal. **72** (1981) 375.
71. C. Louis, J. M. Tatibouet and M. Che, J. Catal. **109** (1988) 354.
72. N. Nakuta, K. Tohji and Y. Ugadawa, J. Phys. Chem. **92** (1988) 2853.
73. C. T. J. Mensch, J. A. R. van Veen, van Wingerden and M. P. van Dijk, J. Phys. Chem. **92** (1988) 4988.
74. J. A. van Veen and P. A. - J. M. Hendricks, Polyhedron **5** (1986) 75.
75. T. Tanaka, H. Yamishita, R. Tsuchitani, T. Funabiki and S. Yoshida, J. Chem. Soc. , Faraday Trans. I **84** (1988) 2987.

76. J. Wong, F. W. Lytle, R. P. Messmer and D. H. Maylotte, Phys. Rev. **B30** (1984) 5569.
77. R. Kozlowski, R. T. Pettifer and J. M. Thomas, J. Phys. Chem. **87** (1983) 5176.
78. G. Bergeret, P. Gallezot, K. V. R. Chary, B. Rama Rao and V. S. Subrahmanyam, Appl. Catal. **40** (1988) 191.
79. B. Moraweck, in *Les Techniques physiques d'étude des Catalyseurs*, ed. B. Imelik and J. C. Vedrine, Technip, Paris (1988) 587.
80. B. Moraweck in *Catalyst Characterization, Physical Techniques for Solid Materials*, ed. B. Imelik and J. C. Vedrine, Plenum, New York (1994) 377.
81. J. C. Vedrine, J. M. Millet and J. C. Volta, Faraday Discuss. Chem. Soc. **87** (1989) 207.

# SELECTIVE CATALYTIC OXIDATION BY HETEROGENEOUS TRANSITION METAL CATALYSTS

R.A. VAN SANTEN
*Schuit Institute of Catalysis, Faculty of Chemical Engineering, Eindhoven University of Technology, P.O. Box 513, 5600 MB Eindhoven, The Netherlands*

## ABSTRACT

**The reaction mechanisms of two transition metal catalysed reactions are discussed: epoxidation of ethylene production and vinylacetate. In addition short reference will be made to methanol and CO oxidation. The surface reactions are related to the corresponding reactions in organometallic complexes. Also a relation between surface science model studies and surface reactivity will be made. Elementary surface reaction steps on surfaces will be highlighted; subsequently the main mechanistic issues in oxygen CH and OH bond activation will be described.**

## 1. Introduction

Three large scale selective oxidation processes are based on heterogeneous metallic catalysis. The epoxidation of ethylene, catalysed by silver, produces oxirane, which is an important intermediate to the manufacture of glycol or polyols. Silver is also used as a catalyst for the oxidative dehydrogenation of methanol to formaldehyde. The third process is the production of vinylacetate by oxidative coupling of ethylene and acetic acid catalysed by palladium. Whereas some of these processes are more than fifty years old, there is still a considerable need to further improve their yields. Only recently it has become possible to formulate a mechanistic basis to these reactions. This will be the subject of this chapter.

The reaction temperatures for these three processes are moderate. Two other important oxidation processes concern the oxidation of $NH_3$ to NO, the Ostwald process and the selective oxidation of $CH_4$ to synthesis gas. These reactions proceed at temperatures higher than 600 °C and are catalysed by noble metals as Pt or Rh. The catalytic surface activates $NH_3$ or $CH_4$. The product distribution of these processes is mainly determined by gas phase radical reactions. Reactions that also should be mentioned are the oxidation of $SO_2$ to $SO_3$ by Pt and the oxidation of CO to $CO_2$. This reaction has been extensively investigated, mainly for fundamental reasons, and occurs for instance in automotive exhaust catalyst systems. Low temperature hydrocarbon conversion catalysis is also of interest in the context of automotive exhaust catalysis.

Here we will focus on the reactivity of transition metal surfaces of relevance to low temperature selective oxidation processes. Another chapter deals with the high temperature oxidation reactions.

As explained in chapter 10, the oxidation mechanism of vinylacetate formation is not yet completely clear. One possibility is that it is related to the homogeneous Wacker reaction[1,2] as will be explained later; the alternative is a surface reaction on large Pd clusters. According to the latter proposal, the large Pd clusters that are found in the reaction mixture activate ethylene to an adsorbed vinyl fragment that reacts with adsorbed acetate to vinylacetate.

The activation of the CH bonds in ethylene will be discussed in section 3. On silver it initiates total combustion, the non-selective reaction in the ethylene epoxidation reaction.

For the ethylene epoxidation reaction[3] comparisons have also been made with homogeneous systems, but main fundamental advances have been due to model catalyst surface science studies.

Recent advances in the understanding of the reactivity patterns of adsorbed oxygenates to transition metal surfaces are providing a molecular basis to a mechanistic description of selective oxidation reactions[4].

This lecture will discuss the mechanism of selective oxidation based on information from surface science studies as well as on the reactivity of organometallic complexes. We will first describe the elementary steps of the corresponding catalytic reaction cycles and then discuss in some detail essential reaction steps.

## 2. Reaction mechanisms

### *2.1. Ethylene epoxidation by silver (3)*

The epoxidation reaction proceeds at 250 °C over a Ag catalyst. The selectivity of the catalyst strongly depends on catalyst composition and the presence of chlorine containing hydrocarbons in the gas phase (~ppm).

The Ag particles are supported on a wide porous $\alpha$-$Al_2O_3$ support and the catalyst is promoted by addition of alkali, especially Cs.

The kinetics of the reaction consists of two parallel reactions and a consecutive reaction:

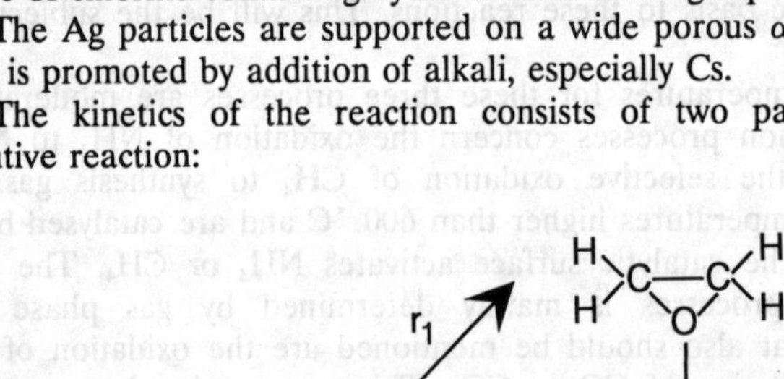


(1)

Total combustion of ethylene can occur by a direct competitive combustion of ethylene, $r_2$, or a consecutive reaction of epoxide, $r_3$.

Ethylene epoxide itself has been demonstrated to be converted with a very low rate by silver. However the ethylene epoxide molecule can become isomerized to the aldehyde by acidic protons on the support. The aldehyde has a high rate of combustion by silver to $CO_2$ and $H_2O$. The role of alkali is amongst others the suppression of the presence of acidic protons on the support and hence to reduce the rate of the consecutive reaction $r_3$.

The moderating chlorine containing hydrocarbons, e.g. vinylchloride are combusted by silver and chlorine becomes deposited on silver. Chlorine containing molecules have to be added continuously to the reactant feed, because chlorine is removed from it by a reaction with ethylene.

Chlorine adsorbed to the silver surface enhances the initial rates ratio $r_1/r_2$. There have been many different proposals concerning the mechanism of epoxide formation. All of them have in common that adsorbed oxygen species of different reactivity are proposed to be present on the silver surface and one species is proposed to give the epoxide upon contact with ethylene.

According to one early proposal oxygen can be adsorbed in the molecular or dissociated form to silver. Only molecular oxygen is proposed to give the epoxide, atomically adsorbed oxygen gives total combustion of ethylene. Since 6 atoms of oxygen will burn one ethylene molecule to $CO_2$ and $H_2O$, 3 adsorbed $O_2$ molecules can produce 6 ethylene epoxide molecules from 6 ethylene molecules. A seventh ethylene molecule then has to be used to remove the 6 oxygen atoms left on the silver surface. According to this proposal the maximum ethylene selectivity of the reaction can never exceed 6/7. Modern epoxidation catalysts, however, exceed this value.

Another proposal, now generally accepted, is based on the strong dependence of the selectivity on silver surface oxygen coverage. The selectivity of the epoxidation reaction is found to have its optimum value when the $O_{Surf.at}/Ag_{Surf.at} \approx 1.0$. This implies a surface composition close to AgO. Since only half of the oxygen can reside on the external surface, the other half of the oxygen atoms has to be located in subsurface positions. At this high oxygen concentration, the O-Ag bond energy is low and O behaves electrophilic. There is no barrier for oxygen insertion into the ethylene $\pi$ bond (figure 1).

Total combustion proceeds by activation of the C-H bond of ethylene, as has been demonstrated using deuterated ethylene. At low surface coverage adsorbed oxygen atoms behave nucleophilically and attack the slightly positively charged ethylene hydrogen atoms. Empty vacancy sites next to adsorbed oxygen are needed to stabilize the resulting $C_2H_3$ species.

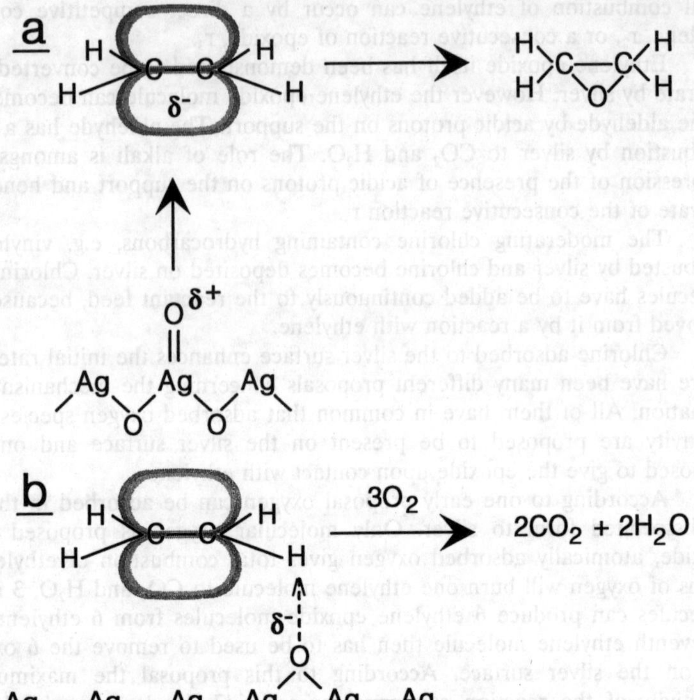


**Figure 1**

At the reaction temperature the surface will only be partially covered by oxygen. This would give a low selectivity. Chlorine adsorbs strongly to Ag. It can take over the role of subsurface oxygen. It will reduce the O-Ag bond energy of the oxygen atoms that share also a bond with Ag (see figure 1).
Whereas in the epoxidation reaction molecularly adsorbed $O_2$ does not play a role, the question of the importance of molecularly versus atomically adsorbed oxygen in substrate activation is a recurring motive in selective oxidation.

Two examples are known for molecularly adsorbed $O_2$ to play a role in oxidation. One example is the oxidation of $SO_2$ by molecularly adsorbed $O_2$ on silver[5]. The other example is the activation of $NH_3$ (6) on Mg, Zn or Cu by molecularly adsorbed $O_2$. In the case of low temperature $NH_3$ activation, convincing occurrence of a transient intermediate with molecular $O_2$ has been proposed.

*2.2 Methanol oxidation*

The methanol oxidation reaction catalysed by silver, proceeds at a temperature (~500 °C) such that the adsorption equilibrium of adsorbed oxygen is shifted to the gas phase. Hence oxygen can only be adsorbed on defects of silver gauze. Methanol adsorbs only weakly to silver, but will dissociate by a reaction with adsorbed oxygen.

$$CH_3OH + O_{ads} \rightarrow \overset{H}{O}_{ads} + CH_3O_{ads} \qquad (2)$$

The adsorbed methoxy will desorb as formaldehyde.

*2.3. Vinylacetate from ethylene and acetic acid*

In the liquid phase a mechanism can be proposed that is analogous to the Wacker reaction. This is shown in figure 2.

The reaction concerns a redox cycle between $Pd^{2+}$ and $Pd^0$. In the Wacker reaction ethylene reacts with $H_2O$ instead of acetic acid, and Pd oxidation is catalysed by $Cu^+$. $Cu^+$ is a catalyst for $Pd^0$ oxidation, $Cu^+$ has only to change one valency, whereas oxidation of Pd requires two electron transfer and hence is more difficult.

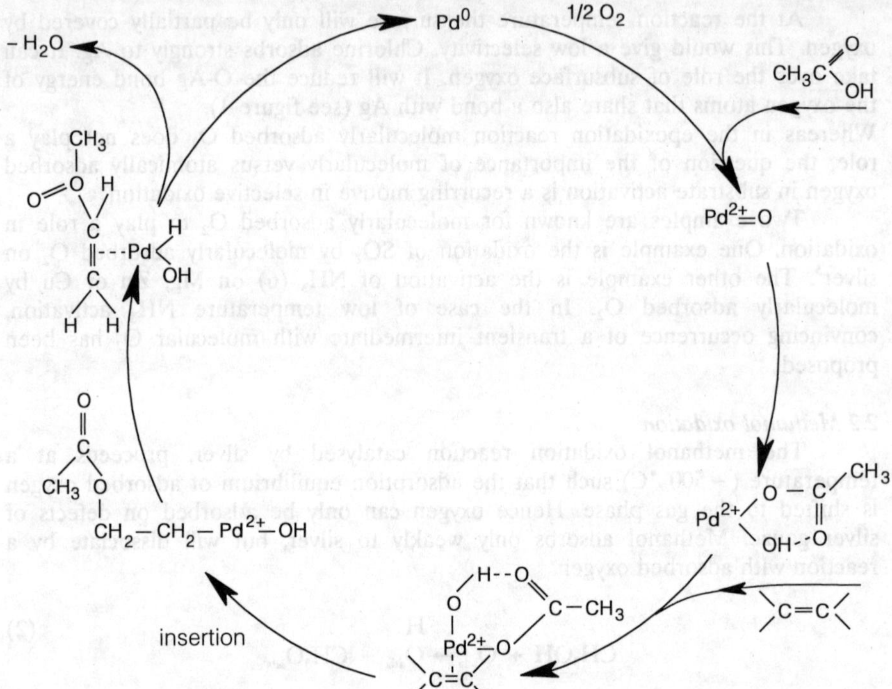


The vinyl acetate oxidation-redox cycle
( analogue Wacker )

Figure 2

The essential difference between the acetoxylation reaction and ethylene epoxidation is that insertion occurs by a nucleophilic attack between a negatively charged acetate ion and ethylene, with formation of ethyl acetate coordinated to the $Pd^{2+}$ ion.

In the epoxidation reaction oxygen addition to the ethylene $\pi$ bond occurs by a electrophilic attack, without stabilization of one of the ethylene carbon atoms. Interestingly it has been found that also in a homogeneous catalytic reaction epoxidation occurs by such a complex with electrophilic oxygen[4]: with Ag in a threevalent state.

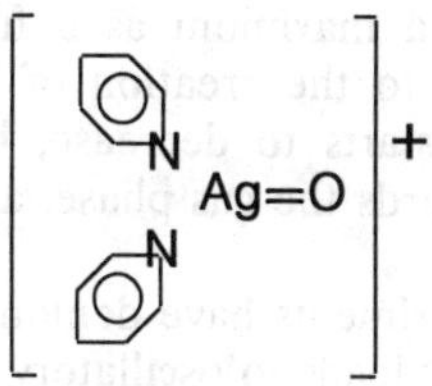

(3)

After the acetate insertion step, the ethyl acetate intermediate undergoes $\beta$ C-H cleavage to vinyl acetate. Bond breaking of this CH bond is facilitated by the presence of the near polar C-O ester bond.

The other important steps are reductive elimination of acetic acid and oxidation of $Pd^0$ to $Pd^{2+}$. Except for the reoxidation step of $Pd^0$ (in acetoxylation $Cu^{+1}$ is not applied), the other elementary steps are well known in organometallic chemistry. The reduced Pd forms metal clusters as in a metallic catalyst. Oxygen can dissociate on these clusters. The non-selective total oxidation of ethylene to acetic acid or combustion of acetic acid probably occurs on these metallic Pd particles. However, according to Moiseev (see chapter 10) the Pd surface also plays a role in the vinyl acetate reaction step.

Whereas Ag metal without adsorbed O will not activate CH bonds, the more reactive Pd metal can activate the CH bond of ethylene.

In the absence of hydrogen ethylene adsorption to Pd will lead to ethane formation by self hydrogenation and formation of carbonaceous residues in a reaction formally represented as:

$$2\ H_2C = CH_{2_{ads}} \rightarrow H-C \equiv CH_{ads} + H_3C - CH_3 \tag{4}$$

Adsorbed acetylene is readily converted into $CH_{ads}$ or $C_{ads}$ and $H_{ads}$ species. The partially hydrogenated 'C', species will readily react with adsorbed oxygen to produce CO or $CO_2$.

Acetic acid interacts more weakly with palladium metal, but can become activated by coadsorbed oxygen as we will discuss later.

*2.4. CO oxidation*

The recombination of adsorbed O atoms and CO to give $CO_2$ occurs readily on most transition metals. The slow step is the dissociation of adsorbing $O_2$ into adsorbed oxygen atoms. At low temperatures the reaction order is negative in CO, implying that CO is Major Adsorbed Reaction Intermediate (MARI). This high surface coverage suppresses $O_2$ dissociation, for which at least two neighbouring vacant sites are needed.

The reaction rate shows a maximum as a function of temperature. The increase in reaction rate is due to the creation of surface vacancies. At higher temperatures the reaction rate starts to decrease, because then the adsorption equilibrium of CO is shifted towards the gas phase, and reaction becomes positive in the CO pressure.

Single crystal surface experiments have demonstrated an interesting surface reconstruction phenomenon[9], that leads to oscillatory behaviour at low pressures.

The (100) surface of Pt reconstructs in a vacuum to a more stable surface in which the surface layer has the more dense (111) packing. The rate of $O_2$ dissociation on this surface is extremely low. CO, however adsorbs with a high rate, but destabilizes the (111) overlayer, so that the more reactive (100) layer is reformed. On this surface the $O_2$ dissociates rapidly. The reaction between $CO_{ad}$ and $O_{ad}$ occurs rapidly and the weakly adsorbing $CO_2$ desorbs. The surface free of adsorbate reconstructs to the stable (111) layer and the process repeats itself.

## 3. The reactivity of transition metal surfaces for oxidation reactions

### *3.1. Oxygen activation*

The reactivity of small hydrocarbons with adsorbed oxygen has been extensively investigated by a number of groups[4,10,11]

The oxygen molecule will dissociate on clear transition metal surfaces below room temperature. In order to accommodate the oxygen atoms large surface atom ensembles are required. Oxygen dissociation can become suppressed by the presence of adsorbed oxygen or other blocking coadsorbates. Hence the rate of dissociation of oxygen dissociation rapidly declines with surface coverage.

Because of the low reactivity of the noble metal surfaces as well as the group IB metals coadsorption of oxygen often has a promoting effect on hydrocarbon reactivity. It assists the dissociation of CH bonds. Due to their completely filled d-valence electron band, the IB metals, Cu, Ag and Au are least reactive.

With respect to oxygen the reactivity of Ag is intermediate between that of Cu and Au. On Ag as well as Cu adsorbed $O_2$ dissociates below room temperature, however $O_2$ will not dissociate on Au. The $O_{ads}$-Au interaction is too weak to overcome the $O_2$ bond energy. Cu has a higher reactivity than Ag. Whereas for instance NO will not dissociate on Ag, dissociation of NO on Cu occurs readily. Because of the high interaction energy with oxygen, and the resulting high temperature of oxygen desorption, catalytic dissociation of NO however is only possible at high temperatures in the absence of oxygen. Strongly adsorbed oxygen atoms will block surface ensemble sites necessary for NO dissociation.

Of recent interest is the reduction of NO by hydrocarbons[12] in excess oxygen catalysed by Cu or Co containing zeolites. In these cases no zerovalent metals are present, but metal-oxo-complexes. NO is oxidized to $NO_2$.

An intermediate formed by recombination with NO then probably reacts in a consecutive reaction with hydrocarbon to $N_2$.

The high reactivity of most of the transition metals with respect to oxygen has as a consequence that oxidation catalysis will usually proceed on transition metal oxides instead of the metals. The catalytically active metals, silver and platinum remain bulk metals also during the catalytic oxidation reaction.

On reactive metal surfaces oxygen coadsorption can prevent decomposition of adsorbed molecules. On Mo (110) coadsorbed oxygen has been found to deactivate the metal surface and to stabilize adsorbed molecules[13].

Also oxygen adsorption may lead to facetting of catalyst particles. Those surfaces will be stable during the catalytic reaction that have their lowest surface energy in the presence of oxygen. This usually is not necessarily the same surface that is most stable in the absence of adsorbed oxygen.

*3.2. The activation of CH and OH bonds*

Oxygen atoms adsorbed at low surface coverage to transition metal surfaces as, Ag, Pd or Rh are nucleophilic and act as Lewis basic atoms.

Whereas acetic acid will mainly desorb molecularly from transition metals and at relatively low temperature, in the presence of coadsorbed oxygen acetate formation occurs readily.

$$CH_3COOH + O_{ads} \rightarrow CH_3COO_{ads}^- + OH_{ads} \tag{5}$$

Such Brønsted acid-Lewis base oxygen reactions are quite common. On Ag it has been shown that CH bonds, that are not activated by the clean metals, will react in the presence of oxygen:

$$HC \equiv CH + O_{ads} \rightarrow HC \equiv C^{(-)} + OH_{ads} \tag{6}$$

Earlier we mentioned that the total combustion of ethylene is initiated by an analogous reaction to eq.7. In the presence of adsorbed oxygen also the methyl group of acetic acid may become activated, providing a pathway for the total combustion of acetic acid.

The CH bonds of the $CH_3$ group in propylene are also easily activated, because of resonance stabilization of the allyl formed upon CH cleavage.

Propylene will react with $O_{ad}$ on Cu to acrolein[12] illustrating the preference of CH bond breaking in propylene versus oxygen insertion into the C-C $\pi$ bond. At low oxygen coverage this reaction will compete with propylene oligomerization, at higher oxygen coverages consecutive reaction of acrolein will occur that will lead to total combustion.

On the less reactive Ag surfaces it appears that with propylene oxygen insertion into the C-C $\pi$ bond competes with methyl CH activation. The CH bond of propylene epoxide reacts so rapidly, that propylene epoxide is only observed as a product when contact times are used of the order of msec[15].

In case Ag is alloyed by Au, the oxygen concentration on the Ag surface is decreased. This enhances the nucleophility of the adsorbed oxygen atoms. It is found that the selectivity for ethylene epoxide formation is decreased, but that propylene now forms acrolein with high selectivity[16]. The reaction of aldehydes or alcohols with atomically adsorbed oxygen can lead to acetate formation. On the transition metals acetate decomposition will occur upon heating[17]. At lower temperature CO and $O_{ads}$ formation occurs. At higher temperature the acetate tends to decarboxylate. Coadsorbed oxygen tends to suppress the low temperature decomposition path because of site blocking.

Only atomic oxygen adsorbed to silver gives epoxide formation by reaction with ethylene. On Pd or Rh aldehydes or ketones will be formed. In the presence of excess oxygen the main reaction will be total oxidation. The analogous homogeneous reaction of ethylene on $Pd^{2+}$ is the Wacker reaction.

The steps in the homogeneous phase that lead to the aldehyde formation are:

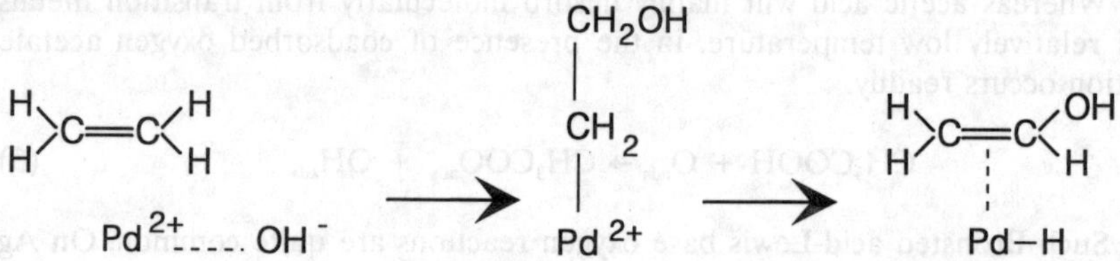

(7)

The reaction is initiated by insertion of $OH^-$ into adsorbed ethylene. Aldehyde formation results by consecutive isomerization of the vinyl alcohol, that is formed after $\beta$ CH cleavage. According to mechanism 7, intermediate I has a strong $\sigma$ metal-carbon bond comparable to that in adsorbed ethyl. The next step in intermediate I is $\beta$ C-H bond cleavage. These reaction paths have to be contrasted with epoxide formation, that does not require stabilization of ethylene and is preferentially formed via electron-deficient oxygen species. This is illustrated by the following typical homogeneous epoxidation reaction:

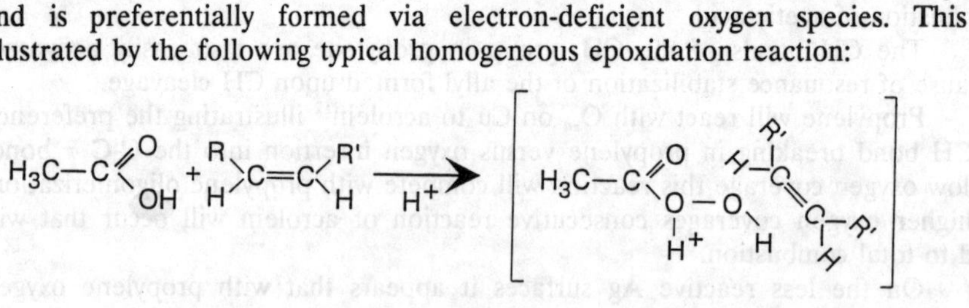

(8)

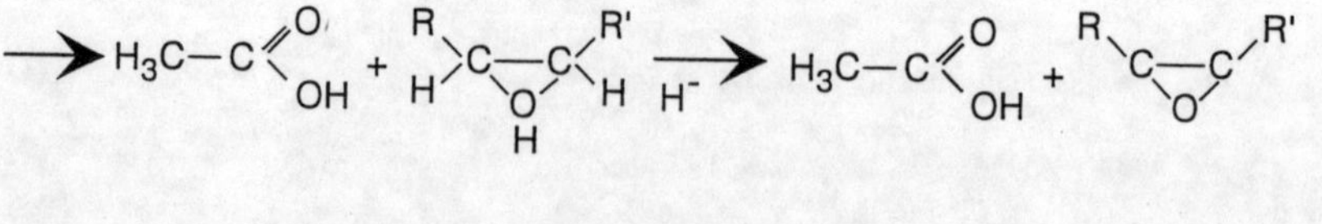

In reaction scheme eq.8 the proton is the catalyst that intermediates oxygen insertion from acetic peroxide into ethylene. Epoxidation occurs via intermediate formation of the electron deficient species II.

Also epoxide formation may occur upon reaction of ethylene with $MnO^{3+}$ stabilized in a porphyrin[18] or a $TlO^{+}$ species as present in $Tl_2O_3$[19]. Again illustrating the need for electron deficient oxygen.

Two factors contribute to the uniqueness of Ag. On Ag the CH bond will not be readily activated, whereas on the transition metal with a partially filled d-valence electron band CH bond activation will readily occur. In addition at high surface coverage the oxygen-metal bond energy has weakened, so that the weakly exothermic epoxidation reaction can occur.

Ethylene when adsorbed to a transition metal surface can have the $\pi$ or di-$\sigma$ adsorbed state:

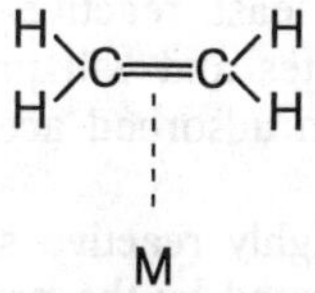

$\pi$ adsorbed ethylene

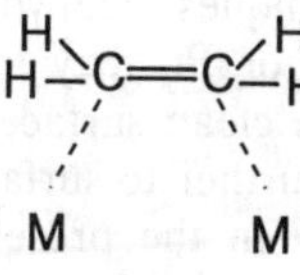

di-$\sigma$ adsorbed ethylene

di-$\sigma$ adsorbed ethylene occurs preferentially on metals with spatially extended d-valence orbitals, as Pd or Pt.

The intermediate III[20]

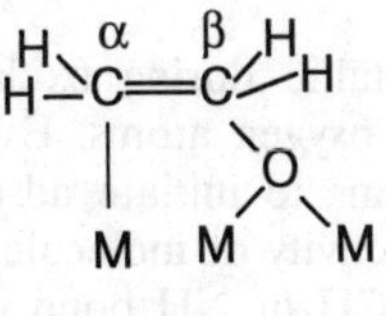

intermediate III,
metallocycle

can be considered the surface analogue of intermediate I. $\beta$ CH cleavage and subsequent hydrogen addition to the $C_\alpha$ atom will produce the aldehyde. This appears to be the preferred alternative next to fragmentation on transition metals. In case no hydrogen activation occurs intermediate III could in principle give epoxide formation by formation of a $C^\alpha$-O bond.

Formation of a metallocycle as intermediate III has been proposed to occur from adsorption of alcohols to transition surfaces[10,17]. On Rh its decomposition gives $CH_4$ and CO. Aldehydes, however have been shown to be CO $\eta_2$ adsorbed. Their decomposition leads to surface carbonaceous residues. On Ag dehydrogenation of the t-butyl alcohol, promoted by adsorbed oxygen has also been shown to give the epoxide. In this case a metallocycle as intermediate III could also be proposed. Epoxide formation is the preferred reaction path because of the absence of activated CH bonds.

Similarly coadsorbed on Ag norbornadiene and isobutylene have been shown to give the epoxide by reaction with atomically adsorbed oxygen[11].

On Ag the low reactivity with respect to CH activation may also result in interesting condensation reactions. Condensation of acetylene has been shown to give benzene.

Condensation of butadiene with surface atom oxygen on Ag has been shown to give ring closure and furfuryl formation[21].

When ethylene adsorbs to a group VIII noble metal, its CH bonds become rapidly activated. This reaction has been extensively investigated with Surface Science techniques. Dehydrogenation of ethylene on the least reactive noble metals as Pd or Pt, may lead to adsorbed vinyl intermediates and will produce acetylene. On clean surfaces and in the absence of hydrogen adsorbed acetylene decomposes further to surface carbidic species.

Clearly in the presence of adsorbed oxygen these highly reactive species will lead to total combustion. Only when a surface is deactivated by the presence of coadsorbed intermediates one may expect that a reactive species as a vinyl species remains stable. Interesting examples by vinyl containing organic metallic complexes are known.

## 4. Summary

Surfaces, mainly metallic during oxidation catalysis, may become more reactive due to coadsorbed oxygen atoms. Especially the OH or CH bonds may react with basic oxygen atoms to initiate adsorbate activation with formation of $OH_{ad}$. However also the reactivity of molecules as $CH_4$ or $NH_3$ may be enhanced, because of cleavage of the CH or NH bond with formation of surface hydroxyls. These are important initiation steps in partial oxidation of $CH_4$ and NO production. The most important selective oxidation reactions catalysed by silver are ethylene epoxidation and methanol oxidation. Cu is a selective oxidation catalyst for the oxidation of propylene to acrolein.

Whereas oxygen adsorbed to Cu is nucleophilic, oxygen adsorbed to Ag can be electrophilic. On transition metals insertion reaction of olefins with atomically adsorbed oxygen leads to aldehydes or ketones.

## References


1. R. van Helden; C.F. Kohll; D. Medema; G. Verberg and T. Jonkhoff, *Recueil* **87** (1968) 961.
2. P.M. Henry, *J. Am. Chem. Soc.* **94** (1972) 7305.
3. R.A. van Santen and H.C.P.E. Kuipers, *Adv. Catal.* **35** (1988) 265.
4. R.J. Madix, *Adv. Catal.* **29** (1980) 1; N.F. Brown and M.A. Barteau, in *Selectivity in Catalysis*; eds. S.L. Suib and M.E. Davis, *ACS Symposium Series* (1993).
5. J.T. Roberts and R.J. Madix, *J. Am. Chem. Soc.* **110** (1988) 8540.
6. C. Au and M.W. Roberts, *Nature.* **319** (1986) 206; M.W. Roberts, *J. Mol. Catal.* **74** (1992) 11.
7. R.N. Hader; R.O. Wallace and R.W. McKinney, *Ind. Eng. Chem.* **44** (1952) 1508.
8. J.M. v.d. Eijk; Th.J. Peters; N. de Wit and H.A. Colijn, *Catalysis Today*, **3** (1988) 259.
9. G. Ertl, *Surf. Sci.*, **152/153** (1985) 328.
10. X. Xu and C.M. Friend, *J. Am. Chem. Soc.*, **113** (1991) 6779.
11. C. Mukord; S. Hawker; J.P.S. Badyal and R.M. Lambert, *Catal. Lett.*, **4** (1990) 57; S.A. Tan; R.B. Grant and R.M. Lambert, *J. Catal.*, **100** (1986) 383; R.B. Grant and R.M. Lambert, *J. Catal.*, **92** (1985) 364; R.A. Barbrow and R.M. Lambert, *Surf. Sci.* **67** (1977) 489.
12. M. Iwamoto; S. Yokoo; S. Sakai and S. Kagawa, *J. Chem. Soc. Far. Trans. I*, **77** (1981) 1629; Y. Li and J.N. Armor, *Appl. Catal.* **76** (1991) L1; J. Valyon and W.H. Hall, *J. Phys. Chem.* **97** (1993) 1204.
13. Y. Iwasawa, in *Elementary Reaction Steps in Heterogeneous Catalysis*, eds. R.W. Joyner R.W. and R.A. van Santen, *Kluwer* (1993) p. 287-304.
14. J.C. Callahan and R.R. Graselli, *A. I. Ch. E.*, **9** (1963) 755.
15. J.T. Gleaves; J.R. Ebner and T.C. Kuechler, *Catal. Rev. Sci. Eng.* **30 (1)** (1988) 49.
16. P.V. Geenen; H.J. Boss and G.T. Pott, *J. Catal.* **77** (1982) 499.
17. N.F. Brown and M.A. Barteau, *J. Am. Chem. Soc.* **114** (1992) 4258.
18. J.T. Groves and Th. Nemo, *J. Am. Chem. Soc.* **105** (1983) 5786.
19. A. McKillop and E.C. Taylor, *Adv. Organomet. Chem.* **11** (1973) 147.
20. N.F. Brown and M.A. Barteau, *Surf. Sci.* **298** (1993) 6.
21. J.T. Roberts; A.J. Capote and R.J. Madix, *J. Am. Chem. Soc.* **113** (1991) 9848.

# PARTIAL OXIDATION ON NOBLE METALS AT HIGH TEMPERATURES

Lanny D. Schmidt and Marylin Huff
Department of Chemical Engineering and Materials Science
University of Minnesota
Minneapolis, MN 55455

## ABSTRACT

This chapter focuses on the direct catalytic partial oxidation of small alkanes and $NH_3$ to produce chemicals. Processes such as these are now the heart of the chemical industry, and the discovery of new processes will be essential to utilize inexpensive hydrocarbon feedstocks in chemical synthesis.

The discussion centers around four industrial processes: $NH_3$ oxidation to $HNO_3$, the ammoxidation of $CH_4$ to produce HCN, the direct oxidation of $CH_4$ to produce syngas, and the oxidative dehydrogenation of $C_2H_6$ to produce ethylene. All of these processes operate at or above atmospheric pressure over Pt or Rh catalysts with contact times of approximately $10^{-3}$ sec.

These reaction systems are very complex because they are so fast that mass transfer effects dominate the rates and so exothermic that the reaction temperature rises to 800 to 1000°C in the very short residence time ($10^{-3}$ sec) over the catalyst. A third complication in direct oxidation processes is the possibility of homogeneous reaction which leads to poor selectivity, flames and explosions.

These processes can only be managed and understood by consideration of mass and heat transfer along with kinetics. We consider the principles by which these processes operate and the factors which must be considered in designing new direct partial oxidation processes.

## 1. Catalytic Oxidation on Noble Metals

This subject goes back to the origins of chemistry and catalytic reactions. In the early $19^{th}$ century, Sir Humphrey Davy and his assistant Michael Faraday were interested in the problem of explosions in coal mines. It was found that when air was pumped into mines so miners could breathe, explosions occurred frequently and unpredictably. The cause of the explosions was of course methane combustion,

$$CH_4 + 2O_2 \rightarrow CO_2 + 2H_2O, \quad \Delta H^o_{298} = -192 \text{ kcal/mole}, \tag{1}$$

a very fast and very exothermic reaction which proceeds homogeneously as a free radical chain reaction which produces flames and explosions.

People were interested in design of miner's lamps and flame arrestors. At about the same time platinum metal had been discovered, and Davy and Faraday discovered that a Pt wire could be made to glow spontaneously with no homogeneous flame when held in a mixture of $CH_4$ + air or $H_2$ + air if the wire was preheated to ignite the surface reaction[1]. This was the first recorded adiabatic catalytic reactor occurring on a single wire. The principle of operation of this reaction, written down 100 years later, is that a reacting steady state exists in which the heat of reaction of the catalytic combustion reaction, -192 kcal/mole, is sufficient to heat the wire to a temperature where reaction is self sustaining and the process is autothermal[2]. This is the principle of operation of the *autothermal catalytic reactor*, the major reactor type used for carrying out catalytic oxidation reactions.

The first commercially viable process was discovered by Wilhelm Ostwald in Leipzig who used the same principle but now with a woven mesh of Pt gauze across a tube[3-5]. He found that the reaction

$$NH_3 + \tfrac{5}{4}O_2 \rightarrow NO + \tfrac{3}{2}H_2O, \quad \Delta H^{o}_{298} = 54 \text{ kcal/mole}, \tag{2}$$

could be made to proceed with >90% efficiency (selectivity) over Pt with a negligible contribution from the competing reaction

$$NH_3 + \tfrac{3}{4}O_2 \rightarrow \tfrac{1}{2}N_2 + \tfrac{3}{2}H_2O, \quad \Delta H^{o}_{298} = -\ 76 \text{ kcal/mole}, \tag{3}$$

which is even more exothermic and has a more favorable equilibrium constant. Nitric acid is then easily produced by further homogeneous oxidation and hydration of NO. Thus the use of catalysts was established for obtaining highly selective production of a desired product, and the use of the heat of oxidation processes to provide internally the heat necessary to heat the catalyst to the desired operating temperature.

Nitric acid plants quickly became an essential part of the world's technology for fertilizer and explosives manufacturing. Current plants operate under similar conditions, now using Pt-10% Rh gauze catalysts and pressures up to 10 atm[3]. A related and even earlier process was the manufacture of sulfuric acid, made by catalytic oxidation of $SO_2$ to $SO_3$ followed by hydration to $H_2SO_4$,

$$SO_2 + \tfrac{1}{2}O_2 \rightarrow SO_3 \rightarrow H_2SO_4 \tag{4}$$

again a catalytic process first carried out over Pt. However, the process was soon switched to the much cheaper $V_2O_5$ catalyst which is used for sulfuric acid production today[3]. The latter is one of the first application of metal oxides in catalytic oxidation. We shall confine this discussion to noble metals, but oxide catalysts also have wide applications in oxidation, giving different product selectivities and being much cheaper than noble metals. The mechanism of oxide catalysis is much different than on noble metals, noble metals working by dissociative chemisorption of reactants and oxides by reduction-oxidation using lattice oxygen.

There is a natural and very distinct division of oxidation processes between fuel-lean and fuel-rich situations. As indicated in Figure 1, for oxidation of any hydrocarbon in fuel-lean environments, reaction selectivities of hydrocarbons (homogeneously or heterogeneously) favor total oxidation to $CO_2$ and $H_2O$. This is the regime of total combustion which finds wide application in *incineration* of volatile organic compounds (VOC's) and in the *automotive catalytic converter* as major environmental reactions which will have increasing importance in the world economy to reduce pollution.

There is a "no man's land" of oxidation near the total oxidation stoichiometry (5 to 14% $CH_4$ in air) where flames will occur in premixed gases and a narrower region near 9.5% $CH_4$ in air where hydrocarbon-air mixtures become explosive. Obviously, catalytic processes using premixed fuel and oxygen cannot operate in this composition regime.

In excess fuel compositions, oxidation reactions can be made to occur selectively to produce *partial oxidation* products if a suitable catalyst can be found. This is the subject of this chapter, the selective production of chemicals by catalytic partial oxidation on noble metals. We shall consider these processes only with gases and on noble metals at fairly high temperature (>600°C); other chapters will consider liquid phase processes and oxide catalysts, both of which require lower operating temperatures.

The reactions we describe are very fast, so fast that they are always limited by the mass transfer or diffusion rates of reactants to the surface. Mass transfer effects and the fact that these processes are usually operated adiabatically require that *reactor engineering* be considered simultaneously with reaction chemistry. Adiabatic operation is an essential mode of operation in any large oxidation reactor because it is impossible to remove heat fast enough to control the temperature independently. In fact, one usually wants to use reaction heat to heat the reactor to sufficiently high temperatures to obtain high reaction rates.

In this chapter, we focus on catalyst geometries that are conducive to high rates of mass transfer. The processes we will describe can use a monolithic type catalyst. These

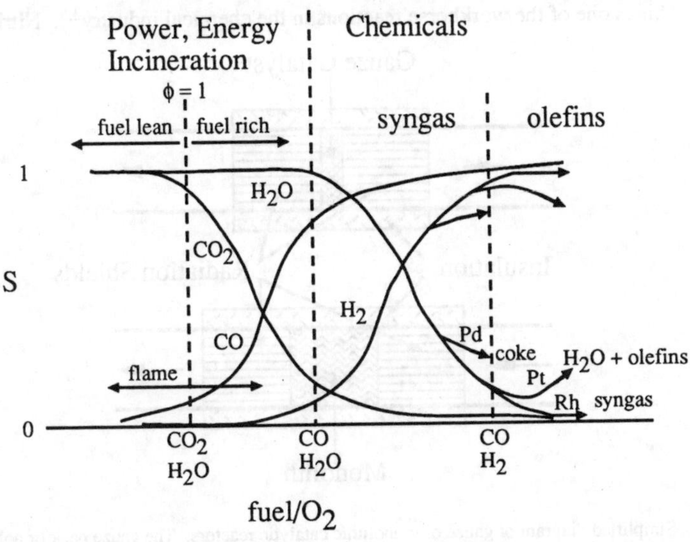


Figure 1: Schematic plot of the selectivities expected versus fuel to oxygen ratio for the oxidation of a hydrocarbon over a suitable catalyst. Total combustion occurs in excess oxygen, syngas in excess fuel, and chemicals possibly in large excess fuel.

reactors are depicted in Figure 2. A pack of several layers of Pt-10% Rh gauze, as used commercially in HCN synthesis and $NH_3$ oxidation[4], is the thin limit of monolithic type catalysts. In later sections, we describe hydrocarbon oxidation over noble metal coated foam monoliths. Like gauze packs, these ceramic supports provide a tortuous path for the reacting gases and lead to high rates of mass transfer. For either gauze or monolithic catalysts, the reaction zone is usually insulated as shown in Figure 2 to better approximate adiabatic operation.

Reaction engineering principles must be understood in considering most chemical processes, but this is especially true of oxidation processes. The student is referred to Chemical Engineering texts such as those by Levenspiel[6] or Fogler[7] for thorough consideration of reaction engineering issues.

## 2. Four Important Industrial Processes

We shall center our discussion around four important industrial reactions for the production of commodity chemicals. These are listed in Table 1.

### *2.1 Nitric acid synthesis*

This is one of the workhorse reactions in the chemical industry[3,5]. Nitric acid is a

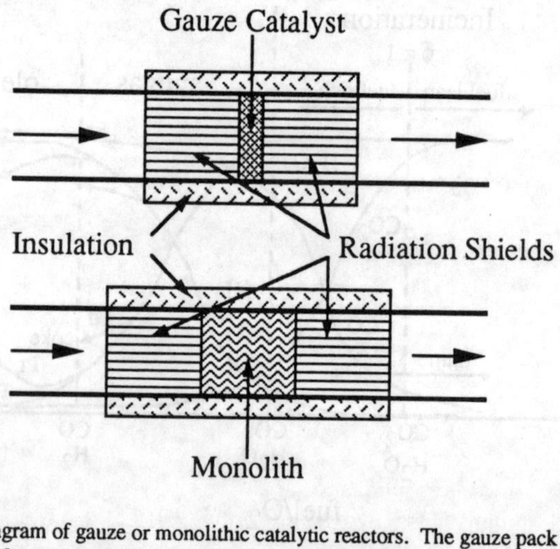


Figure 2: Simplified diagram of gauze or monolithic catalytic reactors. The gauze pack or noble metal coated ceramic foam monolith is placed in the reactor tube between two inert ceramic monoliths which act as radiation shields and reduce the heat loss in the axial direction from the glowing gauze or monolith. The reaction zone is also externally insulated to reduce the radial heat loss and better approximate adiabatic operation.

**Table 1: Industrial Chemical Processes**

| Process | Reaction | Catalyst | Date | Temperature |
|---|---|---|---|---|
| $HNO_3$ synthesis (Ostwald) | $NH_3 + O_2 \rightarrow NO \rightarrow HNO_3$ | Pt-10% Rh gauze | 1900's | 800°C |
| HCN synthesis (Andrussow) | $CH_4 + NH_3 + O_2 \rightarrow HCN$ | Pt-10% Rh gauze | 1950's | 1100°C |
| Syngas generation | $CH_4 + H_2O \rightarrow CO + H_2$ | Ni on $Al_2O_3$ | 1930's | 700°C |
| | $CH_4 + O_2 \rightarrow CO + H_2$ | Rh on monolith | 1990's | 1000°C |
| Ethylene synthesis | $C_2H_6 \rightarrow C_2H_4 + H_2$ | homogeneous | 1940's | 900°C |
| | $C_2H_6 + O_2 \rightarrow C_2H_4 + H_2O$ | Pt on monolith | 1990's | 1000°C |

major industrial acid which is a strong oxidizing agent which produces no liquid residues. It was developed early in the 20th century to prepare nitrates for fertilizers and explosives. The process was developed by Ostwald and has seen only minor modifications over the past 90 years.

The reactor is a Pt-10% Rh woven gauze formed into 10-50 layers forming a catalyst "bed" several millimeters thick and several feet in diameter. In typical operation 10% $NH_3$ in air at 200°C is flowed over this catalyst at up to 10 atm pressure at flow velocities of several meters/second. Exothermic reactions heat the catalyst gauze to 800°C, and under typical operating conditions >90% of the $NH_3$ is reacted with over 90% converted to NO which if further oxidized homogeneously to $NO_2$ and $N_2O_5$ which is then hydrated to $HNO_3$.

As noted previously, the dominant competing reactions in this process can be written as

$$NH_3 + \tfrac{5}{4}O_2 \rightarrow NO + \tfrac{3}{2}H_2O \qquad (5)$$

and

$$NH_3 + \tfrac{3}{4}O_2 \rightarrow N_2 + \tfrac{3}{2}H_2O. \qquad (6)$$

However, the production of $N_2$ also probably occurs through the reaction of product NO with unreacted $NH_3$[8],

$$\tfrac{3}{2}NO + NH_3 \rightarrow \tfrac{5}{4}N_2 + \tfrac{3}{2}H_2O. \qquad (7)$$

This is a series-parallel process which is series in NO and parallel in $NH_3$ as shown in Figure 3. We were quite successful in modeling the performance of industrial nitric acid synthesis reactors using the individual steps such as those listed above from kinetics obtained using Langmuir-Hinshelwood rate forms[9].

*2.2 HCN synthesis*

Andrussow found in the 1950's that the products in the Ostwald process for nitric acid could be completely changed by adding methane and removing some of the oxygen to form not NO but HCN in the overall reaction[3,4]

$$NH_3 + CH_4 + \tfrac{3}{2}O_2 \rightarrow HCN + 3H_2O. \tag{8}$$

This of course competes with ammonia decomposition to $N_2$ (eq. 6) or oxidation to NO (eq. 5) and to methane oxidation to CO and $CO_2$ (eq. 1). These reactions are shown in Figure 3. Andrussow found that using a 1:1:1 feed ratio approximately 90% of the ammonia and $CH_4$ could be reacted and that up to 70% of this formed HCN rather than the above undesired products. Thus, this process can be regarded as a partial oxidation in which the two fuel molecules are coupled to form the C≡N triple bond and form water or as an ammoxidation of methane in which a N atom is added to $CH_4$ in an oxidative environment.

This process also operates adiabatically with a feed temperature near room

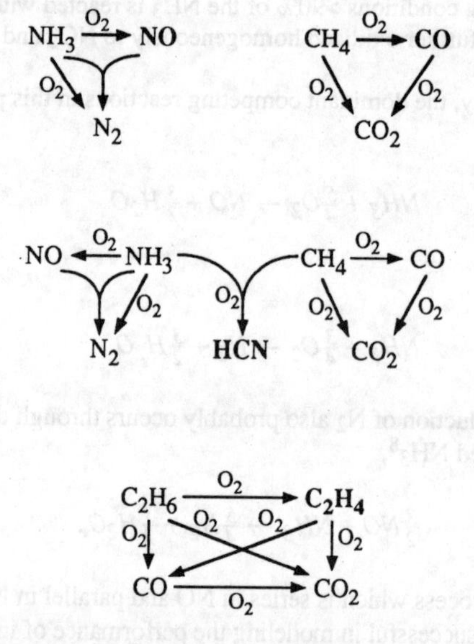


Figure 3: Schematic diagrams for he reaction networks for $NH_3$ oxidation, $CH_4$ oxidation, $CH_4$ ammoxidation, and $C_2H_4$ oxidation. The networks exhibit the series and parallel nature of these partial oxidation reactions. The desired product in each of the systems is highlighted in bold type.

temperature, but the heat of reaction is somewhat higher to yield a gauze temperature of 1100°C, and the process usually operates just above atmospheric pressure.

HCN is a key intermediate in Nylon 66 and in methyl methacrylate, and typical units produce several hundred tons per day using a 1mm thick gauze reactor several feet in diameter[3,5]. Both nitric acid and HCN synthesis reactors operate at extremely short contact times; gas velocities are approximately 1 m/sec, and the catalyst is just over 1 mm thick, so the contact time is approximately $10^{-3}$ sec. We shall see that similar short contact times are also essential in formation of synthesis gas and olefins by direct oxidation.

### *2.3 Production of Syngas*

Synthesis gas, a mixture of CO and $H_2$, is also a widely used chemical intermediate. It is used to produce $H_2$ needed in ammonia synthesis and in hydrotreating of heavy crude oil to produce gasoline. CO is used to produce acetic acid by carbonylation of methanol. Syngas is also used to produce methanol by the reaction

$$CO + 2H_2 \rightarrow CH_3OH \tag{9}$$

over Cu/ZnO catalysts and to produce synthetic diesel fuel by Fischer-Tropsch synthesis

$$nCO + 2nH_2 \rightarrow (CH_2)_n + nH_2O \tag{10}$$

over Co, Fe, or Ru catalysts.

Synthesis gas is now made commercially mostly by the steam reforming of methane[3]

$$CH_4 + H_2O \rightarrow CO + 3H_2, \qquad \Delta H^{o}_{298} = +49 \text{ kcal/mole.} \tag{11}$$

This is a highly endothermic reaction and the reaction must be carried out in a tube furnace with heat supplied from the outside of the tubes to heat each tube to 900°C to attain a favorable equilibrium in this reaction and obtain adequate kinetics. Typically this process is operated at 900°C at pressures of 30 atm with residence times of several seconds.

In a later section, we discuss the direct oxidation of methane to form syngas. The combustion reaction (eq. 1) is in competition with syngas production as shown in Figure 2, but this process shows promise when compared to the endothermic steam reforming reaction (eq. 11).

*2.4 Production of olefins*

The steam cracking of alkanes to olefins is also a very old process which is quite inefficient. Olefins are made by homogeneous pyrolysis of alkanes in a hot tube furnace[10]. With ethane as feed the dominant reaction would be

$$C_2H_6 \rightarrow C_2H_4 + H_2, \qquad \Delta H^o_{298} = +\ 33 \text{ kcal/mole.} \qquad (12)$$

Equilibrium and kinetics require that the reactor be maintained at temperatures greater than ~850°C, and the reaction is run in empty tubes with a contact time of approximately one second. Since these reactions involve gas phase free radicals, many product species are formed, ranging from methane to benzene. One of the products is solid carbon, and a large excess of steam is added with the alkane feed to reduce (but not eliminate) carbon formation. Since carbon formation is inevitable, this carbon must be periodically burned out of the tubes.

Olefins are the major building block of organic chemicals in the chemical industry, and steam cracking is the primary process by which they are formed currently. However, this process is very expensive because of the heat which must be supplied to operate the tube furnace, high separation costs of the complex product mix, and the down time because of carbon formation.

Thus, the reactions listed in Table 1 represent two fairly old oxidation processes which operate quite successfully, and two old endothermic processes which are in need of improvement if the technologies using their products as feedstocks are to be improved.

The intent of this discussion is to describe how the first two operate and to show that it appears possible to replace the endothermic and slow steam reforming and steam cracking processes by direct oxidation processes which offer the promise of much smaller and more efficient reactors.

## 3. Understanding and Designing an Oxidation Process

The standard paradigm in designing a catalytic process is to begin with a bench scale experiment, then scale it up to pilot plane scale and finally construct the full scale industrial reactor according to the sequence

$$\begin{matrix} \textit{bench} \\ \textit{scale} \\ \textit{batch} \end{matrix} \rightarrow \begin{matrix} \textit{bench} \\ \textit{scale} \\ \textit{continuous} \end{matrix} \rightarrow \begin{matrix} \textit{catalyst} \\ \textit{optimization} \end{matrix} \rightarrow \begin{matrix} \textit{pilot} \\ \textit{plant} \end{matrix} \rightarrow \begin{matrix} \textit{full} \\ \textit{scale} \\ \textit{reactor} \end{matrix} \qquad (13)$$

It is our contention that this procedure almost never works for oxidation processes, and that the new paradigm in designing oxidation reaction processes should be simply

$$\underset{reactor}{small} \rightarrow \underset{reactor}{large} \tag{14}$$

Bench scale reactors are frequently worthless in designing a practical reactor process because they fail to capture too many features which are essential in designing the large reactor.

The basic problem is that oxidation reactions are very fast and exothermic so that most oxidation processes are (1) mass transfer limited and (2) heat transfer limited. The former means that it is essentially impossible to measure reaction kinetics accurately: in most experiments one measures only mass transfer rates because these rate coefficients are smaller than reaction rate coefficients. The only way to avoid this problem is to dilute the reactants or to lower the temperature sufficiently to slow the reaction. However, this inevitably results in a regime of conditions where rates are not very meaningful with respect to the operating regime of an industrial reactor.

The second problem is heat transfer effects and temperature nonuniformity in the reactor. In a packed bed microreactor it is essentially impossible to maintain the temperature precisely uniform because exothermic reactions heat the catalyst, particularly near the center of the bed where rates then accelerate and generate even more heat. This obscures rate measurements, and only by severe dilution or lowering temperature can these effects be minimized.

Another consequence of exothermic reactions is the possibility of ignition of the reaction, the phenomenon of the glowing wire. As one begins to heat a reactor, the rate is negligible until the catalyst reaches a temperature where the rate suddenly accelerates, the temperature rises rapidly, and the reaction is suddenly nearly at completion. Thus oxidation reactors exhibit "lightoff" in which multiple steady states occur[6,7]. This is in fact beneficial in oxidation reactors: the reactor is said to become "autothermal" with the heat of reaction providing the heat to sustain the reactor operating adiabatically. Most industrial oxidation reactors operate autothermally and nearly adiabatically. This is a key to their successful operation: fast and exothermic reactions supply their own heat and thus avoid the necessity of external process heat as must be added in steam reforming to produce syngas (eq. 11) and steam cracking to produce olefins (eq. 12).

## 4. Reactor Simulation

Because oxidation reactors involve a complex interplay between reaction kinetics, mass transfer, and heat transfer, is essential in understanding them to develop a model or simulation of the process. In such situations, intuition fails because the experimentalist cannot understand his results, and the theorist cannot interpret his calculations. Only by integrating experimental results with simulations of the processes can these processes be understood and can new processes be created.

One needs a complete model of the process. This comes from solving complete mass and energy balances in the reactor. These can be quite complicated if fluid mechanics must be included, but packed bed reactors can usually be modeled as nearly plug flow reactors

$$u\frac{dC_j}{dz}=\sum_i \nu_{ij}r_i \tag{15}$$

and

$$uC_p\frac{dT}{dz}=\sum_i(-\Delta H_i r_i)-Q(t,z). \tag{16}$$

Equation 15 is the mass balance on species j with concentration $C_j$ and stoichiometric coefficient $\nu_{ij}$ for reaction rate $r_i$ for a fluid flowing with velocity u in a tube versus position z (total length L). Equation 16 is the energy balance on temperature T for heat capacity $C_p$ and $\Delta H_i$ the heat of reaction i. Both gauze and monolithic reactors can be modeled as plug flow reactors.

In a fluidized bed reactor or in any reactor at sufficiently low pressures that diffusion provides rapid mixing (for example, UHV experiments), the mass balance is described as a stirred tank reactor,

$$C_{jo}-C_j=\tau\sum_i \nu_{ij}r_i, \tag{17}$$

which has the energy balance

$$C_p(T-T_o)=-\tau\sum_i \Delta H_i r_i, \tag{18}$$

In the stirred tank reactor the residence time $\tau$ is defined as the reactor volume divided by the volumetric flow rate. In the plug flow reactor, the residence time is L/u, and in any case we are very interested in comparing conversion and selectivity versus residence time $\tau$ for different feed and reactor parameters.

We shall not of course be able to develop these equations in this presentation which are discussed in the texts by Levenspiel[6] and Fogler[7], but the student should be aware of the governing mass and energy balance equations in interpreting experiments and in predicting how experiments suggest operating conditions for an industrial process.

## 5. Mechanism and Kinetics of HCN Synthesis

As we indicated previously, this reaction involves the reaction of $CH_4$, $NH_3$, and $O_2$ (eq. 8) over a Pt-10% Rh surface to yield 70% HCN. We have examined the reactions in this system extensively[11-13]. We measured kinetics of individual steps at low pressure (where mass and heat transfer effects are negligible and the stirred tank reactor equation is applicable. We found that the reaction between $CH_4$ and $NH_3$ without $O_2$ will produce HCN with high efficiency. Thus we suggest that the Andrussow process can be interpreted simply as proceeding through the bimolecular reaction of $CH_4$ and $NH_3$ to form HCN with $O_2$ mainly playing the role of oxidizing the product $H_2$ to provide the heat necessary to compensate for the highly endothermic dehydrogenation reaction and to heat the catalyst to 1100°C.

Figure 4 shows the rates of formation of HCN, $r_{HCN}$, and of $N_2$, $r_{N2}$, the products of this reactions without oxygen. Thus, we assume that the real reaction steps are

$$CH_4 + NH_3 \rightarrow HCN + 3H_2 \tag{19}$$

and

$$NH_3 \rightarrow \tfrac{1}{2}N_2 + \tfrac{3}{2}H_2 \tag{20}$$

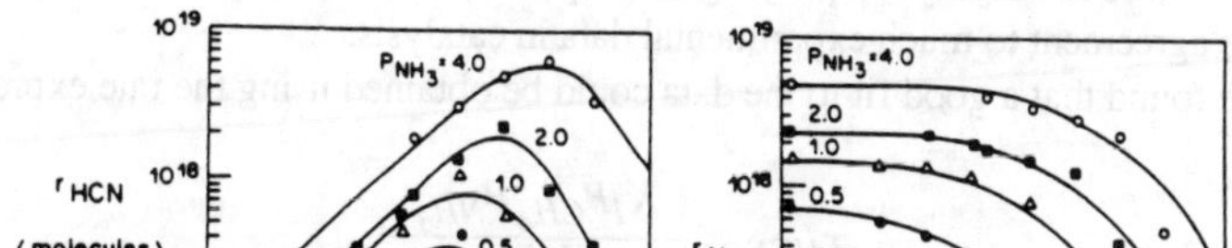

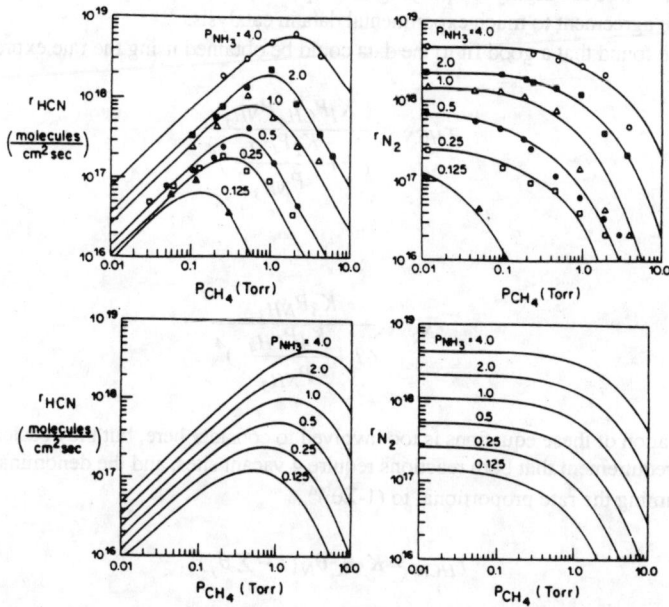


Figure 4: Upper panels: Measured kinetic of the reactions between $CH_4$ and $NH_3$ in the absence of $O_2$ over Pt foils. Rates of HCN synthesis, $r_{HCN}$ (left), and $N_2$ formation, $r_{N2}$ (right), versus $P_{CH4}$ at 1450 K for various $P_{NH3}$ as indicated. Lower panels: Calculated rates of HCN (left) and $N_2$ (right) formation versus $P_{CH4}$ for the same conditions as used in the upper panels. The calculations use Langmuir-Hinshelwood kinetics as listed in eq. 22-23.

which suggests that the mechanism is one of the bimolecular reaction (eq. 19) competing with the unimolecular decomposition of $NH_3$ (eq. 20). It is seen that for pure $NH_3$ the only reaction is formation of $N_2$ (eq. 20) and that, as the $CH_4$ pressure is increased, the HCN increases linearly with $P_{CH4}$ (the rate is first order in $CH_4$ in this regime). Then near a 1/1 ratio, both the HCN and $N_2$ begin to decrease rapidly (the rate is proportional to $P_{CH4}^{-4}$). It is seen that the maximum selectivity to HCN

$$S_{HCN} = \frac{r_{HCN}}{(r_{HCN} + r_{N_2})} = 0.9, \tag{21}$$

and that this occurs when $P_{CH4} = P_{NH3}$. This experiment shows that these low pressure experiments do produce predominantly HCN.

We next attempted to fit the data to a Langmuir-Hinshelwood type rate expressions[14]. These expressions are the ones usually used to fit experimental data in surface reactions. The model assumes that (1) adsorbates are confined to a monolayer, (2) all adsorbates are competitive, and (3) that adsorbate properties are independent of the coverages. These are highly simplifying assumptions, but they are found to give remarkable agreement to much experimental data in catalysis.

We found that a good fit to the data could be obtained using the rate expressions

$$r_{HCN} = \frac{K_1 P_{CH_4} P_{NH_3}}{\left(1 + \frac{K_2 P_{CH_4}}{P_{NH_3}}\right)^4} \tag{22}$$

and

$$r_{N_2} = \frac{K_3 P_{NH_3}}{\left(1 + \frac{K_4 P_{CH_4}}{P_{NH_3}}\right)^4} \tag{23}$$

The derivation of these equations is too involved to consider here, but they basically come from the requirement that both reactions require 4 vacant sites, and the denominators come from assuming the rate proportional to $(1-\Sigma\theta_j)^4$

$$r_{HCN} = K_5 \theta_C \theta_N (1 - \sum_j \theta_j)^4. \tag{24}$$

In all of these expressions the K's are groupings of reaction and equilibrium rate coefficients. It is seen in Figure 4 that these expressions give an excellent fit to the data over a wide range of pressures and pressure ratios. They suggest (but do not prove) that we have the correct mechanism.

These kinetics show that in excess $CH_4$ the rates are strongly blocked by some species (as the 4th power of its coverage), and we initially thought this species was carbon from methane decomposition in excess $CH_4$.

We next used ultrahigh vacuum (UHV) experiments on Pt and Rh single crystals to try to identify the species responsible for the slowing down of $NH_3$ decomposition to allow HCN to form[15,16]. We did this by temperature programmed desorption (TPD), a very versatile and simple technique in which one heats the surface with adsorbates and follows the desorbing produces with a mass spectrometer.

However, we cannot examine this reaction in UHV because neither $CH_4$ or $NH_3$ will react at the low temperatures (<300K) necessary to attain significant adsorption at $10^{-8}$ Torr. However, we were able so simulate the HCN synthesis reaction, not by using a mixture of $CH_4$ and $NH_3$, but instead by using methyl amine, $CH_3NH_2$. This molecule contains the C-N single bond and we can use it to determine the decomposition of C-H, N-H, and C-N bonds. Dissociation of the C-N bond would not lead to HCN, but dehydrogenation would yield HCN

$$CH_3NH_2 \rightarrow HC{\equiv}N + 2H_2. \tag{25}$$

In fact when we adsorb a monolayer of methyl amine on Pt, the only products we observe are $H_2$, HCN, and cyanogen ($C_2N_2$). Cyanogen is the product which must form if H has evaporated to leave only CN which must dimerize to evaporate. The other surprising feature of these experiments is that $C_2N_2$ does not desorb in UHV until 1000 K. In other words, the CN species is very strongly adsorbed on Pt (more strongly than O atoms or any other common adsorbates).

Therefore we are led to the conclusion that the HCN reaction occurs, not because carbon blocks $NH_3$ decomposition but because the surface nitrile CN species blocks the surface against $NH_3$ decomposition and allows HCN to form.

We finally used rate expressions such as those described in section 4 to attempt to model the performance of the atmospheric pressure reactor[14]. The flow of gases through the gauze reactor should be nearly plug flow ( a "tube" several mm long and with a diameter equal to the distance between gauze wires which is 0.1 mm), so the equations to be solved are the plug flow mass balances (eq. 16) with now the concentrations expressed in partial pressures. We solved the approximately 10 equations (one for each species) with in initial conditions corresponding to the partial pressures of the feed gases. We solved for the exit concentrations and compared these results to those observed in the experimental reactor. We observed very good agreement between model and predicted concentrations for all species, especially with the variation with feed conditions, a strong test of the model.

Thus we conclude that we can describe the performance of the HCN reactor exceedingly well. Further, we believe we understand the mechanism in terms of an oxidative dehydrogenation of $CH_4$ and $NH_3$ which produces CN that blocks the surface against excessive $NH_3$ decomposition.

## 6. The Simplest Oxidation Reaction: $H_2+O_2$

One of the problems in interpreting reactions is the multiple products which are formed and the multiple reactions which occur. We must handle many reactions to describe processes of practical interest.

We next consider the simplest example of a catalytic oxidation process which is the oxidation of $H_2$

$$H_2 + \tfrac{1}{2}O_2 \rightarrow H_2O, \qquad \Delta H^o_{298} = 57 \text{ kcal/mole}, \tag{26}$$

on noble metals. Although, this process has little industrial importance, it is a prototype of more realistic reactions. Furthermore, this reaction is a subset of all oxidation reactions involving H and O species.

We know more about $H_2$ and $O_2$ adsorption properties than almost any species except CO, and therefore many of the elementary steps are available from the surface science literature. We stress, however, that we know of no experiments in which the kinetics of this reaction have been measured except under UHV conditions; in all other experiments only the rates of adsorption or mass transfer rates of $H_2$ or $O_2$ can be measured because this reaction is almost invariable mass transfer limited.

The elementary steps in this reaction are probably

$$H_2 \leftrightarrow 2H_s, \tag{27}$$

$$O_2 \leftrightarrow 2O_s, \tag{28}$$

$$O_s + H_s \leftrightarrow OH_s, \tag{29}$$

$$OH_s + H_s \leftrightarrow H_2O_s, \tag{30}$$

$$H_2O_s \leftrightarrow H_2O, \tag{31}$$

$$2OH_s \leftrightarrow H_2O_s + H_s. \tag{32}$$

All of these steps are reversible, so the process must involve 12 distinct elementary steps.

Here we see the problem with this simplest of oxidation reactions: it *must involve at least 12 elementary steps*. All of the adsorption and desorption steps are well characterized in the literature. However, the surface reactions steps, eq. 29, 30, and 32, were not well known until recently. They all require some spectroscopic characterization of the surface coverage of species to measure the kinetics of these steps.

We have recently used laser induced fluorescence (LIF) of desorbing OH (in the gas phase) and laser induced desorption of product water (measured on a microsecond time scale) to characterize the elementary steps of the surface reaction steps[17,18].

TABLE 2: RATE PARAMETERS FOR $H_2$ OXIDATION

| Reaction | Pt $k_o$ $torr^{-1}s^{-1}$ | Pt Ea kcal/mol | Rh $k_o$ $torr^{-1}s^{-1}$ | Rh Ea kcal/mol |
|---|---|---|---|---|
| $H_2(g) \rightarrow 2H$ | $7.5 \times 10^4$ | 0 | $2.25 \times 10^5$ | 0 |
| $2H \rightarrow H_2(g)$ | $5.0 \times 10^{12}$ | 18 | $5.0 \times 10^{12}$ | 18 |
| $O_2(g) \rightarrow 2O$ | $1.25 \times 10^3$ | 0 | $3.5 \times 10^3$ | 0 |
| $2O \rightarrow O_2(g)$ | $5.0 \times 10^{12}$ | 52 | $5.0 \times 10^{12}$ | 70 |
| $H_2O(g) \rightarrow H_2O$ | $5.0 \times 10^4$ | 0 | $7.4 \times 10^4$ | 0 |
| $H_2O \rightarrow H_2O(g)$ | $1.0 \times 10^{13}$ | 10.8 | $1.0 \times 10^{13}$ | 10.8 |
| $OH(g) \rightarrow OH$ | 0 | 30 | 0 | 30 |
| $OH \rightarrow OH(g)$ | $1.5 \times 10^{13}$ | 48 | $8.1 \times 10^{11}$ | 34 |
| $H + O \rightarrow OH$ | $1.0 \times 10^{15}$ | 2.5 | $7.0 \times 10^{12}$ | 20 |
| $OH \rightarrow H + O$ | $1.0 \times 10^8$ | 5 | $1.0 \times 10^{13}$ | 5 |
| $H + OH \rightarrow H_2O$ | $9.0 \times 10^{16}$ | 15 | $3.0 \times 10^{17}$ | 8 |
| $H_2O \rightarrow H + OH$ | $1.8 \times 10^{13}$ | 37 | $5.0 \times 10^{14}$ | 37 |
| $2OH \rightarrow H_2O + O$ | $1.0 \times 10^{15}$ | 12.3 | $4.0 \times 10^{15}$ | 15 |
| $H_2O + O \rightarrow 2OH$ | 0 | 31 | 0 | 63 |

These results are shown in Table 2. Listed are the reaction steps of the above sequence (eq. 27-32) and the preexponential factor and activation energy of each reaction. We assume that adsorption steps are of the form

$$r_{aj} = k_a(T)P_j(1 - \Sigma\theta) \tag{33}$$

desorption steps of the form

$$r_{dj} = k_{dj}\theta_j, \tag{34}$$

and reaction steps of the form

$$r_R = k_R\theta_j\theta_{j'}. \tag{35}$$

Here the k's have Arrhenius forms

$$k_j = k_{oj}\, exp(E_j / RT) \tag{36}$$

and surface concentrations are represented by coverages $\theta_j$ in monolayers. We assume that the total coverage cannot exceed one monolayer so that $\Sigma\theta_j<1$. We also assume that adsorption requires a vacant site so that adsorption rates have a factor $1-\Sigma\theta$.

Thus we have the rate expressions of the mechanistic steps for this reaction. We can use the activation energies of these steps from Table 2 to construct the "potential energy surface" over which molecules must travel in the reaction. This is shown in Figure 5. The vertical scale is energy in kilocalories per mole while the horizontal axis is the "reaction coordinate" over which molecules pass in going from $H_2 + O_2$ at the left to $H_2O$ on the right. The gaseous reactants and product have different energies by the enthalpy of the reaction or the heat of formation of water which is -57 kcal/mole.

It is evident that this is a "downhill" or exothermic process as are all oxidation processes. The largest barrier on Pt is the step $OH_S + H_S \rightarrow H_2O_S$ (eq. 30), which has a

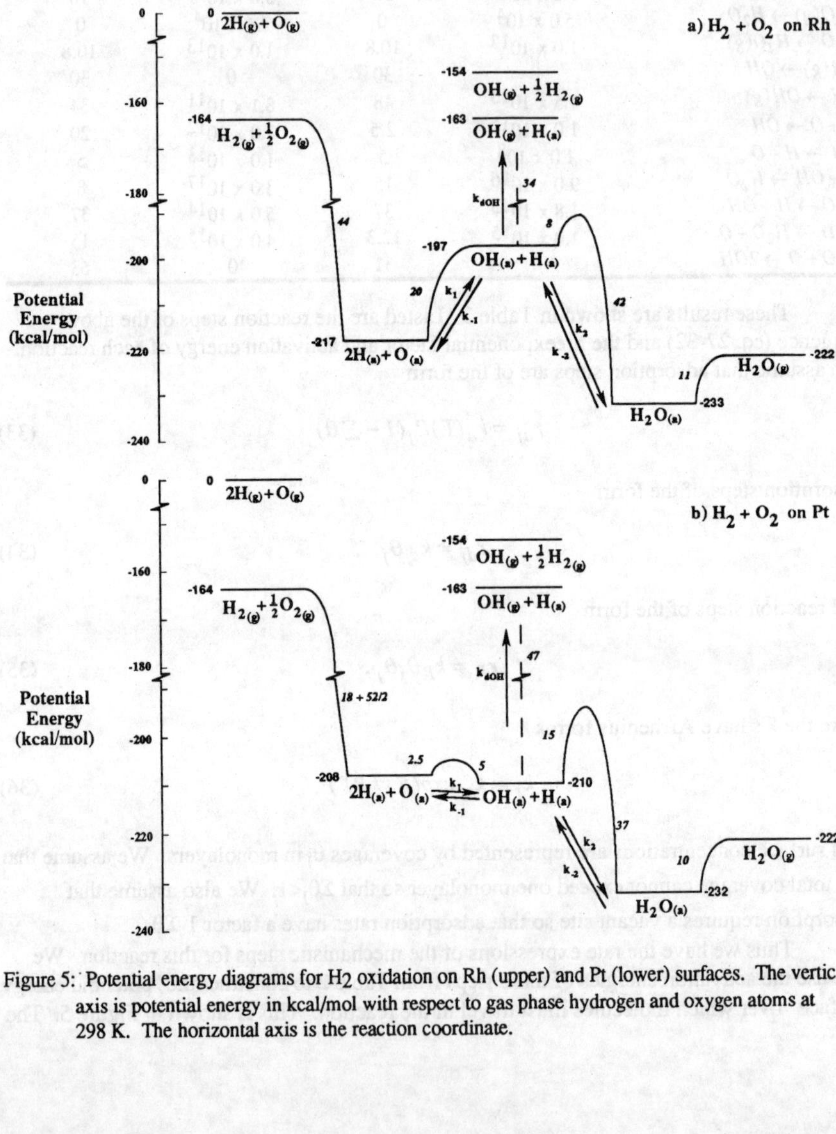


Figure 5: Potential energy diagrams for $H_2$ oxidation on Rh (upper) and Pt (lower) surfaces. The vertical axis is potential energy in kcal/mol with respect to gas phase hydrogen and oxygen atoms at 298 K. The horizontal axis is the reaction coordinate.

barrier of 15 kcal/mole. On Rh the steps $O_s + H_s \rightarrow OH_s$ (eq. 29) and $OH_s + H_s \rightarrow H_2O_s$ (eq. 30) both have significant barriers so that the total barrier in forming water on Rh is 28 kcal/mole. We shall see later that this difference is important in explaining the products in the partial oxidation of $CH_4$ on Pt and Rh.

## 7. Hydrocarbon Oxidation

### *7.1 Syngas by Direct Oxidation of Methane*

The next simplest fuel after $H_2$ is $CH_4$. We now consider the direct oxidation of $CH_4$ over Rh and Pt coated monolithic catalysts to form syngas [19,20]

$$CH_4 + \tfrac{1}{2}O_2 \rightarrow CO + 2H_2 \qquad \Delta H^{o}_{298} = -8.5 \text{ kcal/mol} \tag{37}$$

as compared to the current industrial process of steam reforming $CH_4$ over Ni catalysts to form syngas (eq. 11).

The symbols in Figure 6 represent experimental data for methane oxidation over a

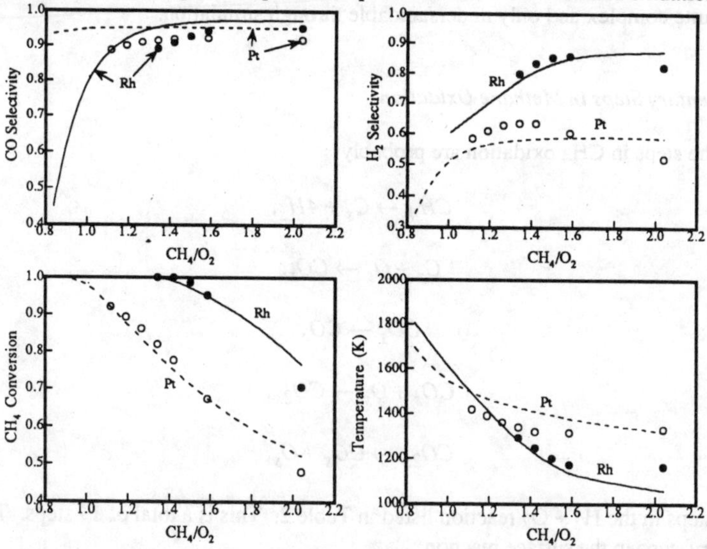


Figure 6: CO selectivity, $H_2$ selectivity, $CH_4$ conversion, and reaction temperature for $CH_4$ oxidation in air as a function of the fuel to oxygen ratio over a 80 ppi 9.8 wt.% Rh foam monolith (closed symbols, solid lines) and a 50 ppi 11.6 wt.% Pt foam monolith (open symbols, dashed lines). The symbols are experimental data points and the curves are model predictions. In all cases, the $CH_4$ and air mixture was premixed and heated to 460°C.

Rh monolith (filled symbols) and a Pt monolith (open symbols). The Pt and Rh catalysts behave similarly for the CO selectivity, but Rh yields a much higher selectivity to $H_2$. This change in reaction products also affects the methane conversion and reaction temperature.

Syngas production by direct oxidation of $CH_4$ (eq. 37) has several advantages over the current steam reforming process (eq. 11). (1) CO and $H_2$ are produced in the correct stoichiometric ratio for downstream processes including $CH_3OH$ synthesis (eq. 9) and Fischer-Tropsch synthesis (eq. 10). (2) The oxidation reaction (eq. 37) is exothermic and the process can operate autothermally. This affords a large savings in production costs compared to the endothermic steam reforming reaction (eq. 11) which must be heated to 700 to 900°C to obtain high yields of syngas. (3) As shown in Figure 6, the direct oxidation process (eq. 37) over a Rh catalyst obtains a high syngas yield (~100% $CH_4$ conversion, > 90% selectivity to CO and $H_2$) in < 10 milliseconds compared to the 1 second reaction time required in steam reforming (eq. 11) with nearly the same syngas yield.

These last two points offer an enormous opportunity to reduce the cost of producing syngas by a large factor. The cost of syngas production currently accounts for 60-70% of the total cost of either $CH_3OH$ synthesis from natural gas or hydrocarbon production via Fischer-Tropsch synthesis. The success of this process hinges on the interactions between reaction kinetics and fast mass and energy transport. This makes the process quite complex and only understandable through simulation.

### *7.2 Elementary Steps in Methane Oxidation*

The steps in $CH_4$ oxidation are probably

$$CH_4 \rightarrow C_s + 4H_s, \tag{38}$$

$$C_s + O_s \rightarrow CO_s, \tag{39}$$

$$CO_s \rightarrow CO, \tag{40}$$

$$CO_s + O_s \rightarrow CO_{2s}, \tag{41}$$

$$CO_{2s} \rightarrow CO_s + O_s, \tag{42}$$

plus the steps in the $H_2 + O_2$ reaction listed in Table 2. This is a total of 19 steps. The elementary steps in the surface reaction

$$CO + \tfrac{1}{2}O_2 \rightarrow CO_2 \tag{43}$$

TABLE 3: RATE PARAMETERS FOR $CH_4$ OXIDATION

| | Pt | | Rh | |
|---|---|---|---|---|
| Reaction | $k_o$ $torr^{-1}s^{-1}$ | Ea kcal/mol | $k_o$ $torr^{-1}s^{-1}$ | Ea kcal/mol |
| $CH_4(g) \rightarrow C + 4H$ | $5.0 \times 10^4$ | 10.3 | $3.0 \times 10^4$ | 5 |
| $CO(g) \rightarrow CO$ | $3.21 \times 10^5$ | 0 | $1.91 \times 10^5$ | 0 |
| $CO \rightarrow CO(g)$ | $1.0 \times 10^{13}$ | 30 | $4.0 \times 10^{13}$ | 31.6 |
| $CO_2(g) \rightarrow CO + O$ | 0 | 36 | 0 | 26 |
| $CO + O \rightarrow CO_2(g)$ | $1.0 \times 10^{15}$ | 24 | $1.0 \times 10^{12}$ | 25 |
| $CO \rightarrow C + O$ | $1.0 \times 10^{11}$ | 44 | $1.0 \times 10^{11}$ | 40 |
| $C + O \rightarrow CO$ | $5.0 \times 10^{13}$ | 15 | $5.0 \times 10^{13}$ | 15 |
| (plus those for $H_2 + O_2$) | | | | |

are well known on Pt and Rh from the surface science literature, although there are in fact many disagreements between experimental determinations of many of these parameters.

Table 3 lists the rate coefficients for these reactions in methane oxidation[21]. The notations and units are the same as for $H_2$ oxidation, and the entries for this reaction are not repeated here. We have used these parameters the develop a computer simulation of methane oxidation over noble metal coated monoliths. The results of this simulation are shown as the curves in Figure 6. Obviously, the model does an excellent job of predicting the data when kinetics, heat transfer, and mass transfer are all considered.

We can use these activation energies to construct a potential energy surface over which these molecules must pass in creating gaseous $H_2$, $H_2O$, CO, and $CO_2$. This is shown in Figure 7. As with $H_2$+$O_2$, the gaseous species energies are obtained from thermodynamic data. Note that the right hand portion of these curves are simply taken from Figure 5 for $H_2 + O_2$.

It is evident that the energy surfaces are very similar on Pt and Rh. The major difference comes in the $H_2 + O_2$ portion of the diagram, and this explains in fact why Rh is a superior catalyst to Pt in producing $H_2$. The barrier to form $H_2$ compared to $H_2O$ is *lower* on Rh, while the barrier to form $H_2O$ is lower on Pt. Restated, the adsorbed $H_s$ atoms on Rh have a lower barrier to dimerize

$$2H_s \rightarrow H_2 \qquad (44)$$

rather than wait for reaction with adsorbed oxygen which forms $OH_s$ and then $H_2O$ which quickly desorbs.

### 7.3. *Ethylene by Oxidative Dehydrogenation of Ethane*

Oxidation reactions become even more complex when we consider the oxidation of larger hydrocarbons. Ethane is the next simplest hydrocarbon, but it can undergo several oxidation reaction including combustion

$$C_2H_6 + \frac{7}{2}O_2 \rightarrow 2CO_2 + 3H_2O, \;\; \Delta H^o_{298} = -\ 348 \text{ kcal/mol,} \qquad (45)$$

partial oxidation to syngas

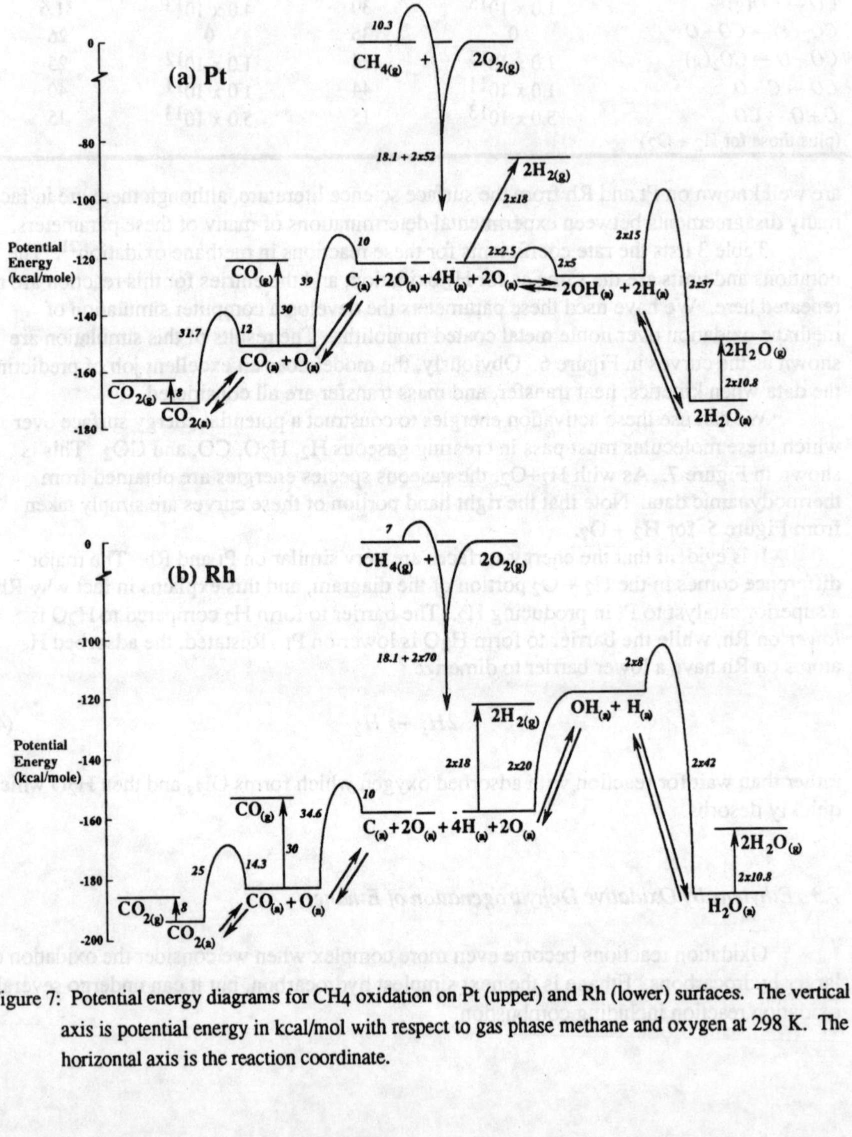


Figure 7: Potential energy diagrams for $CH_4$ oxidation on Pt (upper) and Rh (lower) surfaces. The vertical axis is potential energy in kcal/mol with respect to gas phase methane and oxygen at 298 K. The horizontal axis is the reaction coordinate.

$$C_2H_6 + O_2 \rightarrow 2CO + 3H_2, \qquad \Delta H^o_{298} = -\ 33\ \text{kcal/mol}, \tag{46}$$

and oxidative dehydrogenation

$$C_2H_6 + \tfrac{1}{2}O_2 \rightarrow C_2H_4 + H_2O, \qquad \Delta H^o_{298} = -\ 25\ \text{kcal/mol}. \tag{47}$$

Since all of these reactions (eq. 45-47) are exothermic, they can provide the heat for endothermic reactions to take place including the thermal dehydrogenation of ethane to form ethylene (eq. 12).

We have examined the partial oxidation and oxidative dehydrogenation of ethane over Pt and Rh coated monolith[22] in an autothermal reactor operating at ~1000°C near atmospheric pressure with residence time $\tau$ of 1 to 10 milliseconds. Figure 8 shows

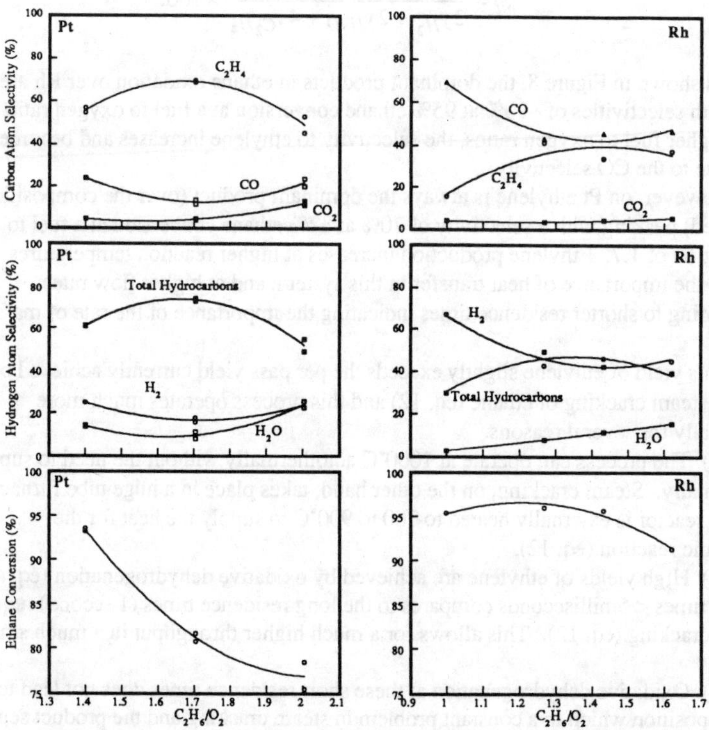


Figure 8: Carbon selectivity, hydrogen selectivity, and ethane conversion for ethane oxidation in $O_2$ over a 2.3 wt. % Pt (left panels) and a 4.0 wt. % Rh (right panels) foam monolith catalyst as a function of the fuel to oxygen ratio. The reactants flow at a total flow rate of 4.5 slpm (with 20% $N_2$ present for calibration purposes) at a total pressure of 1.4 atm.

selectivities and ethane conversion as a function of the reactant composition over both Pt and Rh catalysts.

For these plots, the conversion and selectivities have been defined on a carbon atom or hydrogen atom basis. This implicitly accounts for any mole number change due to reaction. For CO, $CO_2$, and $C_2H_4$ the selectivity is defined as (assuming CO, $CO_2$, and $C_2H_4$ are the only products containing carbon)

$$S_i = \frac{c_i y_i}{y_{CO} + y_{CO_2} + 2y_{C_2H_4}} \times 100 \tag{48}$$

where $y_i$ is the molar flow rate of species i in the product stream and $c_i$ is the number of carbon atoms in species i. The $H_2$ and $H_2O$ selectivities are defined similarly as

$$S_i = \frac{2y_i}{2y_{H_2} + 2y_{H_2O} + 4y_{C_2H_4}} \times 100. \tag{49}$$

As shown in Figure 8, the dominant products in ethane oxidation over Rh are CO and $H_2$ with selectivities of ~70% at 95% ethane conversion at a fuel to oxygen ratio of 1.0. At higher fuel to oxygen ratios, the selectivity to ethylene increases and becomes comparable to the CO selectivity.

However, on Pt ethylene is always the dominant product (over the composition range tested) peaking with a selectivity of 70% at 82% ethane conversion at a fuel to oxygen ration of 1.7. Ethylene production increases at higher reaction temperatures indicating the importance of heat transfer in this system, and at higher flow rates corresponding to shorter residence times indicating the importance of the rate of mass transfer.

This yield of ethylene slightly exceeds the per pass yield currently achieved by industrial steam cracking of ethane (eq. 12) and this process operates much more economically for several reasons.

(1) The process can operate at 1000°C autothermally without the need to supply heat externally. Steam cracking, on the other hand, takes place in a huge tube furnace where the reactor is externally heated to 700 to 900°C to supply the heat for the endothermic reaction (eq. 12).

(2) High yields of ethylene are achieved by oxidative dehydrogenation (eq. 47) in residence times < 5milliseconds compares to the long residence times (1 second) required by steam cracking (eq. 12). This allows for a much higher throughput in a much smaller reactor.

(3) Oxidative dehydrogenation at these short residence times does not lead to carbon deposition which in a constant problem in steam cracking and the product separation should be more straight forward.

*7.4 Elementary Steps in $C_2H_6$ Oxidation*

We believe that the elementary steps in ethane oxidation include

$$C_2H_6 \rightarrow C_2H_{5s} + H_s \tag{50}$$

$$C_2H_{5s} \rightarrow C_2H_{4s} + H_s \tag{51}$$

$$C_2H_{5s} \rightarrow 2C_s + 5H_s \tag{52}$$

$$C_2H_{4s} \leftrightarrow C_2H_4 \tag{53}$$

plus all of the steps in the $H_2 + O_2$ reaction listed in Table 2 as well as all of the additional steps in $CH_4$ oxidation listed in Table 3 except the decomposition of $CH_4$ (eq. 34)[22]. A reaction network describing these probable reaction paths is shown in Figure 9. The rate expressions for these additional steps (eq. 50-53) are less well known and a mechanistic model is still in preparation.

## 8. Summary

We have discussed several catalytic oxidation process, some old, some new. These processes all involve very fast, very exothermic reactions. In all cases very high selectivities to the desired products are obtained at nearly complete reactant conversion in

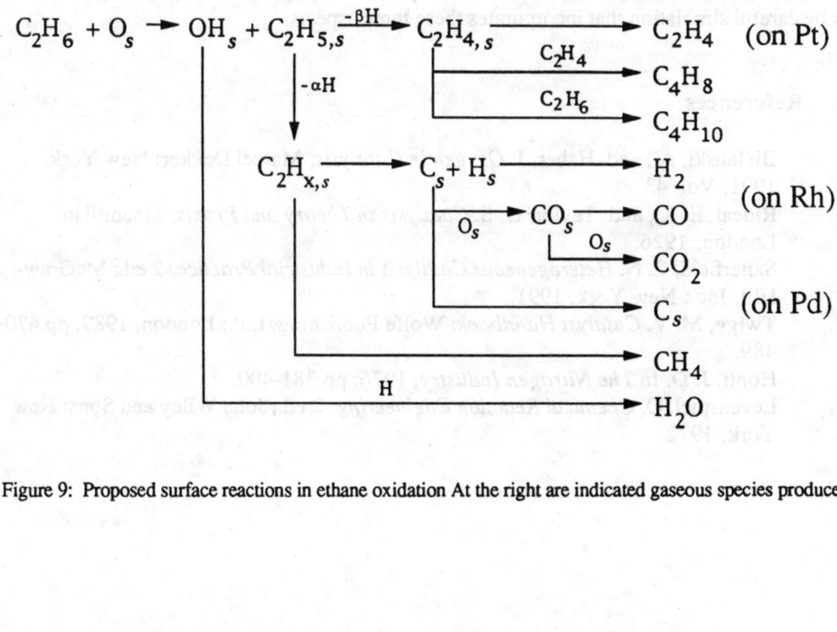


Figure 9: Proposed surface reactions in ethane oxidation At the right are indicated gaseous species produced.

less that 10 milliseconds residence time.

Since these reactions are usually mass transfer limited, for these high conversions to be reached in such short residence times, the rate of mass transfer to and from the catalyst surface must be fast. In fact, at longer residence times, the conversion may decrease because the longer residence time is obtained by reducing the flow rate. At lower flow rates, the rate of mass transfer decreases and the reaction slips into a mass transfer limited regime with a lower conversion. At the longer residence times, the selectivities to the desired products also decrease. This occurs because we are interested in producing intermediate products as shown in Figure 3. At longer residence times, these intermediate products react away. High flow rates not only lead to favorable mass transfer properties, but also reduce the contribution of homogeneous reactions which usually lead to undesired products by thinning the boundary layer over the catalyst surface, scavenging radicals, and preventing excessive gas-phase temperatures.

Obviously, the attainment of high rates of mass transfer is essential in optimizing all of these processes. Unfortunately, mass transfer rates are poorly characterized under these conditions. Correlations exist for mass transfer rates in packed beds which could be applicable to the monolith or gauze structure, but these correlations are strictly for isothermal, nonreacting systems. To further complicate matters, these reactions are occurring so quickly that the important mass transfer characteristics may simply be "entrance effects" that are not well understood at all. Since the gas temperature rises rapidly to the reaction temperature, the gases accelerate as they enter the reactor due to thermal expansion. With all of these complications, no simple correlation can capture the mass transfer characteristics.

Superficially these processes seem simple: small molecules on noble metals. However, due to the interplay between mass transfer, heat transfer, and kinetics, these processes are sufficiently complex that the only way to decipher the reaction steps involved is by careful simulation that incorporates these three aspects.

## 9. References

1. Bielanski, A., and Haber, J. *Oxygen in Catalysis*; Marcel Dekker: New York, 1991; Vol. 43.
2. Rideal, E. K., and Taylor, H. S. *Catalysis in Theory and Praxis*; Macmillan: London, 1926.
3. Satterfield, C. N. *Heterogeneous Catalysis in Industrial Practice;* 2 ed.; McGraw-Hill, Inc.: New York, 1991.
4. Twigg, M. V. *Catalyst Handbook*; Wolfe Publishing, Ltd.: London, 1989, pp 470-489.
5. Honti, J. D. In *The Nitrogen Industry*, 1976; pp 381-400.
6. Levenspiel, O. *Chemical Reaction Engineering;* 2 ed.; John Wiley and Sons: New York, 1972.

7. Fogler, H. S. *Elements of Chemical Reaction Engineering*; Prentice-Hall: Englewood Cliffs, 1986.
8. Pignet, T., and Schmidt, L. D., *Chem. Eng. Sci.* **29** (1974) 1123.
9. Hickman, D. A., and Schmidt, L. D., *Ind. Eng. Chem. Res.* **30** (1991) 50.
10. Song, Y., Velenyi, L. J., Leff, A. A., Kliewer, W. R., and Metcalfe, J. E. In *Novel Production Methods for Ethylene, Light Hydrocarbons, and Aromatics*; L. F. Albright, B. L. Crynes and S. Nowak, Ed.; Marcel Dekkar, Inc: New York, 1992; pp 319.
11. Hasenberg, D., and Schmidt, L. D., *J. Catal.* **97** (1986) 156.
12. Hasenberg, D., and Schmidt, L. D., *J. Catal.* **104** (1987) 441.
13. Hickman, D. A., Huff, M., and Schmidt, L. D., *Ind. Eng. Chem. Res.* **32** (1993) 809.
14. Waletzko, N., and Schmidt, L. D., *AIChE J.* **34** (1987) 1146.
15. Hwang, S. Y., and Schmidt, L. D., *J. Phys. Chem.* **93** (1989) 8327.
16. Hwang, S. Y., Seebauer, E. G., and Schmidt, L. D., *Surf. Sci.* **188** (1987) 219.
17. Williams, W. R., Marks, C. M., and Schmidt, L. D., *J. Phys. Chem.* **96** (1992) 5922.
18. Zum Mallen, M. P., Williams, W. R., and Schmidt, L. D., *J. Phys. Chem.* **97** (1993) 625.
19. Hickman, D. A., and Schmidt, L. D., *J. Catal.* **138** (1992) 267.
20. Hickman, D. A., Haupfear, E. A., and Schmidt, L. D., *Catal. Lett.* **17** (1993) 223.
21. Hickman, D. A., and Schmidt, L. D., *AIChE Journal* **39** (1993) 1164.
22. Huff, M., and Schmidt, L. D., *J. Phys. Chem.* **97** (1993) 11815.

# HIGH TEMPERATURE OXIDATION PROCESSES: OXIDATIVE COUPLING OF METHANE

G.B. MARIN

*Laboratorium voor Chemische Technologie, Schuit Institute of Catalysis, Eindhoven University of Technology, P.O. Box 513, 5600 MB Eindhoven, NL*

## ABSTRACT

**The oxidative coupling of methane is discussed in terms of reaction pathways and catalyst development. The importance of the interplay between chemical kinetics and mass transport phenomena and its consequences for selectivity towards ethane and ethene is highlighted.**

## 1. Introduction

Steam reforming aimed at the production of synthesis gas is the most important commercial process for methane conversion into chemicals and liquid fuels. The conversion of methanol to gasoline (MTG) or olefins (MTO) or the Fischer-Tropsch syntheses to hydrocarbons, such as the Shell Middle Distillate Synthesis (SMDS) use synthesis gas as feedstock. Another important route towards chemicals consists of the hydroformylation or the carbonylation of olefins. In addition to these indirect processes, the direct methane pyrolysis towards ethene and ethyne is also possible, but temperatures higher than 1500 K are required[11,73]. The introduction of dioxygen as reactant, however, makes the direct conversion thermodynamically feasible at much lower temperatures. Thus, recent research efforts were focussed on oxidative routes for direct methane conversion to liquid fuels or chemicals, in particular on the partial oxidation of methane to $C_2$ hydrocarbons, more often referred to as the oxidative coupling of methane.

The oxidative coupling of methane is a stoichiometrically complex reaction, with ethane, ethene, carbon monoxide, carbon dioxide, and water as the major products. The selectivity to the desirable $C_2$ hydrocarbons is hampered by oxidation to CO and $CO_2$. The oxidative coupling of methane is thermodynamically favoured, but the formation of CO and $CO_2$ even more so. Inorganic oxides are usually employed in order to improve the conversion of the reactants and the selectivity for the $C_2$ products. Two modes of operation have been considered, the so-called redox mode and the cofeed mode. In the former mode, a solid oxide is reduced in a reactor by methane which is simultaneously converted to $C_2$ products. Next, the reduced oxide is reoxidized in a regenerator. In the so-called cofeed mode, methane and oxygen are cofed over a catalyst. Temperatures higher than 973 K are required to obtain a reasonable degree of conversion of methane and oxygen and a reasonable selectivity for the $C_2$ hydrocarbons. Atmospheric pressure was used for most of the investigations. Oxygen rather than air has been used in order to allow recirculation of unconverted methane without accumulation of inerts.

Furthermore, the oxygen has to be supplied in substoichiometric amounts in order to allow its total conversion.

Several comprehensive reviews on the oxidative coupling of methane have appeared[1,2,7,9,46,51,53,69] emphasizing the catalytic chemistry. The present contribution is focussed on the kinetics of the reactions involved, both in the absence and in the presence of catalyst with particular emphasis on promoted Li/MgO.

## 2. Coupling in the absence of catalyst

At the temperatures and pressures under which the catalytic coupling of methane is usually carried out, coupling also occurs in the absence of catalyst. Hence, the oxidative coupling of methane in the absence of catalyst has been addressed in several papers[2,5,23,35,50,55,62,92]. The selectivity to $C_2$ products amounts to approximately 50 % at methane conversions lower than 15 % depending on the partial pressures of methane and oxygen. The selectivity for products is defined as the amount of moles of methane converted into $C_2$ products per mole of methane converted. Selectivity not only determines the efficiency with which the carbon of the feedstock is used, but also the reaction heat. Selectivities to ethene exceeding 65% are necessary to maintain the reaction heat below 1200 kJ per mole of ethene produced. The selectivity to CO is higher than that to $CO_2$, in contrast to what is observed for the catalytic reaction. The higher CO selectivity can be an advantage, since CO is a more valuable chemical than $CO_2$. A series of patents assigned to the BP Company[16,40] claims processes consisting of the simultaneous oxidation to $C_2$ hydrocarbons and synthesis gas, CO and $H_2$. A disadvantage of the methane coupling in the absence of catalyst consists of the lower reaction rates. At atmospheric pressure, a ten to hundred fold higher space time is needed to reach the same conversion as in the presence of catalyst. However, this can be compensated for by carrying out the reaction at elevated pressure. It follows from Figure 1 that in the absence of catalyst from 400 kPa on the space-time yield for $C_2$ products, defined as the numbers of moles produced per unit reactor volume per second, rises to a level comparable to that encountered in commercial refining operations. Weisz[91] noted that most space-time yields in the petroleum refining and petrochemical industry fall within a window of 1 to 10 mol $m_{reactor}^{-3}$ $s^{-1}$. The $C_2H_4$ space-time yield in an ethane steam cracking coil, for example, amounts to 12 mol $m^{-3}$ $s^{-1}$ based on the radiation section. The lower limit of the window corresponds to uneconomically high investment costs, while the upper limit corresponds to limitations by physical transport of mass and/or heat. In a highly exothermic process such as the oxidative coupling of methane the latter is limiting. Figure 2 shows the selectivity to the major products as a function of the oxygen conversion. From oxygen conversions of 80% on the product selectivities no longer depend on the total pressure.

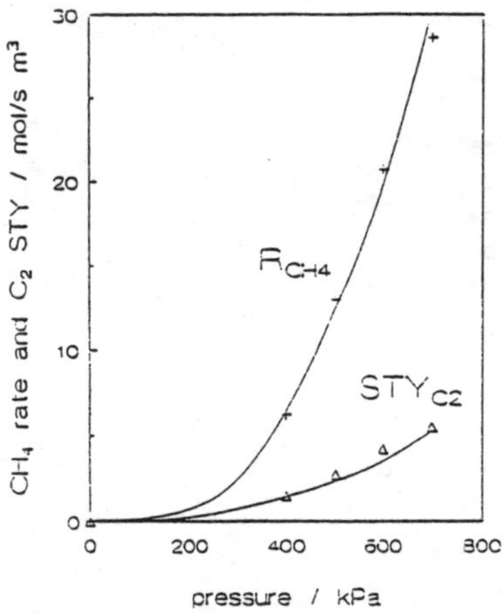


Figure 1 Rate of conversion of methane and spacetime yield of $C_2$ as a function of total pressure. Lines: calculated with kinetic model (Chen et al., 1994 Conditions: $T_{max}$ = 1090 K, $\tau$ = 0.5 s, $CH_4/O_{2,0}$ = 5,0

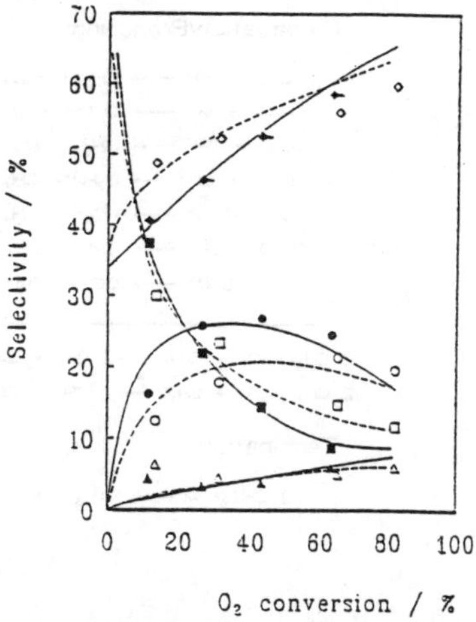


Figure 2: Selectivities for the main reaction products vs. oxygen
Lines: calculated with the model in (ref.Chen et al.)
points: experimental o o $C_2H_4$, □ ■ $C_2H_6$, ◇ ♦ CO, ▵ ▴ $CO_2$. Solid lines & filled points: p = 100 kPa $T_{max}$ = 1100 K. Dashed lines & empty points: pt = 400kPa $T_{max}$ = 1078K $CH_4/O_{2,0}$ = 4,0.

The mechanism of methane oxidation in the absence of catalyst is rather well understood, and mainly derived from combustion chemistry[89,90]. The occurrence of branched chains is the most essential feature in the reaction mechanism. At low temperatures, ca. 700 K, methyl hydroperoxide is the dominating chain-branching species[77], whereas at temperatures around 1500 K, the oxygen atom plays a key role in the chain branching[89]. At the intermediate temperatures at which the methane coupling reactions are carried out the key chain-branching species is hydrogen peroxide[22,23], as illustrated in Figure 3. It is the chain branching which makes it possible to produce large amounts of methyl radicals which couple to ethane in a termination step. The balance between the chain branching and the termination rates beyond the induction period results in a steady state.

Propagation/Branching:

| | $\sigma_1$ | $\sigma_2$ | $\vec{r}$ |
|---|---|---|---|
| $CH_4 + HO_2^{\cdot} \longrightarrow CH_3^{\cdot} + H_2O_2$ | 1 | 1 | 14 |
| $H_2O_2 + M \longrightarrow 2\ OH^{\cdot} + M$ | 1 | 1 | 13 |
| $CH_4 + OH^{\cdot} \longrightarrow CH_3^{\cdot} + H_2O$ | 2 | 2 | 56 |
| $C_2H_6 + CH_3^{\cdot} \longrightarrow C_2H_5^{\cdot} + CH_4$ | 1 | 0 | 3.1 |
| $O_2 + C_2H_5^{\cdot} \longrightarrow HO_2^{\cdot} + C_2H_4$ | 1 | 0 | 3.4 |
| $CH_2O + CH_3^{\cdot} \longrightarrow CHO^{\cdot} + CH_4$ | 0 | 1 | 30 |
| $O_2 + CHO^{\cdot} \longrightarrow CO + HO_2^{\cdot}$ | 0 | 1 | 0.32 |

$$2\ CH_4 + O_2 + C_2H_6 \xrightarrow{I} 2\ CH_3^{\cdot} + C_2H_4 + 2\ H_2O$$

$$2\ CH_4 + O_2 + CH_2O \xrightarrow{II} 2\ CH_3^{\cdot} + CO + 2\ H_2O$$

Termination:

| | | | |
|---|---|---|---|
| $2\ CH_3^{\cdot} + M \longrightarrow C_2H_6 + M$ | | | 12 |

**Figure 3** Typical branched chains towards ethene and CO.
column 1: stoichiometric number for ethene formation
column 2: stoichiometric number for CO formation
column 3: rates (mol $m^{-3}$ $s^{-1}$)of the individual steps at:
$T_{max}$ = 1078 K, p = 400 kPa, inlet methane-to-oxygen ratio 4.0 and $V/F_{CH4,0}$ = 0.020 $m^3$ s $mol^{-1}$.

A number of studies has been devoted to the modeling of the coupling in the absence of catalyst[3,22,23,37,55,92]. Reaction networks were set up on the basis of elementary free-radical steps. The Arrhenius parameters were selected from data bases in the literature which originate mainly from combustion kinetics[82,90]. Chen et al.[22,23] estimated the Arrhenius parameters for the most important reactions in a model consisting of 39 elementary reactions by the regression of experimental data covering a wide range of conditions. The full lines drawn in Figures 1 and 2 are calculated with this model. Figure 4 shows the corresponding typical calculated concentration profiles of the important radicals along the axis of the laboratory reactor used by Chen et al.[23]. The concentrations increase initially and reach a maximum corresponding roughly to the maximum in the axial temperature profile. The radical concentrations are spread over almost 5 orders of magnitude, with the methyl and hydrogenperoxy radicals being the most and the hydrogen atom being the least abundant. The hydrogenperoxy radical is present in high concentrations due to its inactive nature. It is the precursor of hydrogenperoxide, $H_2O_2$, the most important chain-branching agent. The hydrogen atoms, together with the hydroxyl radical are important chain carriers, especially through the hydrogen abstraction from the methane molecule. The methyl radical is also an important chain carrier not only in the chain shown in Figure 3 but also for the dehydrogenation of ethane to ethene.

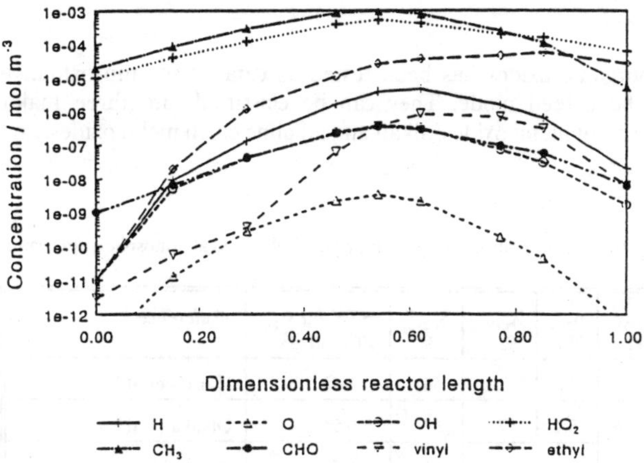


Figure 4 Calculated axial concentration profiles of radicals
$p = 400$ kPa, $T_{max} = 1227$ K, $CH_4/O_{2,0} = 5$, $V/F_{CH4,0} = 0{,}858\ 10^{-2}\ m^3\ s\ mol^{-1}$

## 3. Catalyst development

### *3.1. Criteria*

Several aspects have to be considered in order to evaluate the performance of a catalyst, selectivity being the most important one. The conversion of methane is defined as the amount of moles of methane converted per mole of methane fed. The yield of a coupling product is defined as the amount of moles of methane converted into the product per mole of methane fed and, hence, is equal to the product of the conversion and the selectivity. A negative correlation has been observed between the selectivity for. The corresponding quantities based on the other feedstock, i.e. oxygen, are also important, however. Investment costs, are of course also determining. A distinction can be made between the recirculation section, the separation section and the reactor of the plant. The minimization of the investment cost corresponding to the recirculation requires a high single pass conversion of methane, whereas the minimization of the investment cost corresponding to the separation of the product stream requires a high selectivity. The minimization of the investment costs for the reactor requires a high space-time yield. The space-time yield of a product is defined as the number of moles of the product being produced per unit catalyst mass per second, and can be considered as an average production rate in a reactor which is operated in an integral way.

Finally, a commercial catalyst should show a steady performance over a sufficiently long period of time. In the so-called redox mode the inorganic oxides are reacting stoichiometrically with methane in the reactor and with oxygen in the regenerator. This requires solid circulation rates between the reactor and the regenerator which are uneconomically high with the yields reported to date. The best yields reported to our knowledge with inorganic oxides tested in the redox mode amount to 6.5% for $PbO/\alpha$-$Al_2O_3$ [43] and 5.5% for $Mn/SiO_2$ [80] and are substantially lower than the reported yields on catalysts tested in the cofeed mode.

### 3.2 *Performance*

A large variety of inorganic oxides has been tested as catalyst for the oxidative coupling of methane in the cofeed mode. They can be classified into three major categories: reducible multivalent metal oxides, alkali and alkaline earth metal oxides, and rare earth metal oxides.

Table 1. Optimal conversions, selectivities and space-time yields observed at atmospheric pressure[a]

| Catalyst | T K | $p_{O2}$ kPa | $p_{CH4}$ kPa | $X_{CH4}$ % | $S_{C2+}$ % | STY for $C_{2+}$ $10^{-3}$molkg$^{-1}$s$^{-1}$ | Reference |
|---|---|---|---|---|---|---|---|
| $PbO/\alpha$-$Al_2O_3$ | 1013 | 7.0 | 70 | 6[b] | 56 | 1.6[b] | Hinsen et al.[39] |
| LiCl-$MnO_2$ | 1023 | 2.6 | 5.1 | 47 | 65 | 0.54 | Otsuka et al.[65] |
| Li/MgO | 939 | 4.7 | 7.6 | 43 | 45 | 0.61 | Ito et al.[41] |
| Li/MgO[c] | 1069 | 22 | 78 | 5 | 51 | 6.6 | Couwenberg[26] |
| Li/Sn/MgO | 1123 | 42 | 84 | 42 | 48 | 39 | Couwenberg[26] |
| Na/CaO | 1013 | 7.4 | 67 | 13[b] | 77 | 38[b] | Carreiro & Baerns[20] |
| $K_{0.13}$ $Ca_{0.75}$ $NiO_x$ | 915 | 73 | 27 | 27 | 39 | 0.73 | Dooley et al.[27] |
| Li-$Sm_2O_3$ | 1023 | 2.1 | 5.1 | 37 | 57 | 0.38 | Otsuka et al.[64] |
| CaO-$La_2O_3$ | 1013 | 7.0 | 93 | 12[b] | 78 | 25[b] | Becker & Baerns[10] |
| layered $TiO_2$ | 979 | 2.0 | 5 | 50 | 89 | 0.95 | Chu & Landis[25] |
| $BaTi_{0.75}Mg_{0.25}O_{3-X}$ | 1023 | 2.5 | 7.5 | 29 | 44 | 0.48 | Vermeiren et al.[87] |

a. Balance $N_2$ or He.
b. Not reported explicitly.
c. This data was taken after 216 ks in operation under atmospheric pressure at 1123 K, $p_{O2}$ = 20 kPa, $_{CH4}$ = 80 kPa, and $W/F_{t,0}$ = 1.7 kgs mol$^{-1}$, when the catalyst performance was at a steady state, from 54 ks on. Ten times as much sintered $\alpha$-Al2$O_3$ as the catalyst weight were used to dilute the catalyst bed.

Table 1 summarizes the observed optimal performances as well as the corresponding reaction conditions. The latter are quite diverse, making a direct comparison hazardous. Yields up to 45 % have been reported. These high yields are obtained with strongly diluted feed streams. Increasing the partial pressures of methane and oxygen generally drastically reduces the selectivity at a given conversion and, hence, the yield. Reasonable selectivities can be obtained at conversions as high as 50 %, but again provided the feed stream is diluted. When the dilution is small or absent, selectivities up to 80 % at a conversion between 10 and 15 % have been reported. The corresponding temperature on Li/Sn/MgO was as low as 953 K. Dilution of the feed stream not only would lead to high product separation costs but also to uneconomically high investment costs for the reactor, as can be seen from the corresponding low space-time yields reported in Table 1. The highest space-time yields reported in Table 1 could result in heat removal limitations on industrial scale.

Selectivities between 90 and 100 % at a conversion level of 10 % have been reported by Pereira et al.[68] and Rasko et al.[70] at temperatures lower than 883 K. The catalysts consisted of a mixed oxide with $Ca_{2\text{-}4}NiK_{0.1}O_X$ as stoichiometric composition. The catalyst preparation procedure seems to be critical. Again the feed stream was diluted, but the use of steam as diluent was essential. The space-time yield amounted only to 6 $10^{-5}$ mol $kg^{-1}s^{-1}$. No duplication of these results has been reported to date. It was reported recently by the authors that the high selectivities were due to an artefact: a substantial amount of $CO_2$ was indeed formed but was quantitatively transformed into carbonate on the time scale of the experiments[38].

With the exception of one of the Li/MgO catalysts, the data reported in Table 1 correspond to observations performed shortly after the start-up of the reaction. Catalyst deactivation can occur, however. This is the case for the reducible oxides and the Li/MgO catalysts. The deactivation of $PbO/\alpha\text{-}Al_2O_3$ becomes important after 90 ks and was attributed to the volatization of Pb and sintering. The loss of Cl causes the selectivity over Li-$MnO_2$ to decreases sharply after 7 ks without affecting the conversion. The loss of Li causes deactivation of Li/MgO. However, it follows from Table 1 that a steady-state behaviour can be reached after 54 ks with a sufficiently high space-time yield. Promotion by Sn allows to maintain a constant performance during at least 250 ks[26]. The performance of Na/CaO can be maintained during at least 36 ks. The deactivation behaviour of Li-$Sm_2O_3$ has not been reported, but similar catalysts such as $Sm_2O_3$, Ca-$Sm_2O_3$, and Na-$Sm_2O_3$, are stable during at least 216 ks[45]. It can be expected that Li-$Sm_2O_3$ would show a similar performance. No tests on the deactivation of CaO-$La_2O_3$ nor of the layered $TiO_2$ have been reported to our knowledge. A constant performance during 180 ks was reported for the perovskite listed in Table 1.

## 4. Catalytic routes

It is widely accepted that the primary function of a catalyst for the oxidative coupling of methane consists of the generation of methyl radicals through the abstraction of a hydrogen atom. The possibility that subsequent coupling occurs via combination of two methyl radicals in the gas phase has been demonstrated by isotopic exchange experiments[59,60] and indirectly from the measurement of the methyl radical concentraction in the gas phase[18,19,28,29]. A methyl radical concentration corresponding to 0.006 % of the methane fed was observed over a Li/MgO catalyst[28]. Note that the maximum methyl radical concentration in the absence of catalyst shown in Figure 4 is of the same order of magnitude, which is in line with the similar space time yields observed in both cases. According to Lunsford[54], the coupling of gas-phase methyl radical contributes to the $C_2$ production for at least 28 % on Na/CaO, 40 % on Li/MgO, and 70 % on $La_2O_3$. These values are a lower limit, since the calculations were based on the measurement of radical concentrations in the bulk of the gas phase, i.e., beyond the film surrounding the catalyst pellets. McCarty[57] has shown that important methyl radicals concentration profiles can develop inside the pores of the pellets and in the film.

Next to its obvious thermodynamic function, molecular oxygen regenerates the catalytic sites involved in the abstraction of a hydrogen atom from methane. Figure 5 shows a closed sequence corresponding to the generation of methyl radicals by a Li/MgO catalyst.

| | σ |
|---|---|
| $O_2 + \square + 2O_s^{2-} \rightarrow 4\ O_s^{-}$ | 1 |
| $O_s^{-} + CH_4 \rightarrow OH_s^{-} + CH_3$ | 4 |
| $2OH_s^{-} \rightarrow O_s^{2-} + \square + H_2O$ | 4 |

$$4\ CH_4 + O_2 \rightarrow 4CH_3 + 2H_2O$$

**Figure 5. The production of methyl radicals on a Li/MgO catalyst [Ito et al., 1985].**
**σ: stoichiometric number.**

The nature of the active sites on the catalyst is still a matter of debate. In the closed sequence in Figure 5, the active site is assumed to be the non fully reduced $O_s^-$ based on ESR measurements[41]. Alternative reoxidation mechanisms have been proposed by Sinev and Bychkov[79].

A comparison of Figures 3 and 5 indicates the potential advantages of the catalytic route. The noncatalytic route requires chain carriers such as $HO_2$ and OH which will certainly also carry chains towards CO and $CO_2$ [22]. The catalytic route allows the selective production of methyl radicals only. In order to fully exploit this potential advantage, the interaction of methyl radicals with the catalyst surface which could lead to CO and $CO_2$ through a catalytic route should be minimal. Lunsford[54] has indicated that such a minimal interaction is most probable on inorganic oxides in which the metal ion has only one oxidation state. Lunsford[54] verified experimentally that the reactive sticking probability for methyl radicals on Li/MgO amounted to 1.2 $10^{-7}$ only at 755 K. This is comparable to the reactive sticking probability of alkanes on platinum between 700 and 800 K[81]. The termination of chain carriers such as $HO_2$ and OH by the catalyst surface, on the contrary, should be maximized[75]. Next, the rate of methyl radical generation should be maximal in order to obtain sufficiently high coupling rates without the development of homogeneous branching reactions as shown in Figure 3. This requires according to Figure 5 the generation of oxygen vacancies and of anionic oxygen to be sufficiently fast. The above requirements are met by high-temperature p-type semiconductors[30].

Of course, the consecutive oxidation of the desirable $C_2$ products in particular of ethene also limits the selectivity of the methane coupling at high conversions[36,45]. Hence, the reactive interaction of ethene with the catalyst surface should be minimized. This is in principle possible in view of the higher C-H bond strength in ethene (460 kJ $mol^{-1}$) than in methane (440 kJ $mol^{-1}$)[58]. The corresponding bond strength in ethane (411 kJ $mol^{-1}$) is lower, however. Hence, the abstraction of a hydrogen atom will occur from both methane and ethane. Both routes lead to ethene.

## 5. Catalyst properties

From the previous section it can be understood why the correlation of the catalyst performance and its physico-chemical properties has been studied in particular with respect to surface basicity, if ethene is looked at as a Lewis base, and solid electrical conductivity, if the regeneration of active sites is considered to be important. There are strong indications that basic solids are more appropriate catalysts than acidic solids[8]. A positive correlation between basicity and performance was found for PbO catalysts on different supports[17] and CaO-based catalysts[10]. However, no correlation was observed for the CaO-$La_2O_3$ catalysts, which lead to the conclusion that a certain amount of surface acidity is also required to obtain a maximum $C_2^+$-hydrocarbon selectivity for this type of catalyst[7]. Another important feature of a solid metal oxide relevant to its performance in the oxidative coupling of methane is the oxygen ion conductivity within the solid, or oxygen mobility. Several studies have correlated this property to conversion and selectivity increase with the oxygen mobility[7,67]. Two techniques have been employed to measure oxygen mobility: electrical conductivity measurements and steady-state isotopic transient kinetic analysis (SSITKA). Baerns and coworkers have shown a positive correlation between the selectivity and the oxygen ion conductivity for $CeO_2$-CaO catalysts and promoted $La_2O_3$ catalysts. A similar conclusion was reached by Peil et al.[67] for MgO-based catalysts and $Sm_2O_3$. Dubois and Cameron[30] concluded after a survey of a large number of catalysts that active and selective catalysts are usually high-temperature p-type semiconductors under normal operating oxygen partial pressures. The electrical conductivity not only depends upon the mobility of the charge carriers but also upon their concentration. Dopants effects, e.g., the replacement of $Mg^{2+}$ by $Li^+$, can be explained in these terms[66].

## 6. Chemical kinetics of catalytic coupling

Power law rate equations have been derived for several catalysts, e.g. PbO[39], Na/MgO[42] and Li/MgO[74,86]. Such rate equations allow to investigate the effect of the reactor configuration and operation conditions on the conversions and selectivities that can be obtained in the industrial methane coupling process[76] but not to obtain further understanding of the functioning of the catalysts.

Rate equations taking into account the interaction of the reactants with the catalyst through Eley-Rideal or Langmuir-Hinshelwood steps have also been derived for $Sm_2O_3$ [63], Na/MgO[42], Li/Sn/MgO[44], Na/CaO[52] and Li/MgO[85,86]. In these kinetic models, the role of the gas phase was restricted to mere coupling of methyl radicals or absent.

Clearly, a more detailed kinetic analysis is required in view of the considerations in the previous sections. Several papers indeed report kinetic models consisting of elementary free-radical reactions in the gas phase and of elementary steps involving the catalyst. An overview is given in Table 2. A majority of the models has been built for the Li/MgO catalyst. This overview is limited to models which were validated by experimental data and applied for reactor design. Useful insights in the reaction mechanism and, hence, in the obtainable yields are another important application of such models[48].

Table 2. Kinetic models for the catalytic coupling of methane based on elementary steps

| catalyst | number of homogeneous steps[a] | number of heterogeneous steps | Reference |
|---|---|---|---|
| Li/MgO | 13 | 13 | Forlani et al.[34] |
| Na/CaO | 268 | 10 | McCarty et al.[56] |
| PbO/$\alpha$-$Al_2O_3$ | 183 | 8 | Zanthoff & Baerns[93] |
| Li/MgO | 6 | 7 | Aparicio et al.[4] |
| Li/MgO | 164 | 3 | van der Wiele et al.[86] |
| Li/MgO | 156 | 4 | Shi et al.[78] |
| Li/mixed oxide | > 1000 | 19 | Bistalfi et al.[12] |
| Sn/Li/MgO | 78 | 12 | Couwenberg[26] |

a. One reaction consists of two steps, forward and backward

The approach followed to obtain the Arrhenius parameters of the free-radical reactions has been discussed in the section on the oxidative coupling in the absence of catalyst. Data for the rate coefficients of the heterogeneous steps are rare, in contrast to the abundant data bases for the homogeneous reactions. A preliminary estimation of the Arrhenius parameters for the heterogeneous steps is possible on the basis of chemical rate theories. The preexponential factors can be estimated from the collision theory or transition-state theory[6,15,47,94]. The estimation of activation energies is even more difficult. A linear correlation between the activation energy and the enthalpy of reaction according to the Evans-Polanyi principle[32] is often used[34,57]. Adjustment of the preexponential factors and the activation energies by regression of experimental data remains unavoidable, however.

Despite of the large amount of elementary steps to be considered the number of parameters to be estimated can be reduced by justifiable assumptions. In the model developed by McCarty[57], the site density of active oxygen centers and the activation energy for the hydrogen abstraction from methane were the only parameters estimated by regression of the experimental data. Also, not all the elementary steps are kinetically significant in a given range of reaction conditions. This allows to reduce drastically the number of steps required without compromising with respect to the insight in the reaction mechanism.

Table 3 shows, by way of example, the heterogeneous reactions and the corresponding Arrhenius and van 't Hoff parameters used by Couwenberg[26] in combination with the network of homogeneous reactions proposed by Chen et al.[24] to describe experimental results over a range of conditions shown in Table 4 with a Sn promoted Li/MgO catalyst.

Table 3. Catalytic reaction mechanism and rate coefficients on a Sn/Li/Mgo catalyst[26].

| Reaction | | A | $E_a/\Delta H$ /kJmol$^{-1}$ |
|---|---|---|---|
| $O_2 + 2* \rightleftharpoons 2\,O*$ | (1) | $2.6\ 10^{-3}$ | -11.8 |
| $CH_4 - O* \rightarrow CH_3\bullet - OH*$ | (2) | $1.9\ 10^{9}$ | 123.1 |
| $C_2H_6 - O* \rightarrow C_2H_5\bullet + OH*$ | (3) | $1.4\ 10^{9}$ | 108.1 |
| $C_2H_4 - O* \rightarrow C_2H_3\bullet + OH*$ | (4) | $1.4\ 10^{9}$ | 127.8 |
| $2\,OH* \rightarrow H_2O - O* - *$ | (5) | $6.9\ 10^{9}$ | 152.7 |
| $CH_3\bullet + O* \rightarrow CH_3O*$ | (6) | $3.6\ 10^{6}$ | 0.0 |
| $2\,CH_3O* + 5\,O* \rightarrow 2\,CO_2 + 3\,H_2O + 7*$ | (7) | $\infty$ | 0.0 |
| $CO + O* \rightarrow CO_2 + *$ | (8) | $1.9\ 10^{9}$ | 93.7 |
| $CO_2 + * \rightleftharpoons CO_2*$ | (9) | $5.0\ 10^{-5}$ | -108.4 |
| $4\,HO_2\bullet \xrightarrow{surf} 3\,O_2 + 2\,H_2O$ | (10) | $1.0\ 10^{-2}$ | 0.0 |

Units: $m^3_g\,mol^{-1}$ (1,9), $m^3_g\,m^{-3}_p\,s^{-1}$ (2,3,4,6,8), $mol\,m^{-3}_p\,s^{-1}$ (5), $m^3_g\,m^{-2}_c\,s^{-1}$ (10).

Reactions 2, 3 and 4 are the main source of radicals. Reactions 1 and 5 regenerate the oxygen species which are involved in the production of radicals. At the investigated conditions they are close to equilibrium. The catalyst is also involved in non-selective routes. Tong and Lunsford[83] showed that methyl radicals are oxidized by feeding methyl radicals to a bed with Li/MgO, see reactions 6,7 and 8. Clearly reaction 7 is not elementary, but it occurs on a much smaller time scale than the others and in particular than reaction 6. Hence, the detailed reaction path through which it occurs is not relevant. Van der Wiele et al.[86] showed that carbon monoxide was relatively rapidly converted to carbon oxide over Li/MgO by comparing experiments with and without catalyst and by CO oxidation experiments. It is well known that addition of $CO_2$ lowers the reaction rate of the oxidative coupling of methane[44,88]. This is taken into account by reaction 9. Using the above catalytic reactions together with gas-phase reactions without considering the heterogeneous termination, see reaction 10, did not allow an adequate description of the experimental data. Especially the calculated selectivities at high space times were too low. This can be attributed to an overestimation of the importance of the gas-phase reactions occurring in both the pores of the catalyst and in the interstitial phase. Tulenin et al.[84] suggested that MgO does not only produce but also quenches radicals. This radical quenching has a inhibiting effect on the gas-phase reactions. Therefore the heterogeneous termination reaction 10 suggested by Cheaney et al.[21] was added to the reaction network. The hydrogenperoxy radical is one of the most important chain carriers in the gas-phase reactions.

Table 4. **Conditions covered by the kinetic model of Couwenberg[26] for a Sn/Li/MgO catalyst.**

| | | |
|---|---|---|
| $T$ | / K | 973 - 1023 |
| $p$ | / kPa | 110 - 150 |
| $CH_4/O_2\|_0$ | / mol mol$^{-1}$ | 2 - 12 |
| $C_2H_6/CH_4\|_0$ | / mol mol$^{-1}$ | 0 - 0.1 |
| $CO_2/CH_4\|_0$ | / mol mol$^{-1}$ | 0 - 0.2 |
| $W/F_{t,0}$ | / kg s mol$^{-1}$ | 1 - 10 |
| $X_{CH_4}$ | / % | 2.5 - 15 |
| $X_{O_2}$ | / % | 15 - 75 |

Another important question is related to the source of the methyl radicals. If the chain branching as shown in Figure 3 can be neglected, the major part of the methyl radicals originates from steps involving the catalyst. Shi et al.[78] concluded that at 1 bar without feed dilution, 973 K and a reactor inlet methane to oxygen ratio of two, the methane consumption in the gas phase amounted to 15% of that at the surface of a Li/MgO catalyst. Zanthoff and Baerns[93] estimated that on a $Pb/Al_2O_3$ catalyst about half of the converted methane is activated in the gas-phase at 1 bar, 1020 K and a feed stream composition of $CH_4$:$O_2$:$N_2$ = 10:1:4.

The ultimate $C_2$ yield achievable was estimated by Shi et al.[78]. An ideal catalyst which would only activate methane, but not ethane or ethene was considered. The $C_2$ yield was calculated as a function of the reactants partial pressure $p_{O2}+p_{CH4}$. At a reactant partial pressure of 100 kPa without dilution, a $C_2$ yield of 26% was obtained, in agreement with the maximum yield observed experimentally. This yield limit was caused by the homogeneous oxidation of ethane and ethene. McCarty[57] showed quantitatively that at 1000 kPa, the heterogeneous reactions became relatively less important than that at 100 kPa. At high pressures, the most important overall reaction is the production of formaldehyde and its rapid conversion to CO. The gas-phase oxidation of $C_2$ products alone cannot account for the observed low selectivities[4,78], however, indicating reactive interaction with the catalyst.

## 7. Irreducible transport phenomena

In the preceding section only chemical phenomena were discussed, i.e. no attention was given to the possible effects of transport limitations on the obtainable yields for $C_2$ products. Such effects have indeed been observed[33] and discussed by means of models taking into account both chemical and transport phenomena during reactor simulations[33,72]. Depending upon the details of the reaction network increasing pellet diameters can either lead to a higher or a lower selectivity for $C_2$ products[33,71]. The above reports discussed transport limitations emphasizing the existence of internal concentration profiles of molecules either reactants or products. Clearly, the existence of such profiles induces also concentration profiles of the corresponding radicals which are generated within the catalyst pores[57]. Couwenberg[26] showed that even with internal concentration gradients of molecules small enough to be neglected for all practical purposes, kinetically significant concentration profiles of radicals could exist.

The kinetic network given in Table 3 was used to calculate the concentration profiles in a laboratory fixed bed reactor operated at the conditions summarized in Table 5. Strong concentration gradients for reactive intermediates, e.g. methyl radicals and hydrogen-peroxy radicals, are developed without significant gradients for the stable molecules, such as methane, oxygen, ethane, and ethene. This is shown in figures 6 and 7 where the intraparticle and interstitial concentration of the methyl and the hydrogen-peroxy radicals are plotted versus the axial reactor position and the pellet coordinate. In these figures the zero on the pellet-coordinate axis represents the centre of the catalyst pellet, the gas-solid interface is located at $1.0\ 10^{-4}$ m, and the space between $1.0\ 10^{-4}$ m and $1.5\ 10^{-4}$ m corresponds to the interstitial gas phase. The origin of the axial position axis represents the reactor inlet. The strong concentration gradient for the radicals is caused by the lower time scale for reaction than for transport by diffusion. This is in contrast to oxygen of which the time scale for diffusion is lower than the time scale for reaction and, hence, the concentration gradient can be neglected. In the centre of the pellet, the methyl-radical concentration is high due to their high catalytic production rate. Near the interface the concentration decreases strongly, due to diffusion into the interstitial phase. In the interstitial phase the profile is almost flat because the diffusivity is much higher, but the concentration is approximately one order of magnitude lower than inside the catalyst pellet.

The intraparticle concentration of the surface-terminated hydrogen-peroxy radical is much lower than the interstitial concentration because of the high rate of the heterogeneous termination reaction.

***Table 5:*** *Conditions used during the simulation (Couwenberg et al.[95]).*

| | | |
|---|---|---|
| T | 998 | K |
| p | 135 | kPa |
| $CH_4/O_{2,0}$ | 4.0 | |
| $F_0$ | $1.5\ 10^{-4}$ | mol $s^{-1}$ |
| $W_c$ | $0.375\ 10^{-4}$ | g |
| $d_p$ | $2\ 10^{-4}$ | m |

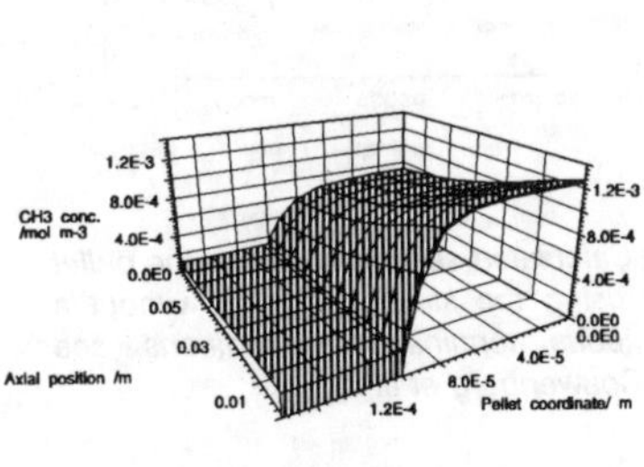


***Figure 6:*** *Calculated concentration profile of methyl radicals on pellet scale and on reactor scale (Couwenberg et al.[95]).*

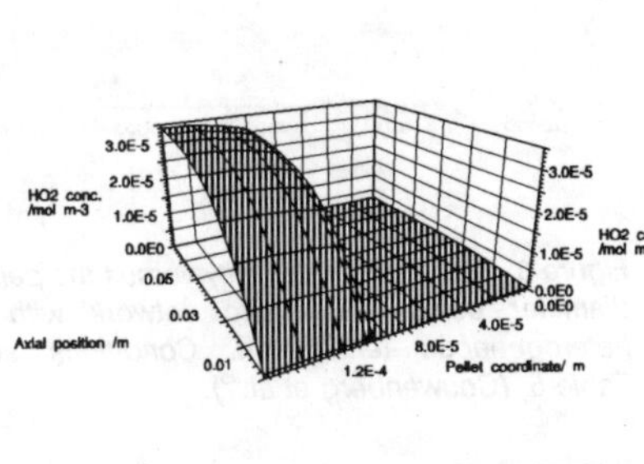


***Figure 7:*** *Calculated concentration profile of $HO_2$ radicals on pellet scale and on reactor scale (Couwenberg et al.[95]).*

The effects of the diffusion limitations of the reactive intermediates on the selectivity were investigated by changing the pellet diameters. Based on the calculated concentration profile of the methyl radical an increase in selectivity is expected, because the volume where the high methyl radical concentration exists is larger when bigger pellets are applied. This would favour their coupling towards ethane, since this is a second order reaction, all other steps being first order with respect to the methyl radical concentration. The results of these calculations presented in figure 8, however, show a decrease in the $C_{2+}$ selectivity with increasing pellet sizes. This is caused by the important contribution of the heterogeneous termination reaction on the termination of the branched chain reactions in the interstitial phase. Increasing the pellet diameter leads to a decrease of the ratio between the gas-solid interfacial surface area and the volume of the interstitial phase. This results in a lower contribution of the heterogeneous termination reaction and thus to higher concentrations of the hydrogen-peroxy radical in the interstitial phase. The higher interstitial $HO_2\cdot$ concentration results in a higher rate of the gas-phase reactions in the interstitial phase and thus in lower selectivities.

This was verified by removing the heterogeneous termination from the kinetic network. The results of the calculations without this reaction are shown in figure 9. It can be seen that the selectivity indeed increases with increasing pellet size. Another significant effect of neglecting the heterogeneous termination is the much lower selectivity, as a result of the increased importance of the non-selective gas-phase reactions. The calculated methane conversion in figures 8 and 9 is constant with increasing pellet diameter, because the intraparticle concentration gradient for the molecules involved in the kinetic network can be neglected. The rate of methane consumption is mainly determined by the rate of heterogeneous methyl radical generation which is only depending on the concentrations of methane, oxygen and carbon dioxide.

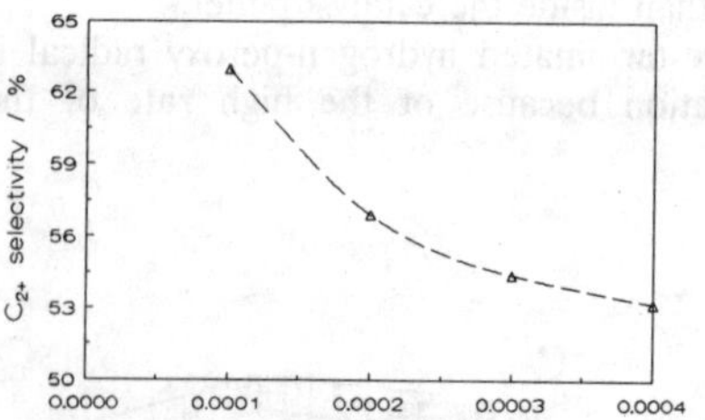


*Figure 8: Calculated selectivity versus the pellet diameter using the kinetic network with a heterogeneous termination. Conditions see Table 5 (Couwenberg et al.[95]).*

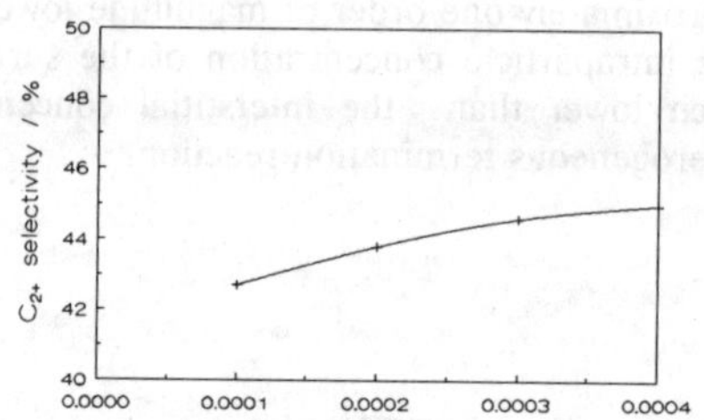


*Figure 9: Calculated selectivity versus the pellet diameter using the kinetic network without a heterogeneous termination. Conditions see Table 5 (Couwenberg et al.[95]).*

## 8. Conclusions

At the temperatures which are required to activate methane either thermally or by means of a heterogeneous catalyst radical chain reactions play an important role. In the absence of catalyst reasonable space time yields are obtained but undesired propagation reactions towards CO are detrimental for the selectivity at which methane is converted into ethane and ethene. One of the major functions of the catalyst consists in increasing the methyl radical concentration and, hence, the selectivity of methane conversion without increasing the concentrations of the chain carriers towards CO. Only limited success has been met so far. None of the presently developed catalysts allows to avoid side reactions of the reaction intermediates or consecutive reactions of the reaction products limiting the obtained yields for ethane and ethene to 20%. The kinetics of the occurring reactions are rather well understood and can be adequally described, however, despite the complexity of the involved chemistry and of the interplay between chemical kinetics and transport phenomena. This should provide ways for further improvement of the process.

**9. Acknowledgement**

The financial support by the Commission of the European Communities in the framework of the Joule programme, subprogramme Energy from Fossil Sources, Hydrocarbons, Contract No. JOUF-0044-C, is acknowledged.

**References**

1. Y. Amenomiya, V.I. Birss, M. Goledzinowski, J. Galuszka and A.R. Sanger, Catal. Review-Science and Engineering, **32** (1990) p. 163.
2. J.R. Anderson, Applied Catalysis, **47** (1990), p. 177.
3. Z.S. Andrianova, A.N. Ivanova, P.E. Matkovskii and G.P. Startseva, Kinetics and Catalysis, **34** (1993) p.344.
4. L.M. Aparicio, S.A. Rossini, D.G. Sanfilippo, J.E. Rekoske, A.A. Trevino, J.A. Dumesic, Ind. Eng. Chem. Res., **30** (1991) p.94.
5. K. Asami, K. Omata, K. Fujimoto, and H. Tominaga, Energy Fuels, **2** (1988) p.574.
6. R.C. Baetzold, and G.A. Somorjai, Journal of Catalysis, **45** (1976) p.94.
7. M. Baerns, Oxidative Coupling of Methane for the Utilization of Natural Gas. Paper presented at Nato-Asi, London, Ontario, Canada (1991).
8. M. Baerns, Basic solids as catalysts for the oxidative coupling of methane. In: Methane conversion by oxidative processes. Fundamental and engineering aspects. (Edited by E.E. Wolf), pages 382-402. Van Nostrand Reinhold, New York (1992).
9. M. Baerns and J.R.H. Ross, Catalytic chemistry of methane conversion. In: Perspective in Catalysis (edited by J.M. Thomas and K.I. Zamaraev), the IUPAC monograph, (1991).
10. S. Becker and M. Baerns, Journal of Catalysis, **128** (1991) p.512.
11. F.G. Billaud, C.P. Guéret, F. Baronnet and J. Weill, Ind. Eng. Chem. Res., **31** (1992) p.2748.
12. M. Bistofli, G. Fornasari, M. Molinari, S. Palmery, M. Dente and E. Ranzi, Chem. Eng. Sc., **47** (1992) p.2647.
13. H. Borchert, Z.L. Zhang and M. Baerns, The effect of oxygen ion conductivity of catalysts for their performance in the oxidative coupling of methane. Paper presented at the American Chemical Society's 1992 National Meeting in San Francisco (1991).
14. M. Boudart, Kinetics of chemical processes; Prentice-Hall; Englewood Cliffs, New Jersey (1968).

15. M. Bowker, I.B. Parker and K.C. Waugh, Applied Catalysis, **14** (1985) p.101.
16. J.H. Bropy, Conversion process. EP 0178855 A2 (1985).
17. W. Bytyn and M Baerns, Applied Catalysis, **28** (1986) p.199.
18. K.D. Campbell, E. Morales and J.H. Lunsford, Journal of American Chemical Society, 109 (1988) p.7900.
19. K.D. Campbell and J.H. Lunsford, Journal of Physical Chemistry, **92** (1988) p.5792.
20. J.A.S.P. Carreiro and M. Baerns, Journal of Catalysis, **117** (1989) p.396.
21. D.E. Cheaney, D.A. Davies, A. Davis, D.E. Hoare, J. Protheroe and A.D. Walsh, Effects of surfaces on combustion of methane and mode of interaction of antiknocks containing metals. 7th Symposium on Combustion, Butterworths Scientific Publications, London, (1959) p.183.
22. Q. Chen, J.H.B.J. Hoebink and G.B. Marin, Ind. Eng. Chem. Res., **30** (1991) p.2088.
23. Q. Chen, P. Couwenberg and G.B. Marin, AIChEJ, **40** (1994) p.521.
24. Q. Chen, P. Couwenberg and G.B. Marin, Catalysis Today, **21** (1994) p.309.
25. P. Chu and M.E. Landis, Selective oxidative coupling. US patent 4,914,252 (1990).
26. P. Couwenberg, PhD Thesis, Eindhoven University of Technology (1995).
27. K.M. Dooley and J.R.H. ROSS, Appl. Catal., **90** (1990) p.159.
28. D.J. Driscoll and J.H. Lunsford, Journal of Physical Chemistry, **89** (1985) p.4415.
29. D.J. Driscoll, W. Martir, J.-X. Wang and J.H. Lunsford, Journal of American Chemical Society, **107** (1985) p.58.
30. J.-L. Dubois and C.J. Cameron, Applied Catalysis, **67** (1990) p.49.
31. J.H. Edwards, K.T. Do, R.J. Tyler, The Oxco process for natural gas conversion via methane oxidative coupling. Proc. Symp. Natural gas conversion, Oslo, (1990) p.489.
32. M.G. Evans and M. Polanyi, Faraday Society, **32** (1936) p.1333.
33. G. Follmer, L. Lehmann and M. Baerns, Catalysis Today, **4** (1989) p.323.
34. O. Forlani, M. Lupieri, V. Piccoli, S. Rossini, D. Sanfilippo, J.A. Dumesic, L.A. Aparicio, J.A. Rekoske and A.A. Trevino, New developments in selective oxidation (edited by G. Centi and F. Trifiro), (1989) p.417. Elsevier, New York.
35. J.W.M.H. Geerts, J.H.B.J. Hoebink and K. van der Wiele, Ethylene from natural gas. Proven and new technology. Paper presented at the American Chemical Society's 1990 National Meeting in Bosten (1990a).
36. J.W.M.H. Geerts, J.M.N. van Kasteren and K. van der Wiele, Catalysis Today, **4** (1989) p.453.
37. J.W.M.H. Geerts, Q. Chen, J.M.N. van Kasteren and K. van der Wiele, Catalysis Today, **6** (1990) p.519.
38. D.M. Ginter, E. Magni, G.A. Somorjai, H. Heinemann, Catal. Lett., **16** (1992) p.197.
39. W. Hinsen, W. Bytyn and M. Baerns, Oxidative dehydrogenation and coupling of methane. Proceedings of the 8th international congress on Catalysis, **3** (1984) p.581.
40. M.J. Howard, The homogeneous partial oxidation of methane-containing paraffinic hydrocarbon. EP 0302665 A1 (1989).
41. T. Ito, J.-X. Wang, C.-H. Lin and J.H. Lunsford, Journal of American Chemical Society, **107** (1985) p.5062.
42. E. Iwamatsu and K.-I. Aika, Journal of Catalysis, **117** (1989) p.416.
43. G.E. Keller and M.M. Bhasin, Journal of Catalysis, **107** (1989) p.9.
44. S.J. Korf, Catalysts for the oxidative coupling of methane. Ph.D. Dissertation, University of Twente, NL (1990).

45. S.J. Korf, J.A. Roos, J.M. Diphoorn, R.H.J. Veehof, J.G. Van Ommen and J.R.H. Ross, Applied Catalysis, **52** (1989b) p.119.
46. O.V. Krylov, Kinetics and Catalysis, **34** (1993) p.1.
47. O.V. Krylov, M.U. Kislyuk, B.R. Shub, A.A. Gezalov, N.D. Maksimova and Yu.N. Rufov, Kinetika i Kataliz, **13** (1972) p.598.
48. J.A. Labinger, Catal. Lett., **1** (1988) p.371.
49. J.A. Labinger and K.C. Ott, Journal of Physical Chemistry, **91** (1987) p.2682.
50. G.S. Lane and E.E. Wolf, Journal of Catalysis, **113** (1988) p.144.
51. J.S. Lee and S.T. Oyama, Catal.Review Science and Engineering, **30** (1988) p.249.
52. L. Lehmann and M. Baerns, J. Catal., **135** (1992) p.467.
53. J.H. Lunsford, Catalysis Today, **6** (1990) p.235.
54. J.H. Lunsford, Methane conversion by oxidative processes. Fundamental and engineering aspects. (edited by E.E. Wolf), (1990) p.3. Van Nostrand Reinhold, New York.
55. J.C. Mackie, Catal.Review Science and Engineering, **33** (1991) p.169.
56. J.G. Mc Carty, Methane conversion by oxidative processes. Fundamental and engineering aspects. (edited by E.E. Wolf) (1990) p.321. Van Nostrand Reinhold, New York.
57. J.G. Mc Carty, A.B. Mc Ewen and M.A. Quinlan, New developments in selective oxidation (edited by G. Centi and F. Trifiro) (1990) p.405. Elsevier, New York.
58. D.F. Mc Millen and D.M. Golden, Annual review physical chemistry, **33** (1982) p.493.
59. C.A. Mims, R.B. Hall, K.D. Rose and G.R. Myers, Catalysis letters, **2** (1989) p.361.
60. P.F. Nelson, C.A. Lukey and N.W. Cant, Journal of Physical Chemistry, **92** (1988) p.6176.
61. T. Norby and A.G. Andersen, Applied Catalysis, **71** (1991) p.89.
62. O.T. Osager, R. Lodeng, P. Soraher, H. Anundskaas and B. Helleborg, Catal. Today, **4** (1989) p.355.
63. K. Otsuka and K. Jinno, Inorg. Chim. Acta, **121** (1986) p.237.
64. K. Otsuka, Q. Liu, M. Hatano and A. Morikawa, Chemistry Letters (1986a) p.467.
65. K. Otsuka, Q. Liu, M. Hatano and A. Morikawa, Chemistry Letters (1986b) p.903.
66. D. Papageorgiou, D. Vamvouka, D. Boudouvas, and X.E. Verykios, Catalysis Today, **13** (1992) p.391.
67. K. Peil, G. Marcelin and J.G. Goodwin, Methane conversion by oxidative processes. Fundamental and engineering aspects. (edited by E.E. Wolf) (1992) p.3. Van Nostrand Reinhold, New York.
68. P. Pereira, S.H. Lee, G.A. Somorjai and B. Heinemann, Catal. Lett., **6** (1990) p.255.
69. M.G. Poirier, A.R. Sanger, and K.J. Smith, The Canadian Journal of Chemical Engineering, **69** (1991) p.1027.
70. J. Rasko, P. Pereira, G.A. Somorjai and H. Heinemann, Catal. Lett., **9** (1991) p.395.
71. S.C. Reyes, C.P. Kelkar and E. Iglesia, Catalysis Letters, **19** (1993a) p.167.
72. S.C. Reyes, E. Iglesia and C.P. Kelkar, Chem. Eng. Sc., **48** (1993b) p.2643.
73. O.A. Rokstad, O. Olsvik, B. Jenssen and A. Holmen, Novel Production Methods for Ethylene, Light Hydrocarbons and Aromatics (L.F. Albright, L. Crynes, S. Nowak, eds.) Marcel Dekker, New York (1992) p.259.
74. J.A. Roos, S.J. Korf, R.H.J. Veehof, J.G. Van Ommen and J.R.H. Ross, Applied Catalysis, **52** (1989) p.131.
75. J. Sanchez-Marcano, C. Mirodatos, E.E. Wolf and G.A. Martin, Catalysis Today, **13** (1992) p.227.

76. J.M. Santamaria, E.E. Miro and E.E. Wolf, Ind. Eng. Chem. Res., **30** (1991) p.1157.
77. N. Semenov, Some problems in chemical kinetics and reactivity. Translated from russian by Boudart, M.; Princeton University Press; Princeton, New Jersey (1958).
78. C. Shi, M. Hatano and J.H. Lunsford, Catalysis Today, **13** (1992) p.191.
79. M.Yu. Sinev and V.Yu. Bychkov, Kinetics and Catalysis, **34** (1993) p.272.
80. J.A. Sofranko, J.J. Leonard and C.A. Jones, Journal of Catalysis, **103** (1987) p.302.
81. G.A. Somorjai, Phil.Trans.R.Soc. London A318 (1986) p.81.
82. W. Tsang and R.F. Hampson, Journal of Physical Chemistry, Reference Data, **15** (1986) p.1087.
83. Y. Tong and J.H. Lunsford, J. Am. Chem. Soc., **113** (1991) p.4741.
84. Y. Tulenin, A. Kadushin, V. Seleznev, A. Shestakov and V. Korchak, Catalysis Today, **13** (1992) p.329.
85. W.-Y. Tung and L.L. Lobban, I&EC Research, **31** (1992) p.1621.
86. K. van der Wiele, J.W.M.H. Geerts and J.M.N. van Kasteren, Methane conversion by oxidative processes. Fundamental and engineering aspects. (edited by E.E. Wolf), pages 259-319. Van Nostrand Reinhold, New York (1992).
87. W. Vermeieren, I.D.M.L. Lenotte, J.A. Martens and P.A. Jacobs, Perovskite-type complex oxides as catalysts for the oxidative coupling of methane. In: Natural gas conversion. (edited by A. Holmen, K.-J. Jens and S. Kolboe) (1991) p.33. Elsevier, Amsterdam, The Netherlands.
88. D. Wang, M. Xu, C. Shi and J.H. Lunsford, Catalysis Letters, **18** (1993) p.323.
89. J. Warnatz, Ber. Bunsenges. Phys. Chem., **87** (1983) p.1008.
90. J. Warnatz, Rate coefficients in the C/H/O system. In: Combustion Chemistry (Edited by Gardiner, W.C.), Springer Verlag, New York (1984).
91. P.B. Weisz, CHEMTECH, (1982) p.424.
92. H. Zanthoff and M. Baerns, Ind. Eng. Chem. Res., **29** (1990) p.2.
93. H. Zanthoff and M. Baerns, Combined kinetics of catalytic and non catalytic reactions in the oxidative coupling of methane. Paper presented at the American Chemical Society's 1992 National Meeting in San Francisco (1991).
94. V.P. Zhdanov, J. Pavlicek and Z. Knor, Catal.Reviews-Science and Engineering, **30** (1988) p.501.
95. P.M. Couwenberg, Q. Chen and G.B. Marin, Plenum, submitted (1994).

# FUEL CELLS

J.A.R. VAN VEEN

*Shell Research B.V. (Koninklijke/Shell-Laboratorium, Amsterdam), P.O. Box 38000, 1030 BN Amsterdam, The Netherlands*

ABSTRACT

The principles and present-day embodiments of fuel cells are discussed. Nearly all cells are hydrogen/oxygen ones, where the hydrogen fuel is usually obtained on-site from the reforming of methane or methanol. There exists a tension between the promise of high efficiency in the conversion of chemical into electrical energy and of very low emissions of noxious compounds, and the enormous difficulty of manufacturing the fuel cells cost-effectively. After three decennia of widespread effort to adapt the fuel cell to terrestrial applications, it is still too early to say whether their large-scale introduction will prove to be viable.

## 1. Introduction

A fuel cell is an electrochemical device in which the chemical energy of the fuels is converted directly into electrical energy, i.e. without being first transformed into heat. A diagram of a fuel cell is shown in Fig. 1. At the fuel electrode, the anode, the fuel is oxidized. In principle, any fuel can be used, but of course a certain reactivity requirement has to be met. From this point of view, hydrogen is the best fuel, and indeed all practical cells to date are based on it. For stationary fuel-cell applications it is often envisaged to be produced through the steam reforming of methane or naphtha, eventually followed by shifting (CO being much less reactive than $H_2$). In the hydrogen case, then, the electrochemical oxidation can be simply written as, assuming an acidic electrolyte:

$$H_2 \rightarrow 2\,H^+ + 2\,e^- \qquad (1)$$

The electrons flow through the external circuit (where they can do their useful work), while protons sustain the current in solution. At the cathode electrons and protons combine again with the oxidizing agent. For the latter duty virtually always oxygen is taken (usually from air), and the reduction reaction can be written as:

$$O_2 + 4\,H^+ + 4\,e^- \rightarrow 2\,H_2O \qquad (2)$$

The overall reaction, $2\,H_2 + O_2 \rightarrow 2\,H_2O$, corresponds exactly to the direct combustion of hydrogen. However, in contrast to the energy conversion in a heat engine, which is subject to the Carnot principle, the available energy can, in principle, be completely transformed into electrical energy. Thermodynamically we have:

$$\Delta G^\circ = -nFE^\circ \qquad (3)$$

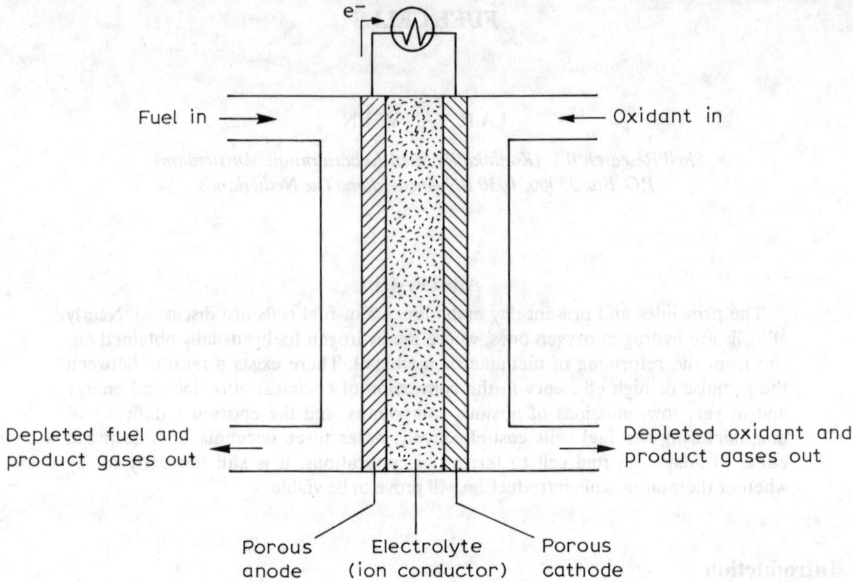


Fig. 1. Schematic illustration of a fuel cell.

where $\Delta G^o$ is the standard free energy change of the cell reaction, $n$ the number of electrons involved (two for hydrogen), $F$ the Faraday constant (96500 C mol$^{-1}$), and $E^o$ the standard cell voltage. It is because of this promise of a very high conversion efficiency that people have been coming back to the fuel cell. Nowadays, the promise of very low production of environmentally objectionable compounds, e.g. of zero $NO_x$ emission, is seen as being quite as important, with Japan leading the way. Both these aspects, in fact, played their part in assuring fuel cells a prominent place in the energy scenario's discussed at the UNCED [1], Rio de Janeiro, 1992. Fuel-cell systems are also characterized by very low noise levels.

The electrolyte within the cell can be of various types. In low-temperature fuel cells one employs aqueous acid or base, either as such or immobilized in a matrix, or a solid polymer membrane. At intermediate temperatures one goes for a molten salt, and for high-temperature (about 1000°C) applications a ceramic membrane is used. Indeed, most fuel cells are characterized by the type of electrolyte upon which they are based. In this contribution we will discuss, however briefly, alkaline fuel cells (AFC), phosphoric-acid fuel cells (PAFC), polymer electrolyte fuel cells (SPFC), direct methanol fuel cells (DMFC), the one exception here because it is distinguished by the fuel it converts, molten-carbonate fuel cells (MCFC), and solid-oxide fuel cells (SOFC). The low-temperature cells, being easier to engineer, are closest to full commercialization, but the higher-temperature ones promise higher overall efficiencies.

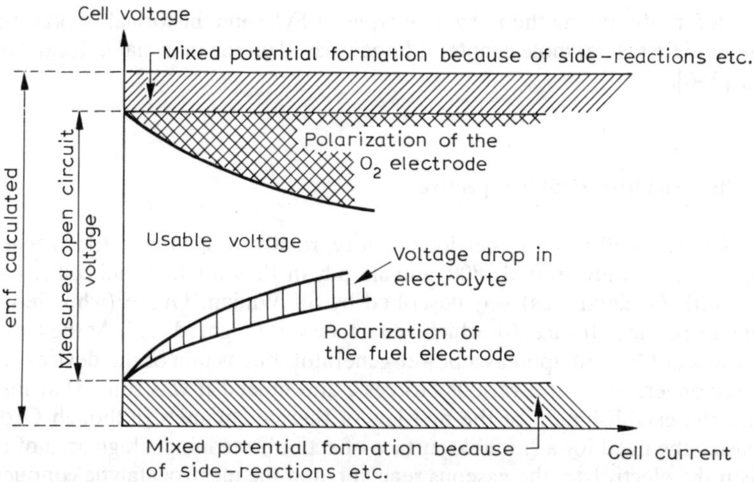


Fig. 2. Schematic representation of the processes leading to loss of cell voltage when current is drawn.

The measured efficiency is always lower than the calculated one, for a variety of reasons. To begin with, even at open-circuit conditions the measured cell voltage is often, certainly in the case of low-temperature cells, less than the thermodynamic value due to, for example, side-reactions or the presence of impurities leading to the formation of a mixed potential. When current is drawn, "polarization" occurs at both electrodes (see Fig. 2) as the electrode reactions are not infinitely fast, i.e. it takes some energy to make them go at an appreciable rate. Polarization can be caused by obstruction of electron transfer between the electrode and the reacting molecule (charge-transfer polarization), by limitations in mass transfer (concentration or diffusion polarization), or by coupled homogeneous reactions being slow (reaction polarization). Also, the electrical resistance of the electrolyte produces a voltage drop proportional to the cell current ($iR$-drop).

To obtain a useful fuel cell, then, polarization must be kept at a minimum — while of course assuring a sufficiently long life (at least five years). This can be done by choosing an electrolyte of good conductivity and, above all, by accelerating the electrode reactions. Especially in low-temperature cells the performance of both anode and cathode can be enormously improved by incorporating very active catalysts. In high-temperature cells, on the other hand, it is materials science that is in the focus of attention. But in all cases it is of paramount importance to have as long or large a three-phase boundary as possible: it is only where catalyst, electrolyte, and reactant meet that the electrochemical conversion takes place with any rate at all, and this calls for the use of porous electrodes that bring the reactant and electrolyte together while not creating unduly long diffusion paths. This is a nice engineering problem.

Before discussing the individual types of FC, some historical background will be provided first. The whole chapter is largely based on the material that can be found in refs. [2–6].

## 2. An ultrabrief historical perspective

The fuel cell has a rather long history, relatively speaking. Electrochemistry began in earnest only around 1800, and already in 1839 the first fuel cell (hydrogen/oxygen with Pt electrodes) was described by Sir William Grove (who went on to become Lord Chief Justice, for which, in fact, he was knighted . . . ). At the time, however, it was not looked upon as a power generator, but as one of the devices showing the interconversion of the various "forces", and indeed it was somewhat lost sight of after the establishment of the first law of thermodynamics (although Grove did emphasize the need for a "notable surface of action", meaning a high area of contact between the electrolyte, the gaseous reactant, and the electrocatalytic conductor, cf. above). The former aspect, however, was much canvassed at the turn of the century, with the emphasis shifting from hydrogen to a much more practical fuel like coal, and this was mainly due (i) to the great improvements in the electrical industry, and (ii) to the conversion efficiency of coal into electricity being well below 10% at the time. Despite the hopes of Ostwald and Jacques ("Think of a smokeless London!") that the 20th Century would become the Age of Electrochemical Combustion, the fuel cell did not make it, defeated for the moment by the low reactivity of common fossil fuels and by the emergence of the internal combustion engine.

The story restarts with Sir Francis Bacon (who *was* knighted for his fuel-cell work), who from 1933 onwards pioneered the cells that were eventually to be employed in the Apollo flights (1960s), after a concentrated development effort by the Pratt & Whitney division of United Technologies, Inc. These were $H_2/O_2$ fuel cells, employing an alkaline electrolyte (KOH), sintered Ni being the anode, and porous lithiated NiO the cathode. It is fair to say, that space travel has played a very important role in keeping fuel cells alive, and indeed $H_2/O_2$ cells are still being used in the Challenger flights (and cf. below). Large-scale terrestrial applications have still to materialize, although they have been predicted to be imminent for some time now.

In the late 1950s, early 1960s the first polymeric acidic membrane fuel cell ($H_2/O_2$, Pt-based electrodes) was developed by General Electric, and applied in the Gemini flights. This type of cell was abandoned for a time in favour of the phosphoric-acid cell (PAFC), but has recently resurfaced. The PAFC was developed in an attempt to convert carbonaceous fuels directly, but again these turned out to be too unreactive despite the relatively high temperatures that can be reached with phosphoric acid (around 200°C), and so one has returned to hydrogen here as well. The use of polytetrafluoroethylene (PTFE, "Teflon") as a wet-proofing agent was very important, in that it allows almost all of the high-surface-area Pt in an electrode to be actually contributing to the activity.

At about the same time, work on the molten-carbonate fuel cell was started by Ketelaar en Broers in the Netherlands. They employed Ni-based electrodes similar to Bacon's. Not much later, the solid-oxide fuel cell was introduced, where the doped zirconia electrolyte was adapted from the "Nernst glower" of 1900. Here again, hydrogen is the fuel of choice.

Still in the 1960s, the direct methanol fuel cell came to the fore as an attractive candidate for such low-power applications as vehicle traction. Methanol was widely seen as the best compromise between reactivity, handleability and cost. Such oil majors as Exxon and Shell were involved in this effort for a time, but the promise has still to be redeemed.

We will skip the last two decades, and turn instead to a slightly more detailed description of the various types of fuel cell in their more or less present-day embodiments. Apart from the DMFC, all FCs are effectively $H_2/O_2$ cells. It should be realized that they have only about 1 V/cell, so that stacks have to be built to obtain useful voltages, but the technologies involved will not be discussed. It is worth pointing out, however, that this modularity makes FC systems very flexible.

## 3. Low-temperature fuel cells

### *3.1. Alkaline fuel cells (AFC)*

As described in Section 2, the AFC was the first fuel cell to be actually applied. It is indeed the best functioning cell to date and it is commercially available. The use of an alkaline electrolyte entails several advantages: (i) it is much less corrosive than its acidic counterpart, enhancing cell life, (ii) it allows the use of non-noble metals as electrocatalysts for the fuel cell reactions, examples being Raney nickel for hydrogen oxidation and Raney silver (rather than Bacon's Li : NiO) for oxygen reduction, reducing cost, and (iii) practical efficiencies are rather high, around 60%, because the oxygen reduction is substantially more rapid in alkaline than it is in acidic electrolyte (at low temperatures). The electrolyte, then, is aqueous 30–45 wt% KOH because of its high conductivity. The system is advantageously operated at around 80°C, so that the water formed in the reaction can be removed by means of gas recycling. Alternatively, the electrolyte can be circulated to effect the same.

According to Murphy's Law, there should also be drawbacks, and these are: (a) when using air, one has to scrub it to remove the $CO_2$, which would otherwise carbonate the electrolyte with deleterious effect on the cell performance, (b) carbonaceous fuels cannot be used for the same reason, $CO_2$ being the product of the electrooxidation reaction, and (c) AFCs are not very compact (limited power density).

At present, several applications are envisaged for the AFC, apart from their continued use in space travel by UTC. An example is Elenco NV's study of their suitability for transportation purposes (e.g., in city buses), with circulating electrolyte and with the hydrogen fuel being stored on board in liquid form. A similar objective is pursued by Siemens, but in this case submarines are the primary target. Also in the

European space-travel programme (HERMES) an AFC is being developed; here a non-circulating electrolyte will be used, trapped in a matrix. For this application the effect of going to higher temperatures (say, 150°C) is also studied. Efficiencies can be substantially higher than at lower temperatures, but Pt-based electrodes will have to be used instead of the non-noble metal ones (for stability reasons).

As to the electrode structure, in the beginning sintered or pressed powders supported on metal screens were applied, which were rather hydrophilic. Later on, following developments in the PAFC programme (vide infra), PTFE ("Teflon")-bonded electrodes were more and more applied. Because of their unique capability of maximizing the "notable surface of action" in aqueous-electrolyte fuel cells, their structure will now be briefly described in a separate paragraph.

### *3.2. The structure of Teflon-bonded electrodes*

The Teflon-bonded electrode has been designed to maximize the contact between gaseous reactant, catalyst and electrolyte. When reacting gases at non-porous electrodes, only small current densities are observed, and this is due to the low solubility of those gases in the electrolyte, resulting in a diffusion-limited current density of only a few $mA/cm^2$. In practical fuel cells a two orders of magnitude higher value is needed. This can be achieved by using porous electrodes containing high-surface-area catalysts, structured in such a way that catalyst utilization is high, i.e. that the performance is not significantly affected by ohmic and mass-transport effects. Teflon-bonded electrodes appear to offer such a structure.

To make such electrodes one mixes the catalyst, say platinum black or Pt/Carbon, with a stabilized Teflon suspension, usually aiming to have 20–35 %w Teflon in the final electrode. The resulting plastic mix is either applied to a current collector screen and sintered in an inert atmosphere at the softening temperature of the Teflon applied (usually around 325°C), or extensively rolled and then pressed into such a screen. The purpose of the heat or rolling treatment is to cross-link the Teflon particles, so that they will form a continuous network, in which the catalyst particles are held.

When the electrode is in contact with an aqueous electrolyte, the catalyst, being almost invariably hydrophilic, is commonly found to be completely wetted. The Teflon network, being extremely hydrophobic, is generally, though not universally, considered to provide the gas channels. Thus, the working mechanism of Teflon-bonded electrodes is explained by assuming that the catalyst particles form porous (and electronically conducting) agglomerates which, under working conditions, are flooded with electrolyte. The catalyst aggregates are kept together by the Teflon binder, which also provides hydrophobic gas channels. A schematic diagram of a Teflon-bonded electrode is given in Fig. 3.

From the above it is obvious that a well-made electrode should consist of two continuous interpenetrating networks, one formed by electrolyte-filled catalyst agglomerates and the other by gas-filled cross-linked Teflon. Also, the flooded agglomerates should not be too large, which would lead to long diffusion paths for

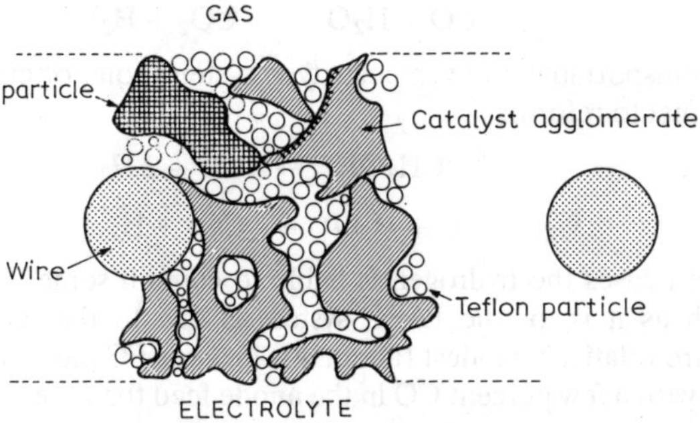


Fig. 3. Schematic representation of a Teflon-bonded electrode [7].

the reactant gas with a concomitant increase in diffusion polarization, nor too small, which would lead to important ohmic losses (the electrode should not be too thick for the same reason). Finally, the Teflon content should be finely tuned, in that too much will make the whole electrode too hydrophobic, preventing the flooding of the agglomerates, and too little will compromise the gas-diffusion characteristics and the structural integrity. These problems can all be satisfactorily solved. (Often an extra porous Teflon layer is applied at the gas side of an electrode to prevent the electrolyte from entering the gas compartment).

### *3.3. Phosphoric-acid fuel cells (PAFC)*

The Grove cell featured aqueous sulfuric acid as the electrolyte, which limits the operating temperature to about 90°C. Since, ceteris paribus, higher temperatures mean higher efficiencies, the switch was later made to phosphoric acid, which can be used at temperatures of up to 210°C. At the latter temperature it actually exists as pyrophosphoric acid, $H_4P_2O_7$.

Both anode and cathode are carbon-supported Pt-based wet-proofed Teflon-bonded electrodes [about 0.5 mg Pt/cm$^2$ (geometric)], while the electrolyte is immobilized in a SiC matrix. The carbon support is preferably a high-surface-area (HSA) graphite, HSA since the Pt dispersion increases (Pt particle size decreases) with increasing carbon surface area, and graphite because it is more corrosion resistant than other types of carbon. Alloying of the Pt is frequently practiced, not only to improve its activity, especially for oxygen reduction (see below), but also for stability reasons. Ternary alloys such as Pt/Co/Cr and Pt/Co/Ga have been mentioned for the cathode; one still has metal leaching in these cases, but the Pt particle growth can be largely suppressed. Alloys like PtPd can be used in the anode; here again Pt dispersion is better maintained, and the poison resistance (CO, $H_2S$) is increased.

As mentioned in the introduction, the $H_2$ feed for a fuel-cell system is often generated through the steam-reforming of, e.g., methane:

$$CH_4 + H_2O \rightarrow CO + 3H_2 \qquad (4)$$

$$CO + H_2O \rightarrow CO_2 + H_2 \quad (5)$$

For transportable and vehicular type systems, one often prefers methanol as the primary fuel to reform:

$$CH_3OH \rightarrow CO + 2H_2 \quad (6)$$

$$CO + H_2O \rightarrow CO_2 + H_2 \quad (7)$$

In both cases the hydrogen is liable to contain some CO, and the technical success, such as it is, of the PAFC system is due to the fact that, although the efficiencies are relatively modest (about 40–45% for FC plus ancillaries), it can cope much better with a few percent CO in the anode feed than the other low-temperature FCs.

The phosphoric-acid utility power concept was developed by United Technologies Corp. (UTC), later International Fuel Cells Corp. (IFC), in the US, leading to the 4.8 MW project in Manhattan. The power plant never came on stream, however, due to several unexpected difficulties largely unconnected with the fuel-cell stacks themselves. The technology was finally licensed to Tokyo Electric Power Co., which resulted in the successful demonstration of an 11 MW utility plant, with an overall efficiency of around 40% (the fuel is natural gas). At the present time, however, PAFCs are expected to be mainly used in such areas as hospitals, paper mills, municipal waste treatment plants and chlor-alkali production.

Some attention is also being paid to PAFC application in the transportation sector, with (reformed) methanol as the fuel, but this would appear to have to take the backseat *vis-a-vis* the polymeric membrane FCs, to which we will now turn.

### *3.4. Solid polymer electrolyte fuel cells (SPFC)*

The solid polymer electrolyte fuel cells use an ion-exchange membrane as electrolyte. Generally a proton exchange membrane is applied and the cells are therefore also referred to as PEMFCs. The Gemini fuel cells of GE contained polystyrene sulfonates. Nowadays one employs such membranes as Nafion, a sulfonated PTFE from DuPont.

The advantage of using a proton exchange membrane is that they can be extremely thin, e.g., 0.05 mm, so that electrolyte ohmic losses are small and current densities can be high. Also, fuel-cell stacks can be very compact, making very high power densities (say, 1 $MW/m^3$) a possibility — indeed, this is the very reason why PEMFCs are so widely considered to have such great potential for application in vehicle traction.

The heart of a PEMFC is the membrane/electrode assembly (MEA), which consists of the PEM, a layer of catalyst (e.g. Pt black or Pt/C) on each side of the membrane, and a gas-porous electrode support material (typically a wet-proofed (Teflon) porous carbon paper or cloth). A MEA is less than a millimeter thick. The catalyst/membrane interface is extremely important in that it determines in large measure the performance of the cell: the problem of the three-phase contact again. The catalyst layer may be impregnated with a soluble form of the polymer electrolyte to produce a more intimate contact between the catalyst and the electrolyte. The MEA is usually bonded together by heating the components under pressure to a temperature at which the polymer softens. Techniques have been developed to emplace the Pt particles exactly there where they are wanted, i.e. at the three-phase boundary, so that high-performance cells do not need more than 0.2 to 0.3 mg Pt/cm$^2$.

The vehicle-traction application is under intense scrutiny, for example, at Ballard Technologies (who have teamed up with Dow as the membrane manufacturer), Los Alamos National Laboratory, and Siemens. The operating pressure is envisaged to be somewhat elevated, 2–5 atm, operating temperatures being of the order of 70–100°C. Many of the present automotive programmes envision using methanol reformed on board of the vehicle as fuel, which requires (i) efficient removal of CO, and (ii) fast load-following of the reformer. Another area requiring attention is the cost (and, indeed, the performance) of the polymer electrolytes. The water/heat management, though not easy, seems to be under control. The major challenge left is whether the PEMFCs can be designed and mass-produced cost-effectively.

### *3.5. Some electrocatalytic aspects of $H_2$ oxidation and $O_2$ reduction*

The efficient conversion of dihydrogen and dioxygen requires electrocatalysts able to adsorb them dissociatively. In acid electrolyte, Pt is the material of choice for both reactions.

In the electrochemical oxidation of dihydrogen, the adsorption step, $H_2 \rightarrow 2\,H_{ad}$, is followed by the Volmer reaction, $H_{ad} \rightarrow H^+ + e$. The rate-controlling step is the dual-site dissociation of the hydrogen molecule. Although this is a non-electrochemical step, the reaction rate is still a function of the potential, because the dihydrogen oxidation reaction is self-poisoned by adsorbed hydrogen atoms, $H_{ad}$, and the hydrogen atom adsorption isotherm is a function of the polarization. At open circuit, the platinum surface is nearly completely covered with $H_{ad}$, and the coverage decreases approximately linearly with the polarization to reach zero at about 0.3 V, where therefore the oxidation current is at its maximum.

The $H_2$ oxidation reaction over Pt is very fast indeed, and the only major complaint here is the sensitivity of Pt for CO. Especially in the case of the PEMFCs, which work at a much lower temperature than the PAFCs, this is a problem: the fuel should be essentially free of CO (<5 ppm). This can in principle be achieved with a selective oxidizer (e.g. 1% Pt/alumina at 125°C with $O_2$/CO about 2/1), if the starting CO level is not too high (a few percent at most). There is as yet no alternative to Pt as an anode catalyst; WC has been a candidate for some time because it is insensitive to CO, but its activity could not be increased to acceptable levels.

The first two steps in the electrochemical reduction of dioxygen over Pt in acid are:

$$O_2 + H^+ + e + Pt = Pt{-}O_2H \qquad (8)$$

$$Pt - O_2H + Pt = Pt{-}O + Pt{-}OH \qquad (9)$$

after which the adsorbed O and OH species are further reduced to $H_2O$. The

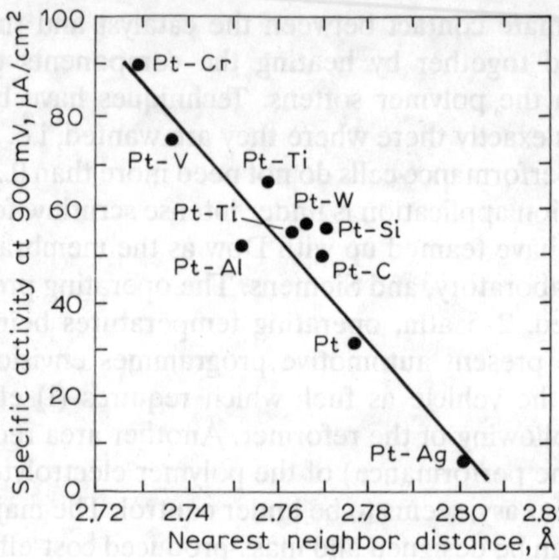


Fig. 4. Specific activity for oxygen reduction vs. electrocatalyst nearest neighbor distance; 100% $H_3PO_4$, 200°C[8].

molecular oxygen reduction reaction is very slow, unfortunately, so that appreciable currents can be drawn only at relatively high polarizations. One also has to guard against strongly adsorbing impurities in the electrolyte, as they impede reaction (9) and, thus, lead to the formation of $H_2O_2$, with a concomitant loss of efficiency (2e vs. 4e reduction), not to mention its corrosiveness. Improvement of the activity can sometimes be achieved through alloying the platinum with other metals. An interesting rationalisation of the effect of alloying is due to Jalan and Taylor[3]: they propose that the activity increases with increasing "fit" between the oxygen molecule and two Pt nearest neighbours (lower energy of activation for the bond-breaking reaction). And in fact, there is a reasonable correlation between specific activity and Pt–Pt nearest neighbour distance (Fig. 4). Of course, this does not prove a direct cause–effect relationship, but it is a pleasing chemical model, which moreover has recently been corroborated in a study of the effect of Pt microstructure on the $O_2$-reduction performance.

Again, alternatives for Pt(-alloys) are not available (in acid), although there has been an active search for them. For a time, chelates supported on carbon were thought to be promising candidates, but their performance is not really good enough, the major drawback being that they tend to reduce $O_2$ to the undesirable $H_2O_2$, instead of to $H_2O$[9].

### *3.6. Direct methanol fuel cells (DMFC)*

In automotive applications, where methanol is often considered to be a very attractive fuel, it would be advantageous to be able to oxidize methanol directly at the anode, thus avoiding the necessity of reforming it first. The product of the reaction being $CO_2$, you need a $CO_2$-rejecting electrolyte, i.e. in general an acid (Shell employed aqueous sulfuric acid, while Exxon rather opted for a carbonate buffer). And once again, the best catalyst is (promoted) Pt.

Very roughly, three reaction steps can be distinguished in the electrochemical oxidation of MeOH over Pt:

$$CH_3OH \rightarrow \gtrdot C-OH + 3H^+ + 3e \quad (10)$$

$$H_2O \rightarrow -OH + H^+ + e \quad (11)$$

$$\gtrdot C-OH + -OH \rightarrow CO_2 + H_2O \quad (12)$$

The general view is that it is the activation of water, step (11), coupled with the strong adsorption of $COH_{ad}$ (or its product, $CO_{ad}$), that is the difficulty here. Many elements and compounds have been evaluated as possible promoters of the activity of Pt, the most promising system being PtRu supported on carbon. The action of Ru is generally discussed in terms of its accounting for reaction (11) — the bifunctional mechanism — but there are also indications that it modifies the bonding between Pt and the methanolic residue (ligand effect). In addition, more complete oxidation of the methanol is observed, thus avoiding the formation of formic acid, and hence methylformiate. The surface coverage of Pt should be relatively low, because the adsorption of methanol requires quite a large ensemble of Pt atoms, which would not be available were the Ru coverage to increase much beyond 0.1 monolayer.

With both the MeOH oxidation and the $O_2$ reduction being on the very slow side, commercialization of the DMFC has always looked a long way off, but the advent of the new generation of PEMFCs has increased hope again. Indeed it seems that it is not so much the slowness of the oxidation reaction that is limiting the possibilities of the PEMDMFC, but the migration of MeOH through the membrane to the cathode, where it interferes with the reduction of dioxygen. So, this is one problem to solve, and one should also have a close look at the stability of the PtRu electrode under operating conditions.

## 4. Molten-carbonate fuel cells (MCFC)

In the case of stand-alone units, there is a preference for the higher-temperature fuel cells, because of the higher quality of the waste heat. Industry's high heat-to-electricity utilization limits the advantages of FCs in any case. Nevertheless, molten-carbonate and solid-oxide fuel cells are considered to be competitive in base-load applications, given a lifetime of at least 40,000 hours. Application in ships and submarines is also envisaged, however.

Of the higher-temperature fuel cells the MCFC is technically the most advanced. Its standard fuel is steam-reformed methane (natural gas) or naphtha. But a coal gasifier + FC can also be an interesting alternative (e.g., to pulverized coal combustion), and at the UNCED conference FCs have been discussed in terms of biomass as the primary fuel[1]. The electrode reactions are:

$$\text{anode:} \quad H_2 + CO_3^{2-} \rightarrow H_2O + CO_2 + 2\,e \qquad (13)$$
$$(CO + CO_3^{2-} \rightarrow 2\,CO_2 + 2\,e)$$

$$\text{cathode:} \quad O_2 + 2\,CO_2 + 4\,e \rightarrow 2\,CO_3^{2-} \qquad (14)$$

The anode catalyst is Ni, the cathode one is NiO, while the electrolyte consists of a $Li_2CO_3/K_2CO_3$ eutectic melt immobilized in a $LiAlO_2$ tile. The cell is operated at about 650°C. Under these conditions the presence of CO in the $H_2$ stream is not problematic. Both the cathodic and anodic reactions are fast, i.e. on a par with the oxidation of $H_2$ over Pt in acid solution (note that $CO_2$ needs to be cofed with $O_2$ to the cathode). Both electrodes are porous with the pores (partially) filled with electrolyte — improved performance, therefore, depends o.a. on the possibility of reducing the electrolyte resistance and of producing a large internal surface area.

The development of MCFCs have only just reached the demonstration phase. Problem areas include: (i) cathode (NiO) dissolution during cell operation — can be counteracted by doping the cathode, adding elements to the electrolyte, and lowering the $CO_2$ partial pressure, (ii) creep; electrolyte loss and migration phenomena; corrosion — the seriousness of these phenomena depends on the seal material and its design, and on the material selection and surface finish for the other cell components, and (iii) compatibility of cell and stack component materials. Single-cell lifetimes have reached the 20,000 hours mark, and people are now studying the behaviour of large stacks (100 kW units).

A hot topic here is so-called "internal" reforming, meaning that the (endothermic) methane reforming reaction is carried out inside the anode compartment (at 650°C), instead of externally (at 800°C), see Fig. 5. Getting the reaction going at 650°C, however, is an as yet unsolved problem, and the "direct" variant suffers also from catalyst deactivation due to electrolyte components evaporating through the anode. Realization of the internal reforming option would boost the efficiency of the MCFC to 60–65%.

## 5. Solid-oxide fuel cells (SOFC)

The solid-oxide fuel cell has the highest operating temperature of all, viz. approximately 1000°C. Because of this high temperature, internal reforming presents no problem, which is good for system simplicity. Power generation efficiency is reasonably high at 55% or so. The all solid-state cells usually show highly stable performance. It is in fact the SOFC that has aroused most industrial interest.

The clear technology leader is Westinghouse Electric Corp., who started out in this field in the mid-Seventies, but there are many other important players (in North America, Japan, Europe). The development of SOFCs is a purely ceramic affair, and the SOFC fabricator maxim's is reputed to be: "They who control the materials, control the technology". Barriers to progress in the larger-scale manufacture are: (i)

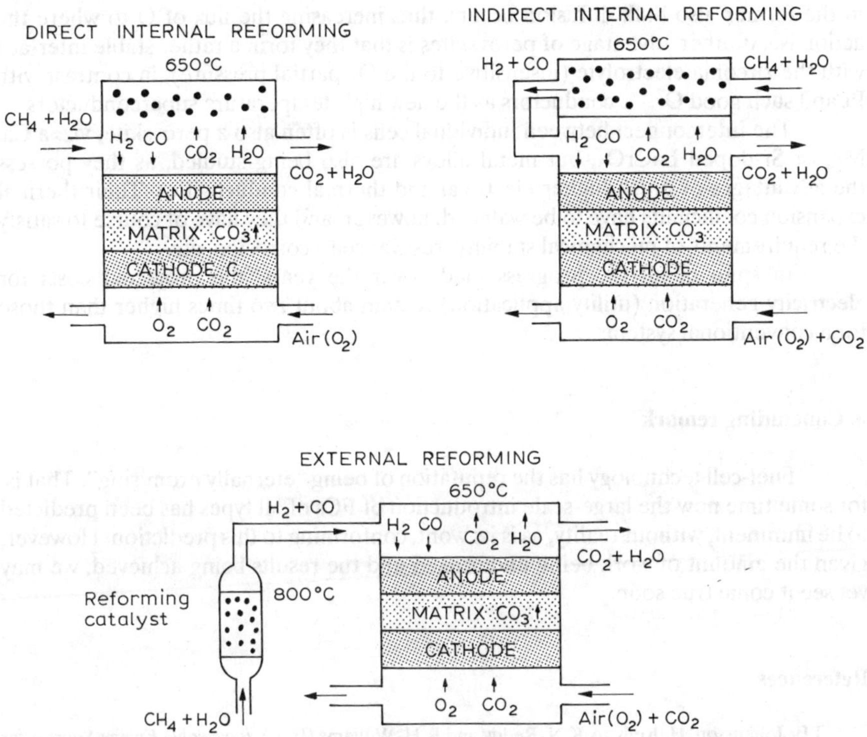


Fig. 5. Type of reforming [10].

assuring that a cell is functioning well (e.g. no cracks in the electrolyte) for a large majority of cells produced, (ii) achieving powder starting materials consistency, and (iii) correct matching of the thermal expansion coefficients of the various cell (and stack) components, to prevent failure after thermal cycling.

The electrode reactions are as simple as you can get:

$$\text{anode:} \quad H_2 + O^{2-} \rightarrow H_2O + 2e \tag{15}$$

$$\text{cathode:} \quad O_2 + 4e \rightarrow 2O^{2-} \tag{16}$$

The anode is a porous Ni-Cermet, fabricated as a mixture of yttria–zirconia and NiO which is converted (reduced) to the conductive cermet in situ within the cell. The electrolyte is an impervious yttria-stabilised zirconia (at least 94% of theoretical density), which is a stable, and reasonably good, oxygen-ion conductor, and a good insulator, at 1000°C. The cathode is a perovskite, e.g. (La,Sr)$MnO_3$. It performs better than, for example, Pt, because in the latter case only surface diffusion of $O_{ad}$ to the electrocatalyst/electrolyte interface can take place (a rate-limiting step), while in the former also bulk diffusion occurs, thus increasing the flux of O to where the action is. Another advantage of perovskites is that they form a rather stable interface with the zirconia electrolyte (insensitive to the $O_2$ partial pressure), in contrast with Pt and such good $O^{2-}$/e conductors as the new high-temperature superconductors.

The interconnect between individual cells is often also a perovskite, viz. a Ca, Mg, or Sr-doped $LaCrO_3$, but metal alloys are also being studied, as they possess the advantages of much higher electrical and thermal conductivities. Their thermal expansion coefficients have to be watched, however, and they have of course to satisfy the requirements of mechanical stability, resistance to corrosion, etc.

In spite of all the progress made over the years, the projected costs for electricity generation (utility application) remain about two times higher than those for a conventional system.

## 6. Concluding remark

Fuel-cell technology has the reputation of being "eternally promising". That is, for some time now the large-scale introduction of FCs of all types has been predicted to be imminent, without reality, as is its wont, conforming to this prediction. However, given the amount of work being undertaken and the results being achieved, we may yet see it come true soon.

## References

1. T.B. Johansson, H. Kelly, A.K.N. Reddy and R.H. Williams (Eds.), *Renewable Energy-Sources for Fuels and Electricity* (Earthscan/Island Press, 1993) (input to the UNCED process).
2. W.J. Albery, *Electrode Kinetics* (Clarendon Press, Oxford, 1975).
3. A.J. Appleby, C.K. Dyer, P.T. Moseley and D.A.J. Rand, *Proceedings of the Third Grove Fuel Cell Symposium, J. Power Sources* **49** (1994).
4. L.J.M.J. Blomen and M.N. Mugerwa (Eds.), *Fuel Cell Systems* (Plenum Press, New York, NY, 1993).
5. D.G. Lovering, *Proceedings of the Grove Anniversary Fuel Cell Symposium, J. Power Sources* **29** (1990).
6. W. Vielstich (Ed.), *Proceedings of a Discussion Meeting on "Fuel cells and their Applications", Ber. Bunsenges. Phys. Chem.* **94**, No. 9 (1990).
7. J. Giner and C. Hunter, *J. Electrochem. Soc.* **116** (1969) 1124.
8. V. Jalan, and E.J. Taylor, *J. Electrochem. Soc.* **130** (1983) 2299.
9. J.A.R. Van Veen and J.F. van Baar, *Rev. Inorg. Chem.* **4** (1982) 293.
10. K. Kishida, in: *Proceedings of a Discussion Meeting on "Fuel cells and their Applications"* (W. Vielstich, Ed.), *Ber. Bunsenges. Phys. Chem.* **94**, No. 9 (1990) 941.

# LIQUID PHASE AUTOXIDATIONS

R.A. SHELDON
*Laboratory for Organic Chemistry and Catalysis,*
*Delft University of Technology, Julianalaan 136*
*2628 BL Delft, The Netherlands*

ABSTRACT

The free radical chain mechanism of liquid phase oxidations with dioxygen is reviewed. The intricate mechanism of the oxidation of substituted toluenes to the corresponding carboxylic acids with the Co/Mn/Br catalyst (Amoco/MC system) is discussed in detail.

## 1. Introduction

Many liquid phase oxidations of organic substrates with dioxygen are known as autoxidations because they are subject to autocatalysis by the initial products of oxidation, alkyl hydroperoxides[1]. The pioneering work of Bäckstrom[2] demonstrated that these processes are radical chain reactions. Criegee[3] made an important contribution in 1939 when he showed that the primary product of the oxidation of cyclohexene by dioxygen is the allylic hydroperoxide (reaction 1). Subsequently it was recognized that the controlled autoxidation of hydrocarbons can be a useful method for preparing oxygenated derivatives. This led to the development, by Hock and Lang[4] in 1944 of what is still the most important industrial process for the production of phenol and acetone, by acid-catalyzed Criegee rearrangement of cumene hydroperoxide derived from the autoxidation of cumene (reaction 2).

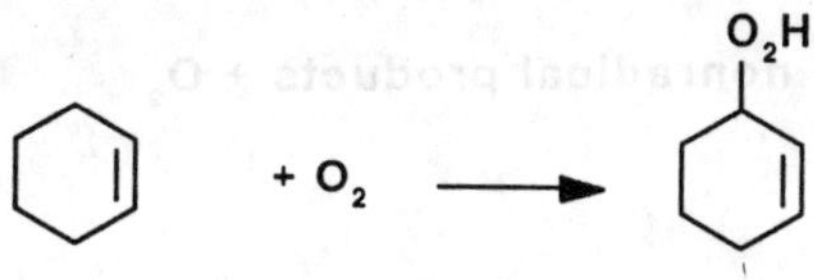


(1)

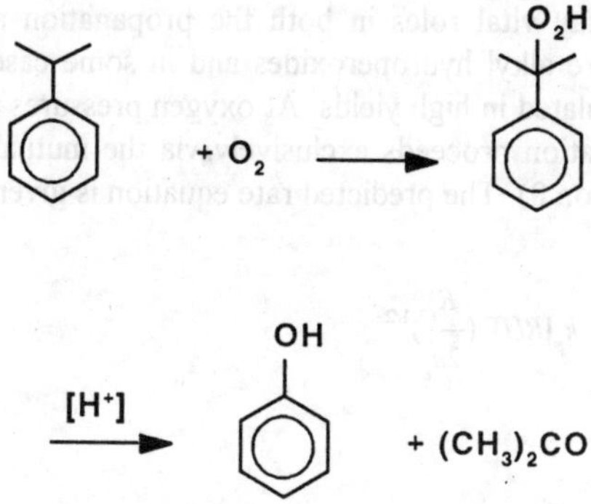


(2)

## 2. Fundamentals of Radical Chain Autoxidations

Liquid phase autoxidations proceed via a free radical chain mechanism described by the general scheme shown in reactions 3-8.

**Initiation**

$$In_2 \xrightarrow{R_i} 2\,In\bullet \qquad (3)$$

$$In\bullet + RH \longrightarrow InH + R\bullet \qquad (4)$$

**Propagation**

$$R\bullet + O_2 \longrightarrow RO_2\bullet \qquad (5)$$

$$RO_2\bullet + RH \xrightarrow{k_p} RO_2H + R\bullet \qquad (6)$$

**Termination**

$$R\bullet + O_2 \longrightarrow RO_2R \qquad (7)$$

$$RO_2\bullet + RO_2\bullet \xrightarrow{2k_t} RO_4R$$

$$\longrightarrow \text{nonradical products} + O_2 \qquad (8)$$

Alkylperoxy radicals play vital roles in both the propagation and termination steps. The primary products are alkyl hydroperoxides and in some cases, e.g. cumene hydroperoxide, they may be isolated in high yields. At oxygen pressures used in practice (1 bar or higher) chain termination proceeds exclusively via the mutual destruction of two alkylperoxy radicals (reaction 8). The predicted rate equation is given by equation 9.

$$\frac{-d[RH]}{dt} = \frac{-d[O_2]}{dt} = k_p[RH]\left(\frac{R_i}{2k_t}\right)^{1/2} \qquad (9)$$

The susceptibility of a particular substrate to autoxidation is governed by the ratio $k_p/(2k_t)^{1/2}$ which is referred to as its **oxidizability**[5]. The oxidizabilities of some typical organic substrates are listed in Table 1.

Table 1. Oxidizability of various organic compounds[a].

| Substrate | $k_p/(2k_t)^{1/2} \times 10^3$ ($M^{-1/2}\,s^{-1/2}$) |
|---|---|
| 2,3-Dimethyl-2-butene | 3.2 |
| Cyclohexene | 2.3 |
| 1-Octene | 0.06 |
| Cumene | 1.5 |
| Ethylbenzene | 0.21 |
| Toluene | 0.01 |
| p-Xylene | 0.05 |
| Benzaldehyde | 290 |
| Benzyl alcohol | 0.85 |

a. Data taken from ref. 5.

### *2.1. Chain Initiation*

Chain initiation is readily accomplished by the deliberate addition of initiators that yield free radicals on thermal decomposition. Typical initiators are aliphatic azo compounds and various peroxides (Table 2). The initiator of choice for a particular autoxidation should have a half-life of about one hour at the temperature of reaction (see Table 2).

Initiation by direct reaction of dioxygen with hydrocarbons is, as noted in Chapter 1, kinetically unfavorable although it has been observed in a few cases, e.g. with indene[6] which forms a highly stabilized radical. When chain initiation is observed in the absence of added initiators it can usually be attributed to the generation of radicals by thermal decomposition of adventitious peroxidic impurities present in the hydrocarbon substrate. In this context it is worth noting that many studies of so-called dioxygen activation in the literature have employed cumene[7] or cyclohexene[8] as the substrate. It is precisely these two substrates that always contain substantial amounts of alkyl hydroperoxide impurities unless they are rigorously purified prior to use[9].

Table 2. Initiators for autoxidations[a].

| Initiator | $E_a$ (kcal.mol$^{-1}$) | Temp. °C for $t_{1/2}$ = 1 hr |
|---|---|---|
| HO-OH | **48** | -- |
| t-BuO-OH | **42** | -- |
| t-BuO-OBu-t | **37** | 150 |
| t-BuO-$O_2$CPh | **34** | 125 |
| Ph$CO_2$-$O_2$CPh | **30** | 94 |
| $CH_3CO_2$-$O_2CCH_3$ | **30** | 85 |
| i-PrC(CN)N=N(CN)Pr-i | **30** | 85 |
| t-BuON=NOBu-t | **28** | 60 |
| t-BuO-$O_2$C$O_2$-OBu-t | **25.5** | 40 |

a. Data taken from ref. 1.

## *2.2. Chain Propagation*

Reaction of the alkyl radical (R·) with dioxygen is, in most cases, diffusion controlled (i.e. $k_2 > 10^9$ $M^{-1}$ $s^{-1}$) and the rate-controlling step in autoxidations is hydrogen transfer from the substrate to the alkylperoxy radical (reaction 6). The rate constants ($k_p$) for this reaction can be roughly correlated with its exothermicity. Oxidations are favorable when the bond that is formed (ROO-H) is at least as strong as that which is broken (R-H). The ROO-H bond is about 90 kcal mol$^{-1}$ which is larger than that for benzylic, allylic and aldehydic C-H bonds (see Table 3).

Table 3. X-H bond energies[a].

| Compound | Energy (kcal.mol$^{-1}$) | Compound | Energy (kcal.mol$^{-1}$) |
|---|---|---|---|
| $CH_3$-H | 103 | Ph$CH_2$-H | 85 |
| n-$C_3H_7$-H | 99 | RCO-H | 86 |
| i-$C_3H_7$-H | 94 | $CH_3$S-H | 88 |
| t-$C_4H_9$-H | 90 | $CH_3$PH-H | 85 |
| $CH_2$=CH-H | 105 | PhO-H | 88 |
| $C_6H_5$-H | 103 | PhNH-H | 80 |
| $CH_2$=CH$CH_2$-H | 85 | ROO-H | 90 |

a. Reproduced with permission from ref. 1.

Propagation rate constants are also dependent on the nature of the attacking alkylperoxy radical. Table 4 compares the propagation rates for the reaction of various substrates with its own alkylperoxy radical with that of the reaction of tert-butylperoxy radicals (cross propagation) with the same substrates.

**Table 4. Rate constants per labile hydrogen for reactions of substrates with their own peroxy radical ($k_p$) and with tert-butylperoxy ($k_p'$) at 30 °C[a].**

| Substrate | $k_p$ ($M^{-1}\ s^{-1}$) | $k_p'$ ($M^{-1}\ s^{-1}$) | $k_p/k_p'$ |
|---|---|---|---|
| 1-Octene | 0.5 | 0.084 | 6.0 |
| Cyclohexene | 1.5 | 0.80 | 1.9 |
| Toluene | 0.08 | 0.012 | 6.7 |
| Ethylbenzene | 0.65 | 0.10 | 6.5 |
| Cumene | 0.18 | 0.22 | 0.9 |
| Tetralin | 1.6 | 0.5 | 3.2 |
| Benzyl alcohol | 2.4 | 0.065 | 37.0 |
| Benzyl acetate | 2.3 | 0.0075 | 307 |
| Benzyl bromide | 0.6 | 0.006 | 100 |
| Benzaldehyde | 33,000 | 0.85 | 40,000 |

a. Data taken from ref. 1.

It is readily apparent that the reactivities of alkylperoxy radicals are strongly influenced by both steric and polar effects. In general rates increase with increasing electron-withdrawing capacity of the α-substituent. Acylperoxy radicals, which possess a strong electron-withdrawing carbonyl group, are considerably more reactive than alkylperoxy radicals, e.g. $PhCO_3\cdot$ is $4 \times 10^4$ times more reactive than $t\text{-}BuO_2\cdot$ toward benzaldehyde as substrate. This difference partly explains the very high rates and long chain lengths observed in the autoxidation of aldehydes. On the basis of bond dissociation energies alone (see Table 3) one would expect aldehydes and alkylaromatics to be autoxidized at roughly the same rate.

### *2.3. Chain Termination*

At reasonable oxygen pressures chain termination occurs exclusively via the self-reaction of two alkylperoxy radicals (reaction 8). The overall rate of autoxidation is governed by both the rate of propagation ($k_p$) and termination ($k_t$) as given by eq. 9. Examination of the rate constants in Table 5 reveals that the lower rates of autoxidation at primary and secondary C-H bonds compared to their tertiary counterparts are due not only to the lower reactivity of the C-H bonds (see Table 3) in the former but also to the significantly higher rates of termination of primary and secondary alkylperoxy radicals. This explains why a fairly reactive hydrocarbon such as toluene [$D(PhCH_2\text{-}H)$ = 85 kcal $mol^{-1}$] has a rather low oxidizability.

Table 5. Approximate rate constants for termination of $RO_2\cdot$ at 30 °C[a].

| $RO_2\cdot$ | $2\,k_t$ ($M^{-1}\,s^{-1}$) |
|---|---|
| $HO_2\cdot$ | $8 \times 10^5$ |
| $RCH_2O_2\cdot$ | $10^7$ |
| $R_2CHO_2\cdot$ | $10^6$ |
| $R_3CO_2\cdot$ | $10^3$ |

a. Data taken from ref. 5.

The mode of decomposition of the intermediate tetroxides is dependent on the structure of the alkyl group. Tetroxides derived from primary and secondary alkylperoxy radicals undergo intramolecular disproportionation to an alcohol and a carbonyl compound (reaction 10). This pathway is unavailable to tetroxides derived from tert-alkylperoxy radicals, which undergo decomposition to dialkyl peroxides and molecular oxygen. In this case further thermolysis of the dialkyl peroxides can afford chain initiating alkoxy radicals.

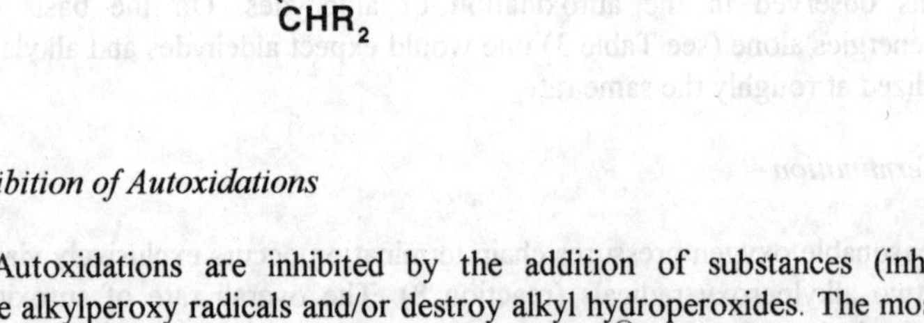

(10)

## 2.4. *Inhibition of Autoxidations*

Autoxidations are inhibited by the addition of substances (inhibitors) that scavenge alkylperoxy radicals and/or destroy alkyl hydroperoxides. The most commonly used inhibitors are substituted phenols, such as Ionol® (2,6-di-tert-4-methyl phenol) and the natural antioxidant vitamin E, which interrupt the autoxidation chain by forming stable phenoxy radicals (reaction 11).

$$RO_2\cdot + ArOH \rightarrow RO_2H + ArO\cdot \quad (11)$$

$$R_2S + R'OOH \rightarrow R_2SO + R'OH \quad (12)$$

$$R_3P + R'OOH \rightarrow R_3PO + R'OH \quad (13)$$

Divalent sulfur and trivalent phosphorus compounds cause inhibition by reducing alkyl hydroperoxides to the corresponding alcohols (reactions 12 and 13). Certain sulfur containing metal complexes, such as zinc dithiocarbamates and dithiophosphates are very effective in the removal of alkyl hydroperoxides.

*2.5. Kinetic Chain Length*

The kinetic chain length (KCL) provides a measure of the efficiency of an autoxidation under given reaction conditions. It is governed by the rates of initiation, propagation and termination and is given by eqn. 14.

$$KCL = k_p[RH]/[2k_tR_i]^{1/2} \quad (14)$$

The yield of alkyl hydroperoxide in a particular reaction is directly related to the kinetic chain length. When the latter is 100, for example, one would expect[9] the alkyl hydroperoxide to be formed in 99% yield, i.e. about 1% hydroperoxide decomposition. Neglect of the correlation of high kinetic chain length with a high yield of alkyl hydroperoxide has led some authors to the erroneous conclusion that a high yield of alkyl hydroperoxide precludes the possibility of chain initiation via metal-catalyzed homolytic decomposition of the hydroperoxide.

## 3. Olefin Autoxidation

In the autoxidation of olefins chain propagation can occur via the usual abstraction mechanism (reaction 15) or via the addition of the alkylperoxy radical to the double bond (reaction 16). Addition can be followed by unimolecular decomposition of the β-alkylperoxyalkyl radical (reaction 17) affording epoxide and an alkoxy radical or by its reaction with oxygen to give polyperoxides (reaction 18).

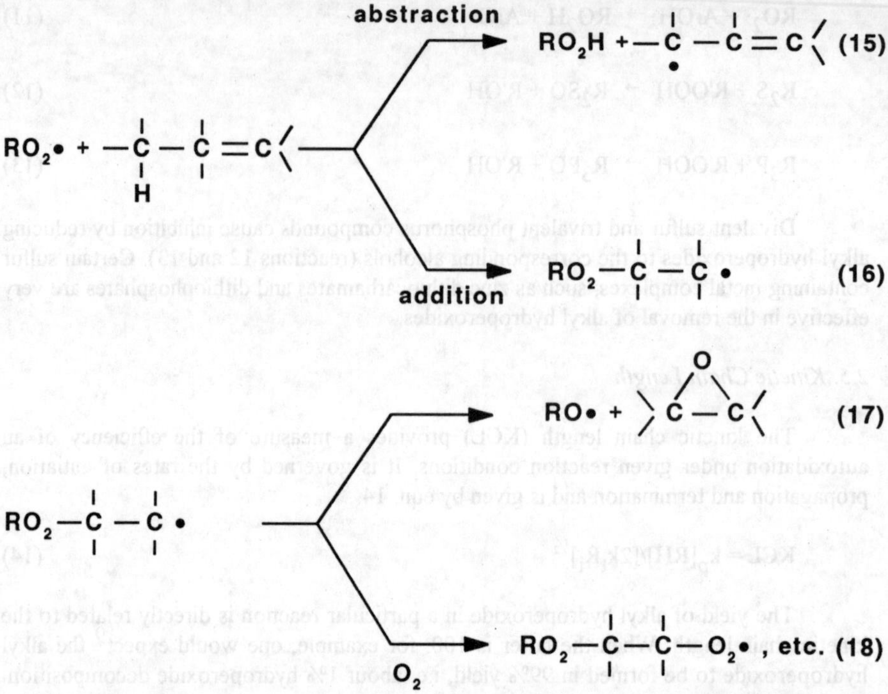


The ratio of addition to abstraction is strongly dependent on the structure of the olefin (see Table 6). Furthermore, the yield of epoxide versus polyperoxide is influenced by the oxygen pressure.

Table 6. Abstraction/addition ratios for selected olefins at 70 °C[a].

| Olefin | Abstraction (%) | Addition (%) |
|---|---|---|
| Propylene | 50 | 50 |
| 1-Hexene | 68 | 32 |
| Cyclohexene | 95 | 5 |
| Cyclooctene | 30 | 70 |
| Isobutene | 17 | 83 |
| 2-Butene | 38 | 62 |
| 1-Butene | 73 | 27 |
| α-Methylstyrene | 0 | 100 |

a. Data taken from ref. 1.

**4. Aldehyde Autoxidation**

The autoxidation of aldehydes is analogous to that of hydrocarbons. Acylperoxy radicals are the principal chain carriers and peroxy acids are the primary products.

$$\mathrm{RCHO} \xrightarrow{\text{Initiation}} \mathrm{R\dot{C}O} \qquad (19)$$

$$\mathrm{R\dot{C}O} + \mathrm{O_2} \longrightarrow \mathrm{RCO_3\bullet} \qquad (20)$$

$$\mathrm{RCO_3\bullet} + \mathrm{RCHO} \longrightarrow \mathrm{RCO_3H} + \mathrm{R\dot{C}O} \qquad (21)$$

As noted earlier the high oxidizability of aldehydes can be mainly attributed to the very high rate of chain propagation (reaction 21).

**5. Cooxidations**

The importance, from a practical viewpoint, of cooxidations of two or more organic substrates cannot be overemphasized. In essence most hydrocarbon autoxidations are, subsequent to the initial stages of reaction, effectively cooxidations of the substrate with reactive secondary products such as alcohols, aldehydes and ketones. Indeed, in many commercial oxidation processes, small amounts of reactive substrates, such as aldehydes and ketones are often added to provide for high initial rates of reaction.

The deliberate addition of small amounts of a second substrate to an autoxidation can sometimes produce dramatic effects. For example, the presence of 3 mol % of tetralin reduces the rate of cumene autoxidation by two-thirds, despite the fact that tetralin is oxidized 10 times faster than cumene[10]. The retardation is due to the higher rate of termination of the secondary tetralylperoxy radicals compared to the tertiary cumylperoxy radicals.

Cooxidations also provide for the possibility of utilizing peroxidic intermediates for additional oxidation processes rather than wasting this active form of oxygen. For example, in the cooxidation of aldehydes and olefins, the acylperoxy radical and the peroxy acid are utilized for the epoxidation of the olefin:

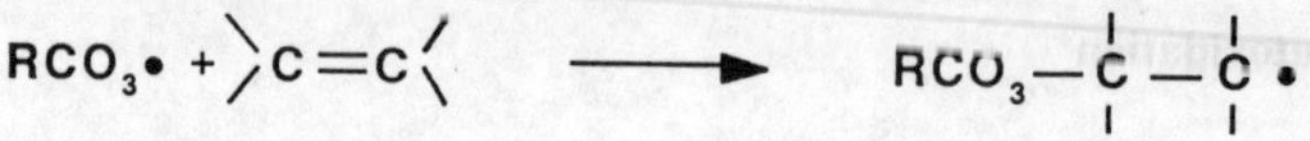

(22)

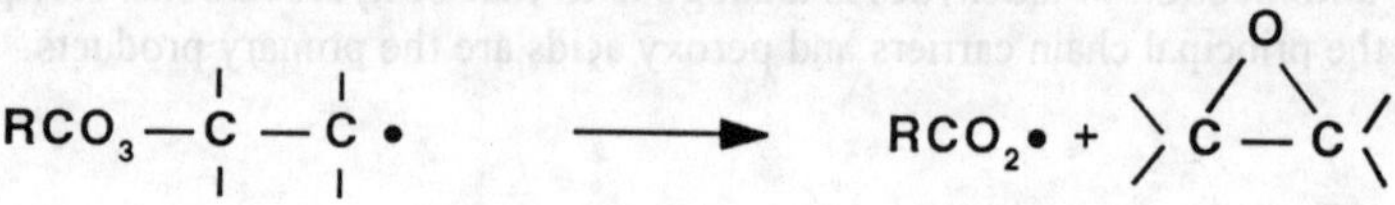

(23)

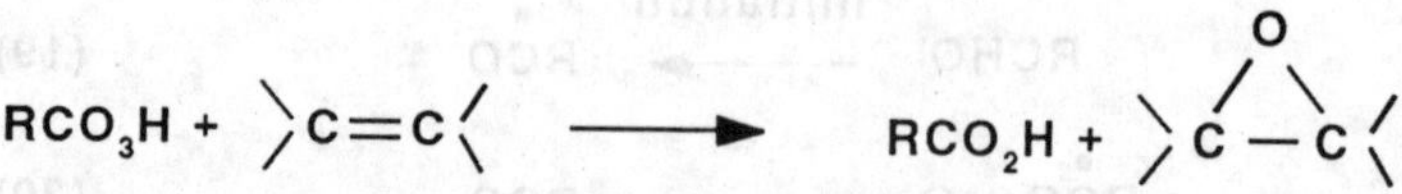

(24)

The cooxidation affords much higher yields of epoxides than those obtained in the autoxidation of the olefin alone, since acylperoxy radicals have a much more favorable addition/abstraction ratio compared to alkylperoxy radicals.

## 6. Gas Phase versus Liquid Phase Oxidation

The primary products of liquid phase autoxidations of hydrocarbons are alkyl hydroperoxides, which in some cases (e.g. cumene) can be isolated in high yields. Gas phase oxidations, in contrast, generally afford carbonyl compounds and or dehydrogenation products. This difference is not due to changes in the fundamental mechanism, but rather to the availability of different pathways for further reaction of the alkylperoxy radicals under gas-phase conditions, i.e. high temperature and low substrate concentration. An illustrative example is the autoxidation of isobutane which has been extensively studied[11,12] and is of considerable importance. The liquid phase autoxidation of isobutane at 125 °C affords a mixture of approximately 75% tert-butyl hydroperoxide (TBHP), 21% tert-butanol (TBA), 2% acetone and 1% isobutyl derivatives[12]. The kinetic chain lengths are rather long and TBHP and TBA are formed via the classical autoxidation mechanism (reactions 25-29).

$$\text{t-Bu}\cdot + O_2 \rightarrow \text{t-BuO}_2\cdot \tag{25}$$

$$\text{t-BuO}_2\cdot + \text{i-BuH} \rightarrow \text{t-BuO}_2\text{H} + \text{t-Bu}\cdot \tag{26}$$

$$2\ \text{t-BuO}_2\cdot \rightarrow \text{t-BuOOBu-t} + O_2 \tag{27}$$

$$2\ \text{t-BuO}_2\cdot \rightarrow 2\ \text{t-BuO}\cdot + O_2 \tag{28}$$

$$\text{t-BuO}\cdot + \text{i-BuH} \rightarrow \text{t-BuOH} + \text{t-Bu}\cdot \quad (29)$$

In the gas phase at 155 °C and at relatively low pressures, in contrast, there is insufficient isobutane to sustain chain propagation (reaction 26). Consequently, the self-reaction of tert-butylperoxy radicals produces tert-butoxy radicals (reaction 28). Since there is also insufficient isobutane to sustain reaction 29 most of the tert-butoxy radicals undergo unimolecular fragmentation to acetone and methyl radicals (reaction 30). The latter react with oxygen to form methylperoxy radicals, which are more reactive in chain termination than tert-butylperoxy radicals. As a result, kinetic chain lengths are short and the principal products are acetone, methanol and TBA:

$$\text{t-BuO}\cdot \rightarrow \text{Me}_2\text{CO} + \text{Me}\cdot \quad (30)$$

$$\text{Me}\cdot + \text{O}_2 \rightarrow \text{MeO}_2\cdot \quad (31)$$

$$\text{MeO}_2\cdot + \text{t-BuO}_2\cdot \rightarrow \text{MeO}\cdot + \text{t-BuO}\cdot + \text{O}_2 \quad (32)$$

$$\text{MeO}_2\cdot + \text{t-BuO}_2\cdot \rightarrow \text{t-BuOH} + \text{H}_2\text{CO} + \text{O}_2 \quad (33)$$

Interestingly, there is a smooth transition from the liquid to gas phase reaction. Thus, an increase in the isobutane pressure in the gas phase oxidation at 100 °C leads to an increase in the yield of TBHP at the expense of TBA, acetone and methanol. At low rates of initiation and 13 bar isobutane a 92% yield of TBHP can be obtained. Similarly, dilution of the liquid phase reaction at 100 °C with the inert solvent, carbon tetrachloride, leads to a simulation of gas phase conditions and the yields of TBA and acetone increase at the expense of TBHP.

At even higher temperatures (300 °C) in the gas phase, the major primary product becomes isobutene, formed via reaction 34. Although reaction 25 is faster than reaction 34 at all temperatures, it is reversible at high temperatures, whereas reaction 34 is not.

$$\text{t-Bu}\cdot + \text{O}_2 \rightarrow \text{Me}_2\text{C=CH}_2 + \text{HO}_2\cdot \quad (34)$$

The scheme outlined above for isobutane applies to all autoxidations. In the liquid phase at relatively low temperatures, the kinetic chain lengths are long and the major products are hydroperoxides. In the gas phase, kinetic chain lengths are short and the major products are carbonyl compounds resulting from thermal fragmentation of intermediate alkylperoxy radicals.

These differences between gas and liquid phase conditions also have implications for oxidations in the presence of metal catalysts. In the liquid phase it is difficult to compete with the ubiquitous free radical chain autoxidation. In the gas phase, in contrast, kinetic chain lenghts are short and intermediate radicals can react via alternative pathways with the metal catalyst. This partly explains why Mars van Krevelen type mechanisms (see Chapter 1) are favored in the gas phase.

## 7. Metal Catalyzed Autoxidations

Variable valence metals, such as cobalt, manganese, copper and iron, catalyze liquid phase autoxidations by promoting the homolytic decomposition of alkyl hydroperoxides into chain initiating radicals via reactions 35 and 36.

$$RO_2H + Co^{II} \rightarrow RO\cdot + Co^{III} + HO^- \quad (35)$$

$$RO_2H + Co^{III} \rightarrow RO_2\cdot + Co^{II} + H^+ \quad (36)$$

Net reaction:

$$2\,RO_2H \xrightarrow{Co^{II}/Co^{III}} RO\cdot + RO_2\cdot + H_2O \quad (37)$$

Since alkylperoxy radicals are strong oxidants they are also capable of oxidizing the reduced form of the catalyst (reaction 38). In this case the metal ion is acting as an inhibitor. Hence, transition metal ions, especially in media of low polarity such as neat hydrocarbons, often behave as autoxidation catalysts at low concentrations and inhibitors at high concentrations. This phenomenon is referred to as <u>catalyst-inhibitor conversion</u>[13]. It manifests itself in the long induction periods often observed in metal-catalyzed autoxidation in nonpolar media.

$$Co^{II} + RO_2\cdot \xrightarrow{ki} RO_2Co^{III} \quad (38)$$

At high cobalt concentrations, Co(II) competes effectively with the substrate RH for the alkylperoxy radicals, obviating chain propagation via reaction 6. Under these conditions termination proceeds virtually exclusively via reaction 38 rather than the self-reaction of two alkylperoxy radicals. A consequence of this is that the expression for the kinetic chain length is different for the two sets of conditions:

Low [Co]

$$KCL = k_p[RH]/[2k_tR_i]^{1/2} \quad (39)$$

High [Co]

$$KCL = k_p[RH]/k_i[Co^{II}] \quad (40)$$

Catalyst inhibitor conversion is observed when the chain length becomes less than unity, i.e. $[Co^{II}] > k_p[RH]/k_i$. In practice an abrupt transition from catalysis to inhibition is generally observed for hydrocarbon autoxidations. In the cobalt-catalyzed autoxidation of neat tetralin at 65 °C, for example, an abrupt transition from rapid reaction to inhibition was observed[14] at a catalyst concentration of approximately 0.1 M.

Catalyst-inhibitor conversion can be circumvented by adding an alkyl hydroperoxide such that $[RO_2H] > [Co]$ or by carrying out the reaction in polar media, such as acetic acid (see later). The use of polar solvents also prevents catalyst deactivation by precipitation as insoluble carboxylate salts from nonpolar media.

## 8. Catalytic Autoxidations of Alkylaromatics

The catalytic autoxidation of toluenes to the corresponding carboxylic acids is of enormous industrial importance, e.g. in the production of benzoic and terephthalic acids by the liquid phase autoxidation of toluene and p-xylene, respectively:

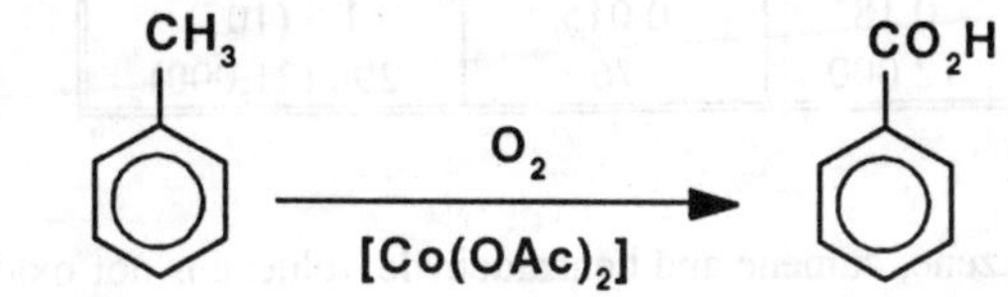


(41)

Temp: 165 °C

Pressure: 10 bar

Conversion: ca. 30%

Selectivity: 90%

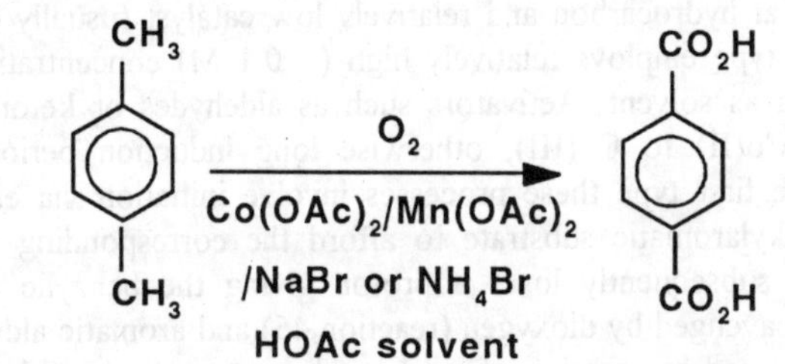


(42)

Temp: 195 °C

Pressure: 20 bar

Conversion: > 95%

Selectivity: > 95%

The exact conditions of the two processes depicted in eqns 41 and 42 are quite different. In order to explain the need for such different conditions, in particular the necessity of the combination Co/Mn/$Br^-$ in HOAc in the Amoco/MC process (reaction 42) we need to delve more deeply into the mechanisms of these fascinating processes.

If one compares the oxidizabilities of toluene, ethylbenzene and cumene one sees that the much lower reactivity of toluene is largely due to the much higher rate of termination of the primary benzyloxy radicals (Table 7). The much higher oxidizability of benzaldehyde, on the other hand, is largely due to a much higher propagation rate.

**Table 7. Comparison of oxidizabilities for aromatic substrates at 30 °C.**

| Substrate | $k_p$ ($M^{-1}s^{-1}$) | $2k_t \times 10^{-6}$ ($M^{-1}s^{-1}$) | $k_p/(2k_t)^{1/2} \times 10^3$ ($M^{-1/2}s^{-1/2}$) |
|---|---|---|---|
| $PhCH_3$ | 0.24 | 300 | 0.014 (1) |
| $PhCH_2CH_3$ | 1.3 | 40 | 0.21 (15) |
| $PhCH(CH_3)_2$ | 0.18 | 0.015 | 1.5 (107) |
| $PhCHO^a$ | 12,000 | 1760 | 290 (21,000) |

a. Measured at 0 °C.

Hence, unlike ethylbenzene, cumene and benzaldehyde, toluene is not oxidized at any appreciable rate by dioxygen in the absence of catalysts. Indeed, Partenheimer showed[15] that when toluene was subjected to dioxygen in acetic acid no reaction occurred, even at 205 °C and 27 bar.

Basically three different types of processes are employed for alkylaromatic autoxidations. The first type, exemplified by the benzoic acid process (reaction 41) involves oxidation of the neat hydrocarbon and relatively low catalyst (usually cobalt) concentrations. The second type employs relatively high (~ 0.1 M) concentrations of cobalt acetate in acetic acid as solvent. Activators such as aldehydes or ketones are usually added to oxidize Co(II) to Co(III), otherwise long induction periods are observed. In contrast to the first type these processes involve initiation via electron transfer oxidation of the alkylaromatic substrate to afford the corresponding radical cation (reaction 43) which subsequently loses a proton giving the benzylic radical (reaction 44). The latter is scavenged by dioxygen (reaction 45) and aromatic aldehydes are the primary products, formed by reaction of the benzylperoxy radicals with Co(II) with simultaneous regeneration of the Co(III) oxidant.

$$ArCH_3 + Co^{III} \longrightarrow [ArCH_3]^{+\bullet} + Co^{II} \quad (43)$$

$$[ArCH_3]^{+\bullet} \longrightarrow ArCH_2\bullet + H^+ \quad (44)$$

$$ArCH_2\bullet + O_2 \longrightarrow ArCH_2O_2\bullet \quad (45)$$

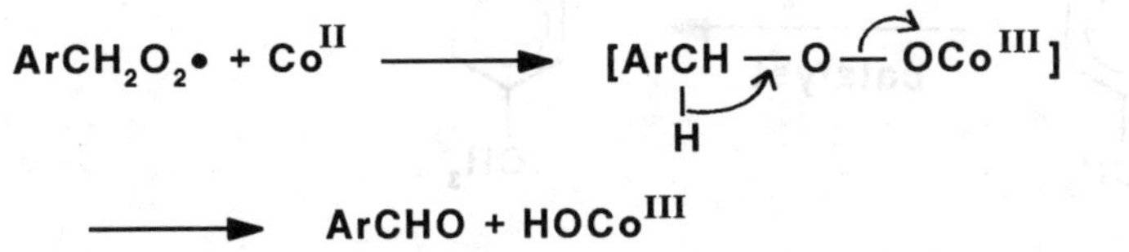

(46)

In the third type, typified by the Amoco/MC terephthalic acid process (reaction 42), lower concentrations of cobalt(II) are employed, in acetic acid solvent, and bromide ion and manganese are added as cocatalysts. The second and third type processes are generally employed with substrates that are more difficult to oxidize.

The metal-catalyzed autoxidation of substituted toluenes can be conveniently divided into two stages. In the initial stage the toluene is oxidized to the corresponding benzaldehyde as discussed above. Since the benzaldehyde is much more reactive than the substrate the reaction soon enters a second stage in which the aldehyde undergoes rapid autoxidation to give the corresponding aromatic percarboxylic acids as key intermediates. Interestingly, Jones[16] showed that the latter, in contrast to alkyl hydroperoxides, oxidize $Mn^{II}$ and $Co^{II}$ via a heterolytic mechanism in acetic acid solution, affording the μ-oxo dimer of $Mn^{III}$ or $Co^{III}$, respectively (reaction 47).

$$ArCO_3H + 2M^{II} \xrightarrow{95\%\ HOAc} ArCO_2H + M^{III}\!-\!O\!-\!M^{III} \quad (47)$$

M = Mn or Co

## 9. Terephthalic Acid from p-Xylene via the Amoco/MC Process

In the oxidation of p-xylene the first methyl group undergoes rapid autoxidation to afford p-toluic acid (reaction 48). The second methyl group is, however, deactivated by the electron-withdrawing carboxyl group, and further oxidation to terephthalic acid (reaction 49) is much slower (toluene is 26 times as reactive as p-toluic acid). It is not surprising, therefore, that the autoxidation of p-xylene to terephthalic acid proved to be a difficult proposition[17].

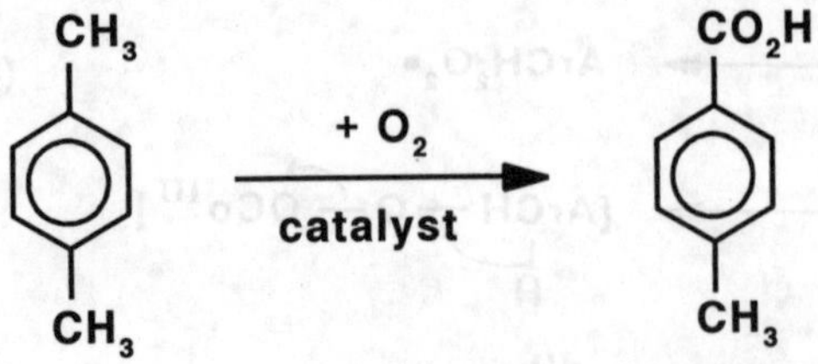


(48)

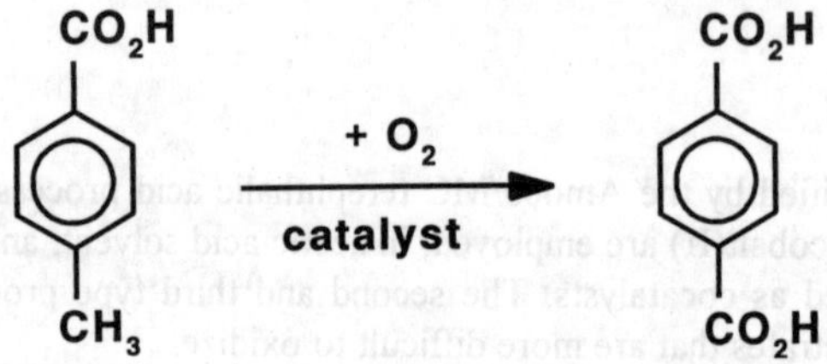


(49)

Two types of processes are used for the industrial oxidation of p-xylene to terephthalic acid (see Chapter 1). In the Eastman Kodak/Toray process a cosubstrate (acetaldehyde or methylethyl ketone) is used in combination with high concentrations of cobalt(III) acetate in acetic acid solvent. The mechanism involves direct reaction of cobalt(III) with the p-xylene substrate as outlined in reactions 43-47. The Amoco/MC process, on the other hand, employs low concentrations of a catalyst cocktail comprising cobalt(II), manganese(II) and bromide ion. As noted by Partenheimer[15] this technology has been successfully applied to the oxidation of about 270 different substrates. Some examples of aromatic di- and tricarboxylic acids that are commercially produced using the Amoco/MC process are shown in Figure 1.

Because of its industrial importance the mechanism of the Amoco/MC process has been extensively studied, notably by Jones[16,18] and by Partenheimer[15,19-21]. A question which inevitably arises in this context is: why the combination Co/Mn/Br in acetic acid? In order to answer this question we need to examine the role of the various catalyst components.

Partenheimer showed[15] that when a solution of cobalt(II) acetate in acetic acid at 113 □C was treated with dioxygen about 1% of the cobalt was converted to the trivalent state. In the presence of a substituted toluene two reactions are possible: formation of a benzyl radical via one-electron oxidation of the substrate or decarboxylation of the acetate ligand (reactions 50 and 51).

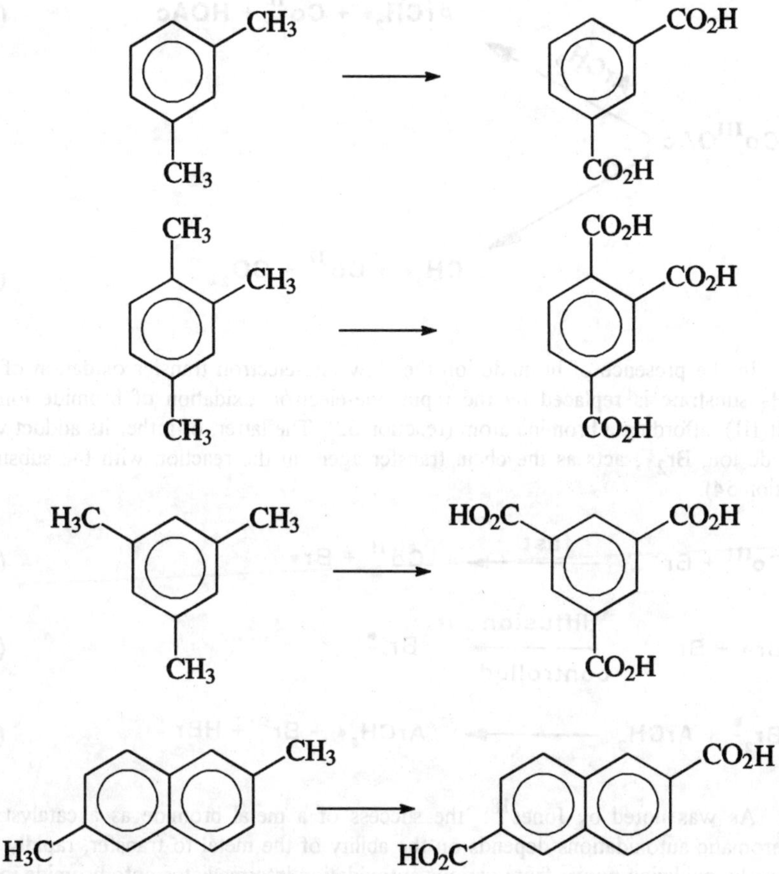


Figure 1. Acids produced by the Amoco/MC process.

Unfortunately, at the temperatures required for a reasonable rate of reaction (> 130 °C) decarboxylation predominates. As noted above, two methods are employed to circumvent this: addition of a cosubstrate which allows for reaction at < 130 °C, or the addition of bromide ion.

$$Co^{III}OAc \xrightarrow{ArCH_3} ArCH_2\bullet + Co^{II} + HOAc \quad (50)$$

$$Co^{III}OAc \longrightarrow CH_3\bullet + Co^{II} + CO_2 \quad (51)$$

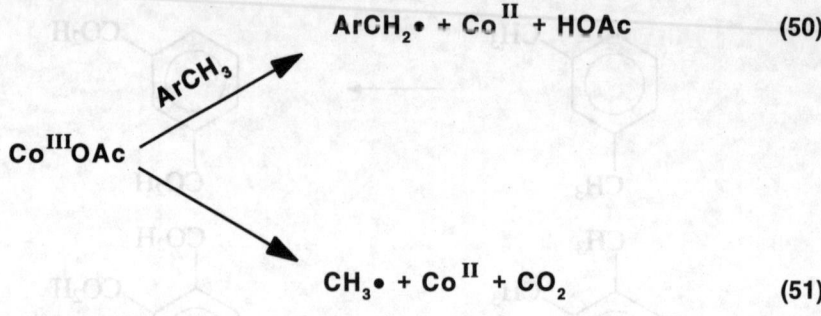

In the presence of bromide ion the slow one-electron transfer oxidation of the $ArCH_3$ substrate is replaced by the rapid one-electron oxidation of bromide ion by cobalt(III), affording a bromine atom (reaction 52). The latter, or rather its adduct with bromide ion, $Br_2^{\dot{-}}$, acts as the chain transfer agent in the reaction with the substrate (reaction 54).

$$Co^{III} + Br^- \xrightarrow{\text{fast}} Co^{II} + Br\bullet \quad (52)$$

$$Br\bullet + Br^- \xrightarrow[\text{controlled}]{\text{diffusion}} Br_2^{\bullet -} \quad (53)$$

$$Br_2^{\bullet -} + ArCH_3 \longrightarrow ArCH_2\bullet + Br^- + HBr \quad (54)$$

As was noted by Jones[18]: `the success of a metal bromide as a catalyst for alkylaromatic autoxidations depends on the ability of the metal to transfer, rapidly and effectively, oxidizing power from various autoxidation intermediates onto bromide ion in a manner which generates bromine atoms'. The fact that no free bromine is observable in this system is consistent with rapid reaction of intermediate bromine atoms with the substrate. Inhibition of the reaction by cupric salts can be explained by the rapid removal of $Br_2^{\dot{-}}$ or $ArCH_2\cdot$ via one-electron oxidation by $Cu^{II}$ (reactions 55 and 56).

$$Cu^{II} + Br_2^{\dot{-}} \rightarrow Cu^{I} + Br_2 \quad (55)$$

$$Cu^{II} + ArCH_2\cdot \rightarrow Cu^{II} + ArCH_2^+ \quad (56)$$

In order to provide an insight into the nature of the catalytic species Jones[18] investigated the reaction of cobalt(II) acetate with m-chloroperbenzoic acid in 95% aqueous acetic acid at 0 °C. The composition of this mixture corresponds reasonably well with that which is formed during $ArCH_3$ autoxidation. He found that $Co(OAc)_2$ was instantaneously oxidized to a μ-oxocobalt(III) dimer. The latter was a very active catalyst and was denoted as $Co^{IIIa}$ (see Figure 2). Within a few minutes at 25 °C this apple green complex reacted with a molecule of water to form an olive green, hydroxyl-bridged dimer. The latter was much less reactive and was denoted as $Co^{IIIs}$. On standing for several days at 25 °C the hydroxyl-bridged dimer reacted with $Co^{II}$ to form a μ-oxo mixed trimer of $Co^{III}$ and $Co^{II}$, denoted as $Co^{IIIc}$, which had previously been identified by Ziolkowski and coworkers[22].

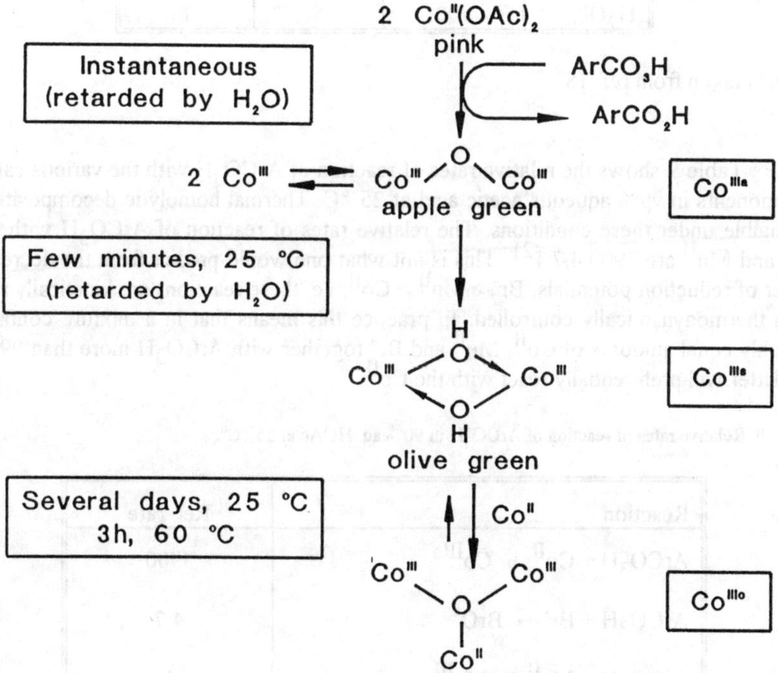


Figure 2. Reaction of $ArCO_3H$ with $Co(OAc)_2$ in HOAc.

Jones[18] subsequently investigated the relative reactivities of the various cobalt(III) species with $Br^-$, $Mn^{II}$ and $H_2O_2$. The active μ-oxodimer, $Co^{IIIa}$ was two to four orders of magnitude more reactive than $Co^{IIIs}$ which was four to five times more reactive than $Co^{IIIc}$ (Table 8). Furthermore, the rate of conversion of $Co^{IIIa}$ to $Co^{IIIs}$ was much higher than the rate of reaction of $Co^{IIIa}$ with $ArCH_3$. In other words, in the absence of $Br^-$ or $Mn^{II}$ the cobalt species that reacts with $ArCH_3$ cannot be $Co^{IIIa}$.

Table 8. Relative reactivities of $Co^{III}$ species[a].

| | $Co^{IIIa}$ | $Co^{IIIs}$ | $Co^{IIIc}$ |
|---|---|---|---|
| $Br^-$ | 30,000 | 4 | 1 |
| $Mn^{II}$ | 6,000 | 4 | 1 |
| $H_2O_2$ | 700 | 5 | 1 |

a. Data taken from ref. 18.

Table 9 shows the relative rates of reaction of $ArCO_3H$ with the various catalyst components in 90% aqueous acetic acid at 25 °C. Thermal homolytic decomposition is negligible under these conditions. The relative rates of reaction of $ArCO_3H$ with $Co^{II}$, $Br^-$ and $Mn^{II}$ are 3900:4.7:1[21]. This is not what one would predict from the decreasing order of reduction potentials: $Br^- > Mn^{II} > Co^{II}$, i.e. these reactions are kinetically rather than thermodynamically controlled. In practice this means that in a mixture containing roughly equal amounts of $Co^{II}$, $Mn^{II}$ and $Br^-$ together with $ArCO_3H$ more than 99% of the latter will preferentially react with the $Co^{II}$.

Table 9. Relative rates of reaction of $ArCO_3H$ in 90% aq. HOAc at 25 °C[a].

| Reaction | Rel. rate |
|---|---|
| $ArCO_3H + Co^{II} \rightarrow Co^{IIIa}$ | 3900 |
| $ArCO_3H + Br^- \rightarrow BrO^-$ | 4.7 |
| $ArCO_3H + Mn^{II} \rightarrow Mn^{III}$ | 1 |
| $ArCO_3H \xrightarrow{\Delta T} ArCO_2\cdot + HO\cdot$ | $10^{-4}$ |

a. Data taken from ref. 21.

Once the active $Co^{IIIa}$ catalyst has been formed by peracid oxidation of $Co^{II}$ its fate is determined by the relative rates of its reaction with the other species present in the reaction mixture, i.e. $Mn^{II}$, $Br^-$ and substrate compared to its rearrangement to $Co^{IIIs}$. As can be seen from the relevant data (Table 10) by far the most favorable reaction is oxidation of $Mn^{II}$ to $Mn^{III}$ which is 940 times faster than conversion of $Co^{IIIa}$ to $Co^{IIIs}$ under these conditions.

Table 10. Relative rates of reaction of $Co^{IIIa}$ in 90% HoAc[a].

| Reaction | Temp °C | Rel. rate |
|---|---|---|
| $Co^{IIIa} + Mn^{II} \rightarrow Mn^{III}$ | 23 | 940 |
| $Co^{IIIa} + Br^- \rightarrow Br\cdot$ | 23 | 84 |
| $Co^{IIIa} \rightarrow Co^{IIIs}$ | 25 | 1 |
| $Co^{IIIa} + ArCH_3 \rightarrow ArCH_2\cdot$ | 80 | 0.03 |

a. Data taken from ref. 21.

In other words, in the mixture containing $Co^{IIIa}$, $Mn^{II}$, $Br^-$ and p-xylene more than 90% of the $Co^{IIIa}$ reacts with $Mn^{II}$ to afford $Mn^{III}$ and there is negligible reaction of $Co^{IIIa}$ with the substrate.

Based on this detailed kinetic analysis of the individual steps we are now able to provide an interpretation of the synergistic effect of the Co/Mn/$Br^-$ catalyst cocktail. With cobalt alone in acetic acid the reaction of cobalt(III) with p-toluic acid is much too slow. Bromide alone is rapidly oxidized by $ArCO_3H$ to afford hypobromite, by a heterolytic mechanism. In the presence of cobalt, bromide ion is oxidized to chain-propagating bromine atoms. Unfortunately, one electron oxidation of acetate ligands, leading to decarboxylation, seriously competes with this process. In the presence of manganese and bromide ion the oxidation of $Mn^{II}$ to $Mn^{III}$ by $ArCO_3H$ is too slow. In contrast, with the three-component system, of cobalt, manganese and bromide, the $Mn^{II}$ is rapidly oxidized by $Co^{IIIa}$ to give $Mn^{III}$ which rapidly oxidizes $Br^-$ to $Br\cdot$. The latter abstracts a hydrogen from the substrate to give the benzylic radical. Because $Co^{III}$ is rapidly removed from the reaction mixture by reaction with $Mn^{II}$ the steady-state concentration of $Co^{III}$ is maintained at a low level, thus preventing undesirable decarboxylation of acetic acid by $Co^{III}$. Decomposition of acetic acid by $Mn^{III}$ is negligible under the reaction conditions (see Table 11).

Table 11. Half-lives of various reactions in 90% HOAc at 100 °C[a].

| Reaction | $t_{1/2}$ (min) |
|---|---|
| $Co^{III}OAc \rightarrow Co^{II} + CH_3\cdot + CO_2$ | 14 |
| $Mn^{III}OAc \rightarrow Mn^{II} + \cdot CH_2CO_2H$ | 790 |
| $Co^{III} + Mn^{II} \rightarrow Co^{II} + Mn^{III}$ | < 0.2 |
| $Mn^{III} + Br^- \rightarrow Mn^{II} + Br\cdot$ | < 0.2 |

a. Data taken from ref. 21.

The use of the Mn/Co/Br⁻ system allows for higher reaction temperatures and lower catalyst concentrations than the bromide-free processes. The only disadvantage is the corrosive nature of the bromide-containing system which necessitates the use of expensive, titanium-lined reactors. The complete mechanism of the autoxidation of substituted toluenes in the presence of the Co/Mn/Br⁻ catalyst, which must surely be considered a work of art, is depicted in Figure 3.

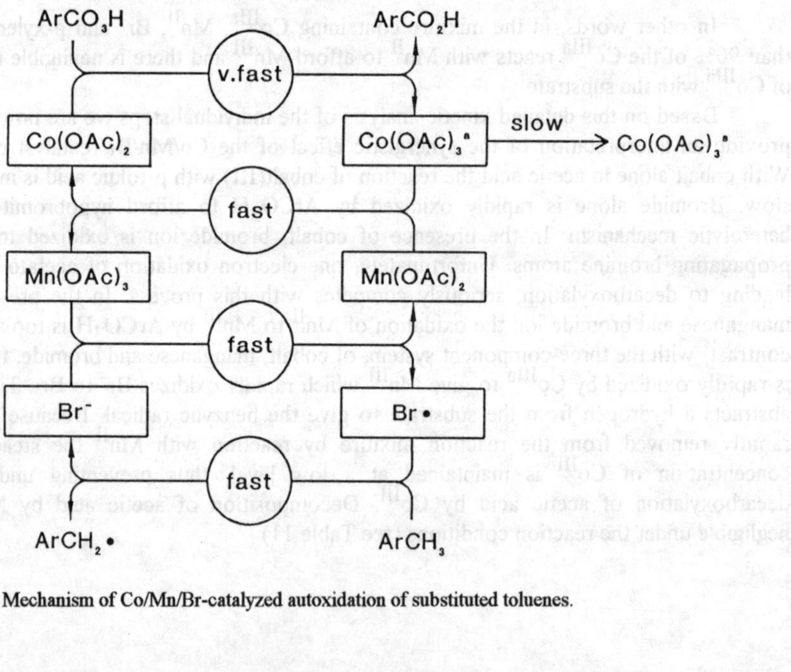


Figure 3. Mechanism of Co/Mn/Br-catalyzed autoxidation of substituted toluenes.

The Amoco/MC catalyst system has also been applied to other types of alkylaromatic oxidations, e.g. the oxidation of m-phenoxyethylbenzene to the pharmaceutical intermediate, m-phenoxyacetophenone (reaction 57)[23].

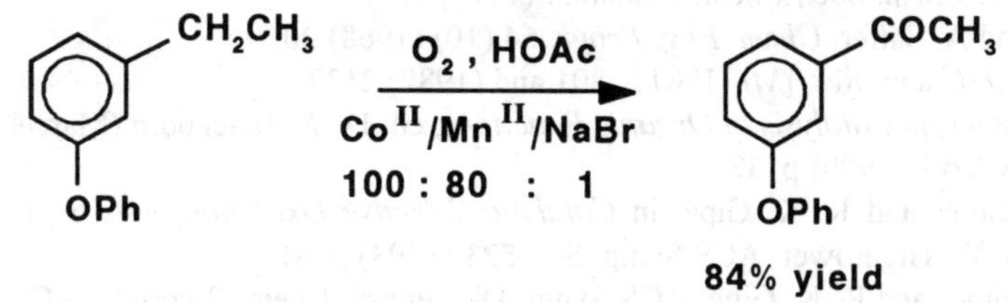


(57)

Indeed, one cannot help but wonder if the full potential of this fascinating and elegant technology has yet been realized.

## 10. Concluding Remarks

Catalytic autoxidations of hydrocarbons in the liquid phase have been around for several decades. Nevertheless, they still constitute a very useful technology for the synthesis of a variety of products. Their utility is likely to be further broadened in the future as they are applied to the manufacture of fine chemicals as replacements for classical oxidations with stoichiometric inorganic reagents. Finally, it cannot be emphasized enough that free radical chain autoxidation is always occurring as a background reaction in any system comprising metal catalysts, dioxygen and hydrocarbon substrates in the liquid phase. Consequently, it is difficult to design conditions in the liquid phase in which alternative pathways, involving dioxygen activation, can predominate.

## References

1. R. A. Sheldon and J. K. Kochi, *Metal-Catalyzed Oxidations of Organic Compounds* (Academic Press, New York, 1981), p. 17.
2. H. L. J. Bäckstrom, *J. Am. Chem. Soc.* **49** (1927) 1460.
3. R. Criegee, H. Pilz and H. Flygare, *Chem. Ber.* **72** (1939) 1799.
4. H. Hock and B. Lang, *Chem. Ber.* **77** (1944) 257.
5. J. A. Howard, *Adv. Free-Radical Chem.* **4** (1972) 49.
6. G. A. Russell, *J. Am. Chem. Soc.* **78** (1956) 1035, 1041.
7. E. W. Stern, *Chem. Commun.* (1970) 736.
8. J. P. Collman, M. Kubota and J. W. Hosking, *J. Am. Chem. Soc.* **89** (1967) 4809.
9. R. A. Sheldon, *Chem. Commun.* (1971) 788.
10. G. A. Russell, *J. Am. Chem. Soc.* **77** (1955) 4583.
11. F. R. Mayo, *Acc. Chem. Res.* **1** (1968) 193.
12. D. E. Winkler and G. W. Hearne, *Ind. Eng. Chem.* **41** (1949) 2597.

13. J. F. Black, *J. Am. Chem. Soc.* **100** (1978) 527.
14. Y. Kamiya and K. U. Ingold, *Can. J. Chem.* **42** (1964) 1027, 2424.
15. W. Partenheimer, *J. Mol. Catal.* **67** (1991) 35.
16. G. H. Jones, *J. Chem. Soc., Chem. Commun.* (1979) 536.
17. R. Landau and A. Saffer, *Chem. Eng. Progr.* **64** (10) (1968) 20.
18. G. H. Jones, *J. Chem. Res. (M)* (1981) 2801 and (1982) 2137.
19. W. Partenheimer, in *Catalysis of Organic Reactions*, ed. D. W. Blackburn (Marcel Dekker, New York, 1990) p. 321.
20. W. Partenheimer and R. K. Gipe, in *Catalytic Selective Oxidation*, eds. S. T. Oyama and J. W. Hightower, ACS Symp. Ser. **523** (1993) p. 81.
21. W. Partenheimer and R. K. Gipe, ACS Symp. Div. Petrol. Chem. Preprints, ACS Meeting, Washington, D.C., Aug. 1992, pp. 1098-1104.
22. T. Szymanska-Buzar and J. J. Ziolkowski, *J. Mol. Catal.* **5** (1979) 341.
23. Jap. Pat. 5867640 (1983) to Nippon Kayaku; CA **99** (1983) 382046.

# HETEROGENEOUS CATALYSIS OF LIQUID PHASE OXIDATIONS

R.A. SHELDON
*Laboratory for Organic Chemistry and Catalysis,*
*Delft University of Technology, Julianalaan 136*
*2628 BL Delft, The Netherlands*

ABSTRACT

**The various types of heterogeneous catalysts for liquid phase oxidations are reviewed. In recent years there is a marked trend towards the use of molecular sieve catalysts, containing redox metal ions incorporated in the framework or metal complexes encapsulated in the micropores, instead of the more traditional metals, metal ions or metal oxides on amorphous supports. These redox molecular sieves and ship-in-the-bottle complexes have many features in common with redox enzymes.**

## 1. Introduction

In choosing a suitable methodology for the (industrial) oxidation of a particular organic substrate there are various options[1], as outlined in Figure 1. The first choice - stoichiometric versus catalytic - is not really a viable option anymore. The use of classical stoichiometric inorganic oxidants is becoming prohibitive. The second choice - liquid versus gas phase - will depend largely on the boiling point and thermal stability of the molecule in question. When gas phase oxidation with dioxygen is technically feasible it will probably be economically more attractive than other options. Examples of the use of gas phase oxidations in the production of fine chemicals are discussed in Chapter 11.

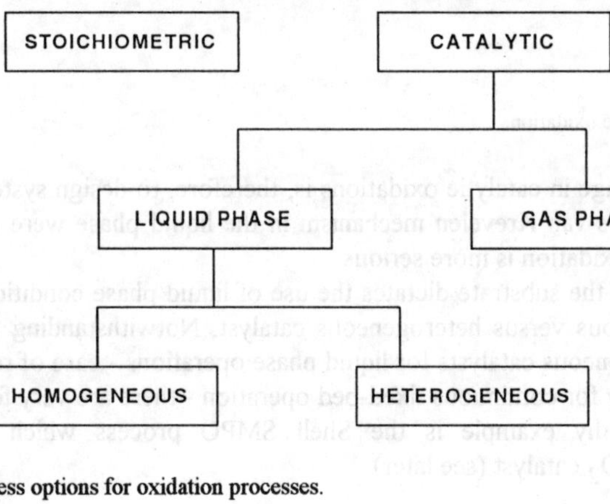


**Figure 1. Process options for oxidation processes.**

In this context it is worth noting, as was already discussed in Chapter 8., that the chance of observing selective, non-classical oxidation of organic substrates with dioxygen is, generally speaking, greater in the gas phase than in the liquid phase. In the gas phase substrate concentrations are much lower and radical chain oxidation is less favorable. Consequently, reactive intermediates, e.g. alkyl radicals, may react with metal oxidants at the catalyst surface leading to a Mars-van Krevelen type mechanism (Figure 2).

**Liquid phase**

$$RH \xrightarrow{\text{initiation}} R\bullet$$

$$R\bullet + O_2 \longrightarrow RO_2\bullet$$

$$RO_2\bullet + RH \longrightarrow RO_2H + R\bullet$$

**Gas phase (Mars-van Krevelen mechanism)**

$$S + M{=}O \longrightarrow SO + M$$

$$2\,M + O_2 \longrightarrow 2\,M{=}O$$

Figure 2. Liquid vs gas phase oxidation.

A major challenge in catalytic oxidations is, therefore, to design systems capable of operating via a Mars-van Krevelen mechanism in the liquid phase were competition from radical chain autoxidation is more serious.

If the nature of the substrate dictates the use of liquid phase conditions the next option is a homogeneous versus heterogeneous catalyst. Notwithstanding the obvious advantages of heterogeneous catalysts for liquid phase operations - ease of recovery and recycling and suitability for continuous fixed-bed operation - there are very few practical examples. A noteworthy example is the Shell SMPO process which employs a heterogeneous $Ti^{IV}/SiO_2$ catalyst (see later).

## 2. Homogeneous versus Heterogeneous Catalysis: Advantages and Limitations

In addition to the obvious disadvantages regarding their recovery and recycling homogeneous catalysts suffer from two other drawbacks which are unique to oxidation processes. First, homogeneous catalysts often undergo deactivation via the formation of μ-oxo dimers or oligomers. Typical examples include the formation of μ-oxo dimers from reactive oxometalloporphyrins (reaction 1) and μ-oxo oligomers from titanyl species (reaction 2). Second, most organic ligands, e.g. porphyrins and Schiff's base ligands, undergo oxidative destruction under oxidizing conditions[2]. This is also a common cause of deactivation of homogeneous catalysts.

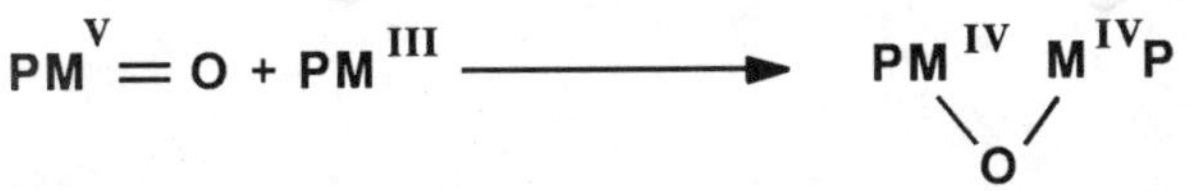


(1)

P = porphyrinato M = Fe , Mn

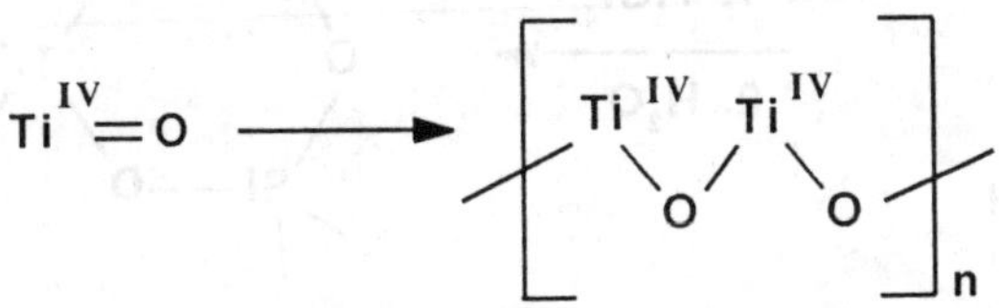


(2)

Both of these problems can, in principle, be circumvented by immobilizing the active catalytic species, e.g. an oxometal (M=O) moiety, in an inorganic matrix such as silica or zeolites and related molecular sieves. For example, the catalyst in the epoxidation step in the Shell SMPO process (Figure 3) is $Ti^{IV}/SiO_2$. The high activity of this catalyst compared to homogeneous titanium(IV) compounds was attributed[3,4] to site-isolation of active monomeric titanyl species in the silica lattice (Figure 3).

If the solution to the problem of deactivation were so simple why are there not many more examples of heterogeneous catalysts for liquid phase oxidations? The answer lies in a fundamental problem associated with the use of heterogeneous catalysts in liquid phase oxidations, namely, leaching of the metal catalyst from the surface. In oxidation processes highly polar molecules are generated, e.g. water, alcohols, carboxylic acids, which can readily solvolyse the metal-oxygen bonds attaching the catalyst to the surface of the support. For example, in the development of the Shell SMPO catalyst many different metal-silica combinations were tested and although several were active only the $Ti^{IV}/SiO_2$ combination was stable towards leaching[3].

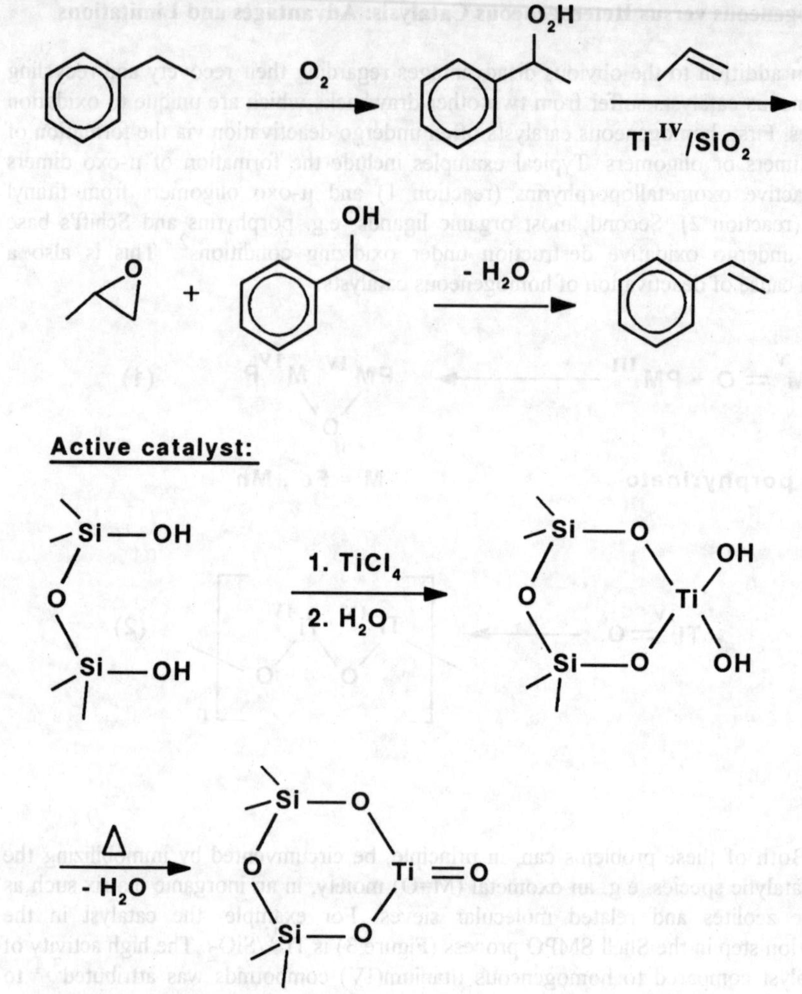


Figure 3. Shell SMPO catalyst.

## 3. Types of Heterogeneous Catalysts

Heterogeneous catalysts for liquid phase oxidations can be divided into four basic types: (a) supported metals (e.g. Pt/C), (b) supported metal ions and complexes, e.g. metal ions on ion exchange resins and metal ion exchanged zeolites, (c) supported oxometal (oxidic) catalysts, e.g. $Ti^{IV}/SiO_2$ and redox molecular sieves (see later) and (d) metal complexes encapsulated in zeolites and related molecular sieves, so-called ship-in-the-bottle complexes.

## 4. Supported Metal Catalysts - Oxidative Dehydrogenation

Platinum- and palladium-catalyzed oxidative dehydrogenations of alcohols with dioxygen in the presence of aqueous alkali were already known in the last century[5]. The noble metal is generally supported on active charcoal. Since dioxygen can be replaced by other hydrogen acceptors these reactions are assumed to involve an oxidative dehydrogenation mechanism:

$$RCH_2OH \xrightarrow{Pt} RCHO + H_2 \qquad (3)$$

$$2\,H_2 + O_2 \xrightarrow{Pt} 2\,H_2O \qquad (4)$$

The exact role of the base in these reactions is still not clear. It is generally thought to be necessary to remove strongly absorbed carboxylic acids, the ultimate products of these reactions, from the catalyst surface.

Platinum and palladium-based catalysts have been widely used[1,5] in the oxidative dehydrogenation of vicinal diols, hydroxy acids and carbohydrates. Some examples are shown in reactions 5-7.

$$CH_3CH(OH)CH_2OH \xrightarrow[\text{Pt-Pb/C or Pt-Bi/C}]{O_2\text{ ; NaOH}} CH_3COCO_2H \qquad (5)$$

$$CH_3CH(OH)CO_2H \xrightarrow[\text{Pt-Pb/C}]{O_2\text{ ; NaOH}} CH_3COCO_2H \qquad (6)$$

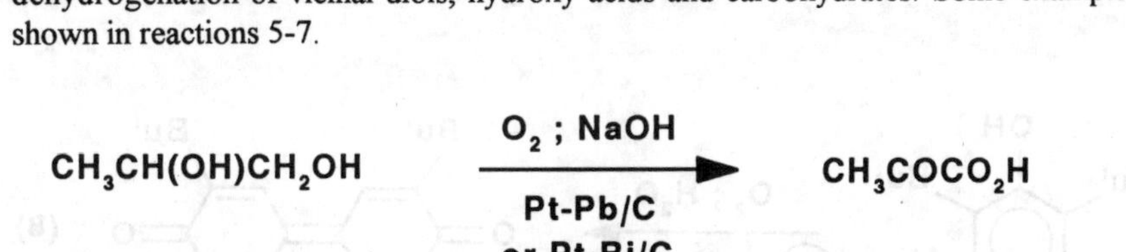

(7)

One reaction that has been the focus of much attention is the catalytic oxidation of D-glucose to D-gluconate. Palladium catalysts exhibit high selectivity to gluconate but the catalyst is rapidly poisoned by dioxygen[6]. However, the simple deposition of bismuth onto palladium-on-charcoal affords a catalyst with excellent activity, stability and selectivity (> 99%)[7].

Notwithstanding the enormous effort that has been devoted to noble metal-catalyzed oxidative dehydrogenations few of these processes have been reduced to industrial practice, largely due to the above mentioned problem of catalyst poisoning.

## 5. Supported Metal Ions and Complexes

A simple means of immobilizing metal ion catalysts is to employ ion-exchange resins as the support[1,5]. For example, weak acid resins exchanged with cobalt(II) ions catalyzed the autoxidation of cyclohexane or cyclohexanone to dibasic acids in acetic acid solvent at 85-105 °C and 5-20 bar[8]. Metal complexes have also been attached to ion exchange resins. For example, a colloidal catalyst prepared by attaching cobaltphthalocyanine tetrasulfonate (CoPcTs) via the anionic sulfonate groups to a styrene-divinyl benzene copolymer containing quaternary ammonium groups, catalyzed the autoxidation of 2,6-di-tert-butylphenol (reaction 8) in aqueous solution at a rate ten times that observed with the soluble CoPcTs catalyst[9].

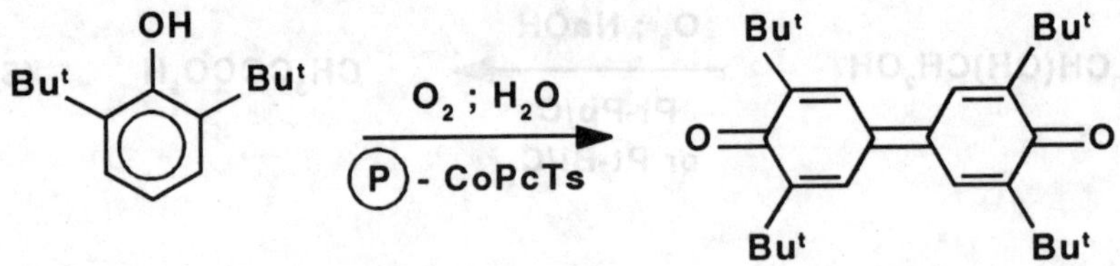


(8)

Transition metal ions supported on ion exchange resins have also been used to catalyze a variety of oxygen transfer reactions with hydrogen peroxide and alkyl hydroperoxides[1,5]. From a practical viewpoint the crucial question is whether these catalysts retain the metal over a long period, i.e. whether or not they are subject to leaching.

Transition metal ions can also be immobilized by ion exchange with zeolites[1,5]. However, the same reservations apply to metal ion-exchanged zeolites as to the metal ions supported on ion-exchange resins discussed above, i.e. their practical utility is crucially dependent on their long term stability towards leaching. In this context it is worth noting that redox molecular sieve catalysts (see later) are probably less prone to leaching than traditional supported metal catalysts.

## 6. Supported Oxometal (Oxidic) Catalysts

Metal oxides are often used as catalysts for hydrocarbon autoxidations[10]. In most cases the metal oxide dissolves in the reaction medium, probably via the formation of a metal carboxylate, to become a homogeneous catalyst. Surprisingly, a cerium oxide catalyst reportedly[11] catalyzes the liquid phase oxidation of cyclohexanone in acetic acid (5-15 bar; 98-118°) without dissolving in the reaction medium. As noted above, although many metal oxide-silica combinations were tried as catalysts for olefin epoxidations with $RO_2H$ only $Ti^{IV}/SiO_2$ displays the unique combination of high activity and true heterogeneity[3,4]. The $Ti^{IV}/SiO_2$ catalyst is, however, not effective for epoxidations with aqueous $H_2O_2$. Interestingly, a $MoO_3$-$Bu_3SnCl$ catalyst, supported on chemically pretreated charcoal, was effective[12] for the epoxidation of olefins with 30% aqueous $H_2O_2$ in isopropyl alcohol at 50°. However, in the epoxidation of cyclohexene the yield of epoxide decreased from 73% to 60% on recycling three times indicating that some leaching probably took place.

In the noble metal-catalyzed oxidative dehydrogenations described in section 4 vicinal diol cleavage is observed only as a minor side reaction. Recently, the selective cleavage of diols with dioxygen in the presence of ruthenium pyrochlore oxide catalysts (mixed oxides of Ru and Pb or Bi) has been reported[13]. Cyclohexane-1,2-diol was selectively oxidized to adipate (reaction 9) in aqueous alkaline medium under mild conditions.

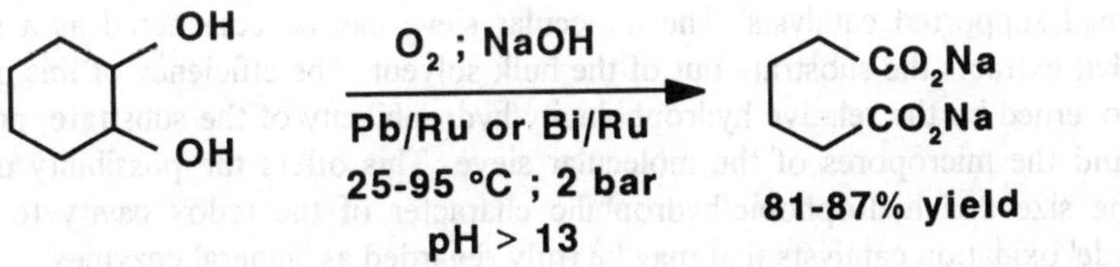


(9)

Although the experimental conditions closely resemble those of the noble metal-catalyzed oxidative dehydrogenations (see section 4) the active oxidant in the ruthenium pyrochlore system is clearly a high-valent oxoruthenium species, i.e. it is a heterogeneous oxidic type catalyst.

## 7. Redox Molecular Sieves - Unique Solid Catalysts for Liquid Phase Oxidations

As noted earlier, two major problems associated with oxidation catalysis by soluble oxometal complexes - the propensity of active oxometal species towards oligomerization to inactive μ-oxo complexes and the oxidative destruction of organic ligands - can, in principle, be circumvented by site isolation of discrete oxometal species in an oxidatively stable inorganic matrix. Unfortunately, attachment of oxometal species to inorganic surfaces, e.g. via Si-O-M bonds, often leads to catalysts that are susceptible to solvolysis by polar molecules, e.g. water, diols and carboxylic acids, present in oxidation reaction mixtures. This leads inevitably to leaching of the catalyst from the surface to form homogeneous systems. Moreover, such catalysts are often deactivated by strong coordination of polar molecules, such as water, thus preventing diffusion of hydrophobic hydrocarbon substrates to the active site.

One approach to isolating redox (oxo)metal species in stable inorganic matrices is via incorporation in the framework of a molecular sieve (zeolite, silicalite, aluminophosphate, etc.). We coined the generic name **redox molecular sieves**[1,14,15] to describe such materials, which have many features in common with redox enzymes and several advantages compared to conventional supported catalysts. Unlike amorphous materials they possess a regular microenvironment having highly homogeneous internal structures with well-defined cavities and channels. They also exhibit enhanced stability towards leaching. A possible explanation is that confinement of the active catalytic species on the curved internal surface of a molecular sieve renders it less accessible to solvolysis by polar molecules. Furthermore, confinement of the active site in channels and cavities of molecular dimensions imparts redox molecular sieves with the additional feature, in common with enzymes, of shape selectivity. Another feature that redox molecular sieves share with enzymes is the possibility of creating a hydrophobic or hydrophilic environment around the active site, by a suitable choice of molecular sieve. Silicalite, for example, contains hydrophobic micropores while aluminophosphates are hydrophilic. This can lead to more pronounced solvent effects than are observed with conventional supported catalysts. The molecular sieve can be considered as a second solvent that extracts the substrate out of the bulk solvent. The efficiency of this process will be governed by the relative hydrophobicity/hydrophilicity of the substrate, product, solvent and the micropores of the molecular sieve. This offers the possibility of `fine tuning' the size and hydrophobic/hydrophilic character of the redox cavity to create `tailor made' oxidation catalysts that may be truly regarded as 'mineral enzymes'.

Various types of molecular sieves are, in principle, amenable to framework substitution by transition metal ions. The basic building blocks are tetrahedral $SiO_4$, $AlO_4$ and $PO_4$ units (Figure 4). Silicalites are examples of all-silica molecular sieves, combination of $SiO_4$ with $AlO_4$ leads to zeolites, $AlO_4$ with $PO_4$ to aluminophosphates (APOs), and the combination of all three building blocks to silicaaluminophosphates (SAPOs).

Silicalites and aluminophosphates are electroneutral materials devoid of cation exchange properties. In zeolites, on the other hand, the extra charge on silicon ($4^+$) compared to aluminium ($3^+$) has to be balanced with a proton (or a cation), which confers ion exchange properties on these materials. Moreover, the proton forms of zeolites are very strong acids.

Molecular sieves are synthesized[16] by allowing the appropriate sol gel to crystallize, usually at temperatures of around 175 °C, in the presence of a template (so-called structure directing agent) which is generally an amine or a tetraalkylammonium salt. Subsequent calcination of the crystalline material at about 500

°C destroys the template, affording a molecular sieve the topology of which is determined by the shape and size of the template. When this so-called hydrothermal synthesis is performed in the presence of transition metal ions this can lead to their incorporation into the framework of the molecular sieve, thus conferring redox properties on the latter.

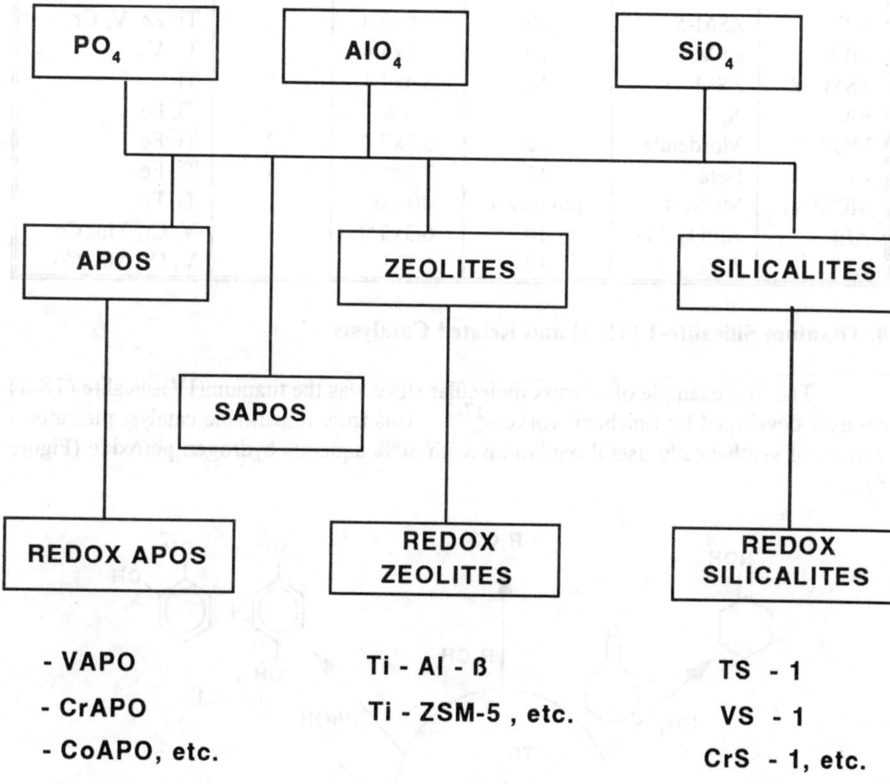


Figure 4. Redox molecular sieves.

Selected examples of molecular sieves, with various pore shapes and sizes, are collected in Table 1. Examples of transition metal ions that have been incorporated into the various structures are also shown. Molecular sieves can have channels that are disposed in a three dimensional array or that are uni- or bidirectional. ZSM-5, for example, has a three dimensional array of intersecting channels while $AlPO_4$-5 and $AlPO_4$-11 have unidirectional channels that do not intersect. The dimensionality of molecular sieves can be of importance with regard to the accessibility, to the substrate, of the active catalytic site.

Table 1. Structural types of molecular sieves.

| Structure type | Isotopic framework structure | Pore structure (ring size) | Pore size (Å) | Dimension | Metals incorporated |
|---|---|---|---|---|---|
| MFI | ZSM-5 | 10 | 5.6x5.4 | 3 | Ti, Zr, V, Cr |
| MEL | ZSM-11 | 10 | 5.1x4.4 | 3 | Ti, V |
| ZSM-48 | ZSM-48 | 10 | 5.4x4.1 | 1 | Ti |
| FAU | X, Y | 12 | 7.4 | 3 | Ti, Fe |
| MOR | Mordenite | 12 | 6.7x7.0 | 2 | Ti, Fe |
| BEA | Beta | 12 | 7.6x6.4 | 3 | Ti, Fe |
| MCM-41 | MCM-41 | unknown | 40-100 | 1 | Ti, Fe |
| AEL | $AlPO_4$-11 | 10 | 6.3x3.9 | 1 | V, Cr, Mn, Co |
| AFI | $AlPO_4$-5 | 12 | 7.3 | 1 | V, Cr, Mn, Co |

## 8. Titanium Silicalite-1 (TS-1) and Related Catalysts

The first example of a redox molecular sieve was the titanium(IV)silicalite (TS-1) catalyst developed by Enichem workers[17-20]. This truly remarkable catalyst mediates a variety of synthetically useful oxidations with 30% aqueous hydrogen peroxide (Figure 5).

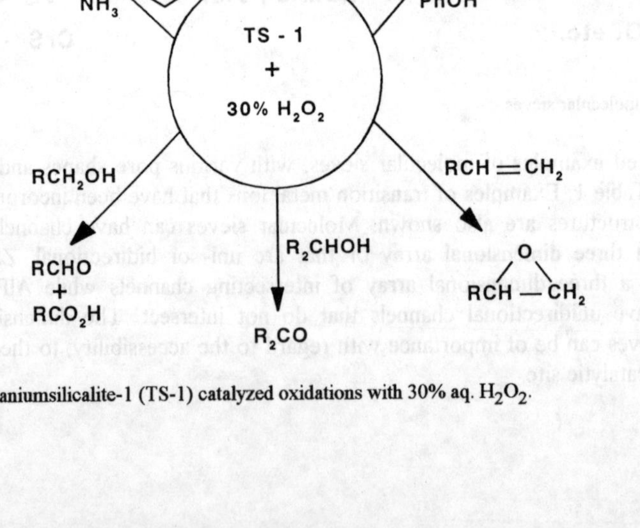

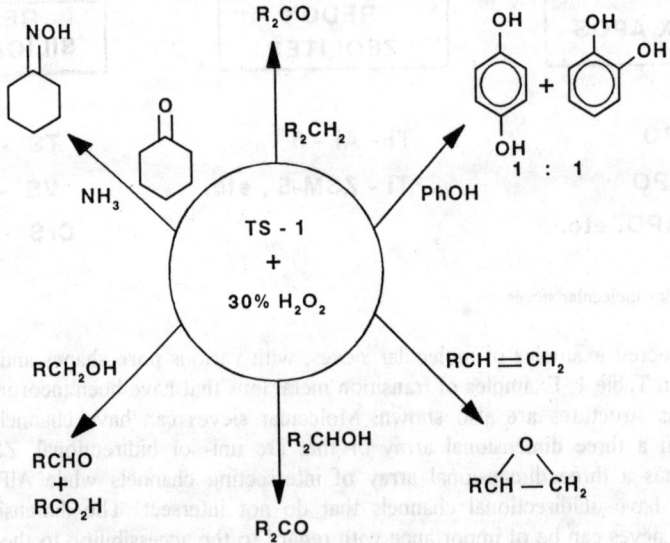


Figure 5. Titaniumsilicalite-1 (TS-1) catalyzed oxidations with 30% aq. $H_2O_2$.

Examples include olefin epoxidation, phenol hydroxylation, cyclohexanone ammoximation with $NH_3/H_2O_2$ and alcohol oxidations to the corresponding carbonyl compounds. The TS-1-catalyzed hydroxylation of phenol to a 1:1 mixture of catechol and hydroquinone has been commercialized by Enichem. Compared to the existing Rhône-Poulenc and Brichima processes, employing a mineral acid and $Fe^{II}/Co^{II}$ catalyst, respectively, the Enichem process gives a higher yield of dihydroxybenzenes (Table 2), i.e. higher or equal selectivity at significantly higher conversions.

Table 2. Comparison of various processes for the hydroxylation of phenol.

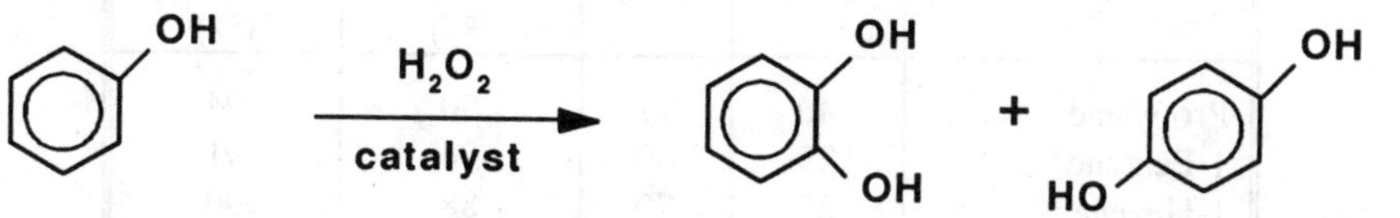


| Process (catalyst) | Rhône-Poulenc ($HClO_4$, $H_3PO_4$) | Brichima ($Fe^{II}/Co^{II}$) | Enichem (TS-1) |
|---|---|---|---|
| Phenol conversion (%) | 5 | 10 | 25 |
| Selectivity on phenol (%) | 90 | 80 | 90 |
| Selectivity on $H_2O_2$ | 70 | 50 | 70 |
| Catechol/hydroquinone ratio | 1.4 | 2.3 | 1.0 |

Similarly, the ammoximation of cyclohexanone to cyclohexanone oxime has commercial potential as a low-salt alternative to the existing technology (Figure 6).

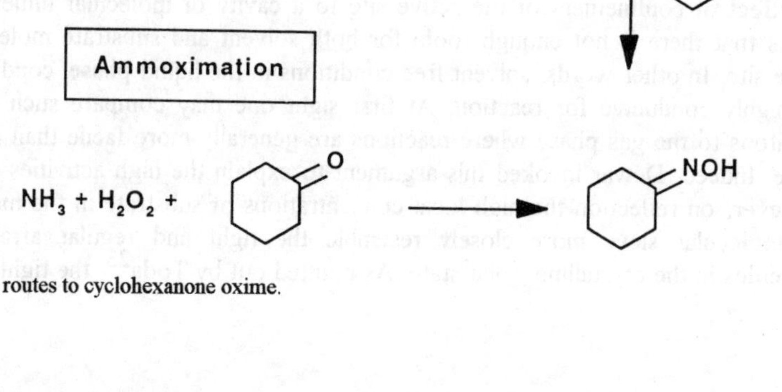


Figure 6. Two routes to cyclohexanone oxime.

The TS-1 catalyst exhibits some quite remarkable activities and selectivities. For example, relatively unreactive olefins, such as ethylene, propylene and allyl chloride, are epoxidized efficiently[21] under very mild conditions in methanol as solvent (Table 3). As a result of the shape selective properties of the catalyst larger, more reactive olefins such as cyclohexene, are epoxidized only slowly, if at all.

Table 3. TS-1-catalyzed epoxidations of olefins with aq. 60% $H_2O_2$ in MeOH[a].

| Olefin | Temp. (°C) | Time (min) | $H_2O_2$ conversion (%) | Epoxide selectivity (%) |
|---|---|---|---|---|
| Propylene | 40 | 72 | 90 | 94 |
| 1-Pentene | 25 | 60 | 94 | 91 |
| 1-Hexene | 25 | 70 | 88 | 90 |
| 1-Octene | 45 | 90 | 81 | 91 |
| Cyclohexene | 25 | 90 | 10 | n.d. |
| Allyl chloride | 45 | 30 | 98 | 92 |
| Allyl alcohol | 45 | 35 | 81 | 72 |

a. Data taken from ref. 21.

From a mechanistic viewpoint it is noteworthy that the TS-1 catalyst contains the same chemical elements, in roughly the same proportions (2% Ti), as the Shell $Ti^{IV}/SiO_2$ catalyst discussed earlier. Yet these two catalysts exhibit strikingly different and complementary catalytic behavior. $Ti^{IV}/SiO_2$ catalyzes epoxidations with alkyl hydroperoxides, such as TBHP, but is uneffective with $H_2O_2$ while TS-1 is effective with aqueous $H_2O_2$ but not with TBHP. The novel activities of both catalysts are assumed to accrue from the site-isolation of monomeric $Ti^{IV}$ centres, which are probably tetrasilanoxytitanium(IV) species $(\rightarrowtail SiO)_4Ti$ rather than titanyl, $(\rightarrowtail SiO)_2Ti{=}0$. one factor influencing the type of reactivity displayed is the relative hydrophobicity of the two materials. The channels of TS-1 are strongly hydrophobic, thus facilitating selective adsorption of the hydrophobic substrate rather than water. A second important factor is the effect of confinement of the active site to a cavity of molecular dimensions. This means that there is not enough room for both solvent and substrate molecules in the active site. In other words, solvent free conditions in the liquid phase, conditions which are highly conducive for reaction. At first sight one may compare such solvent-free conditons to the gas phase where reactions are generally more facile than in the liquid phase. Indeed, Dewar invoked this argument to explain the high activities of enzymes. However, on reflection the high local concentrations of substrate in the micropores of the molecular sieve more closely resemble the tight and regular arrangement of molecules in the crystalline, solid state. As pointed out by Toda[23], the tight and regular packing of molecules in crystals can lead to reaction rates that are higher than the corresponding ones in solution.

The mechanism of TS-1 catalyzed epoxidation of olefins most likely involves oxygen transfer from a coordinated tert-butylperoxy group to the double bond of the olefin (Figure 7). This could be facilitated by a silanoxy ligand (mechanism a) or by a coordinated molecule of methanol solvent, or silanol ligand (mechanism b).

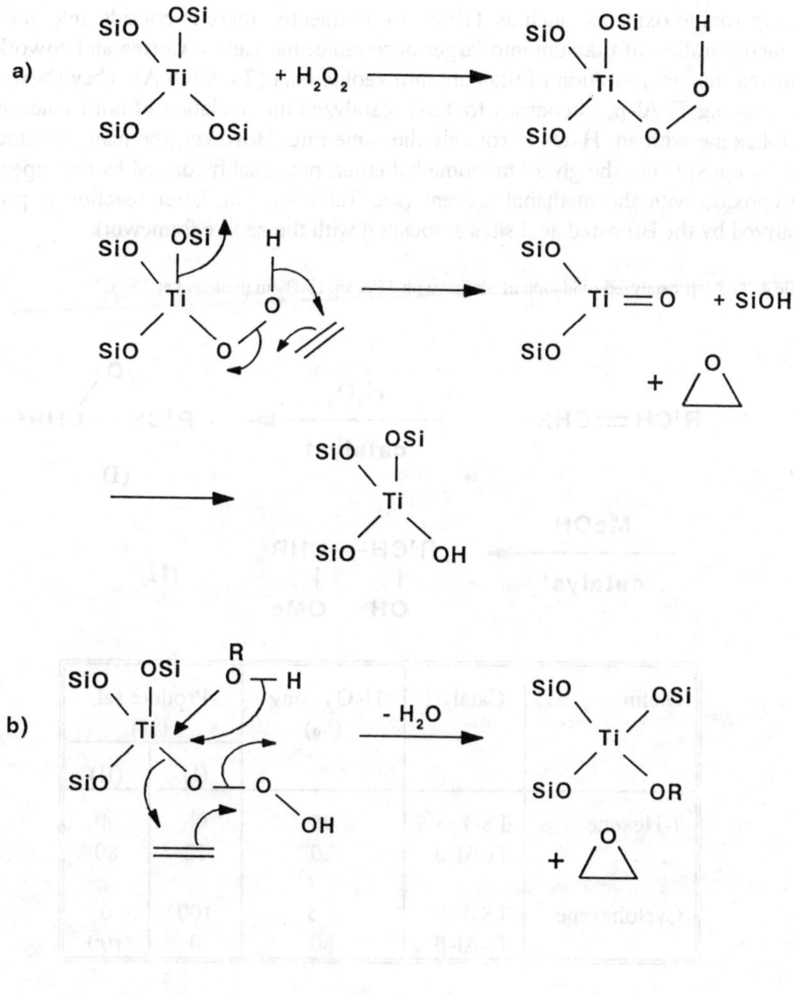


Figure 7. Mechanisms of TS-1-catalyzed oxygen transfer.

Molecular graphics studies[24] show that although the cavity of TS-1 can accommodate a molecule of TBHP severe steric restrictions are imposed on the approach of an olefin to the coordinated tert-butylperoxo group. Thus, TS-1 catalyzes the formation of TBHP from tert-butanol and $H_2O_2$ but it does not catalyze epoxidations with TBHP. Similarly, the lack of reactivity observed with cyclohexene is presumably a result of severe steric constraints imposed on the transition state for oxygen transfer.

In short, TS-1 is an excellent catalyst for oxidations of relatively small substrates with aq. $H_2O_2$. It is not effective, however, with larger substrates and/or typical alkyl hydroperoxide oxidants, such as TBHP. Consequently, there is considerable interest in the incorporation of titanium into larger pore molecular sieves. Corma and coworkers[25] reported the incorporation of titanium into zeolite beta (7.6 x 6.4 Å). They showed that the resulting Ti-Al-β, in contrast to TS-1, catalyzed the oxidation of both 1-hexene and cyclohexene with aq. $H_2O_2$ at roughly the same rate. However, the major product was not the epoxide but the glycol monomethyl ether, presumably formed by ring opening of the epoxide with the methanol solvent (see Table 4). The latter reaction is probably catalyzed by the Brønsted acid sites associated with the zeolite framework.

Table 4. Ti-Al-β catalyzed oxidation of olefins with 35% aq. $H_2O_2$ in methanol at 25 °C[a].

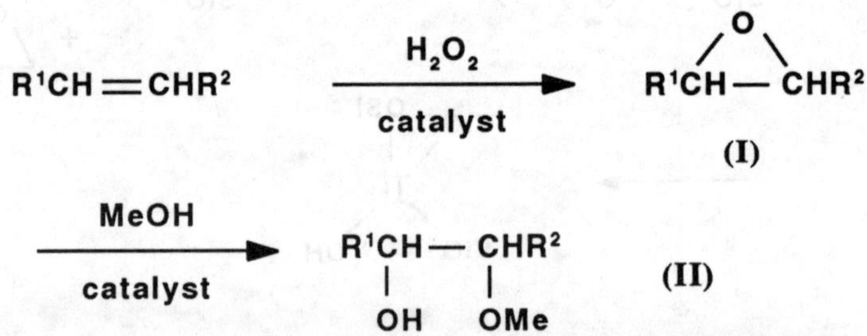


| Olefin | Catalyst | $H_2O_2$ conv. (%) | Product sel. (%) | |
|---|---|---|---|---|
| | | | (I) | (II) |
| 1-Hexene | TS-1 | 98 | 96 | 4 |
| | Ti-Al-β | 80 | 12 | 80 |
| Cyclohexene | TS-1 | < 5 | 100 | 0 |
| | Ti-Al-β | 80 | 0 | 100 |

a. Data taken from ref. 25.

When the Brønsted acid sites of the Ti-Al-β were neutralized by treatment with an aqueous solution of an alkali metal acetate prior to use this led[26] to a dramatic increase in the epoxide selectivity in the oxidation of 1-octene with aq. $H_2O_2$ (Table 5).

Table 5. Effect of alkali metal exchange on selectivity of Ti-Al-β catalyzed epoxidation of 1-octene[a].

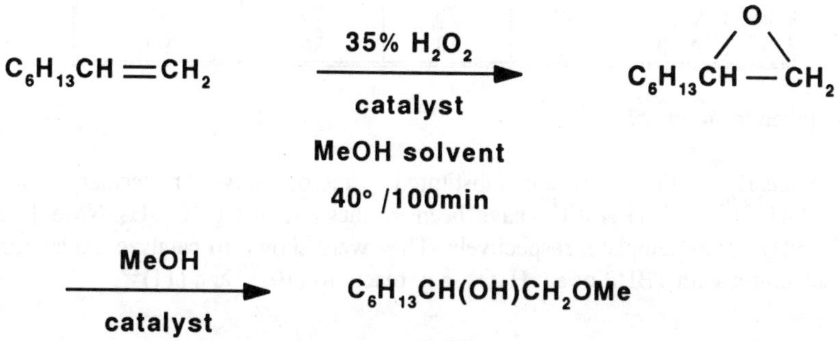


| Catalyst | $H_2O_2$ conv. (%) | Selectivity (%) | | $H_2O_2$ |
|---|---|---|---|---|
| | | Epoxide | Glycol ether | Efficiency (%) |
| TS-1 | 95 | 76 | 24 | 98 |
| Li-TS-1 | 85 | 98 | 0 | 94 |
| Ti-Al-β | 48 | 0 | 97 | 97 |
| Li-Ti-Al-β | 31 | 87 | 5 | 89 |
| Na-Ti-Al-β | 22 | 84 | 6 | 99 |
| K-Ti-Al-β | 25 | 63 | 0 | 74 |

a. Data taken from ref. 26.

The best results were observed with lithium-exchanged Ti-Al-β. Rather surprisingly, even TS-1 (which should not be acidic) gave higher epoxide selectivities when pretreated with lithium acetate (see Table 5). The Ti-Al-β was also an effective catalyst for the epoxidation of 1-octene with TBHP in trifluoroethanol as solvent[27]. Here again significantly higher selectivities were observed when the catalyst was pretreated with an alkali metal acetate prior to use (Table 6)[27].

Table 6. Ti-Al-β catalyzed epoxidation of 1-octene with TBHP in 2,2,2-trifluoroethanol at 90 °C[a].

| Catalyst | Conv. (%) | | Epoxide sel. (%) | |
|---|---|---|---|---|
| | TBHP | 1-octene | on TBHP | on 1-octene |
| Ti-Al-β | 58 | 47 | 30 | 38 |
| Li-Ti-Al-β | 40 | 38 | 95 | 100 |
| Na-Ti-Al-β | 15 | 13 | 86 | 99 |

a. Data taken from ref. 27.

Similarly, the titanium-substituted mesoporous molecular sieves, Ti-MCM-41[28,29] and Ti-HMS[29] have been synthesized, using $[C_{16}H_{33}NMe_3]^+$ and $C_{12}H_{25}NH_2$ as the template, respectively. They were shown to catalyze oxidations of bulky substrates with TBHP or aq. $H_2O_2$, e.g. reactions (10)[28] and (11)[29].

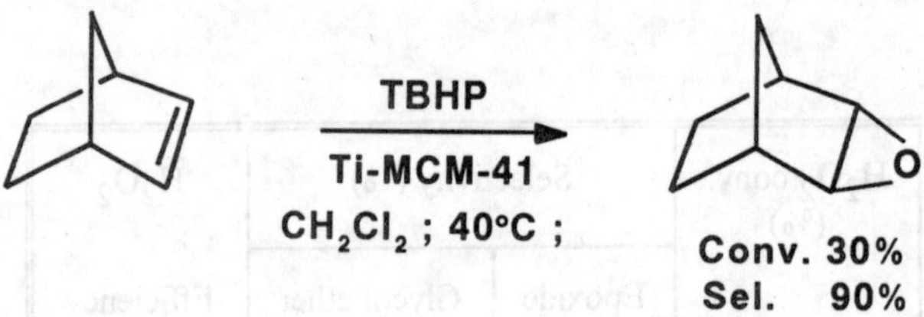


(10)

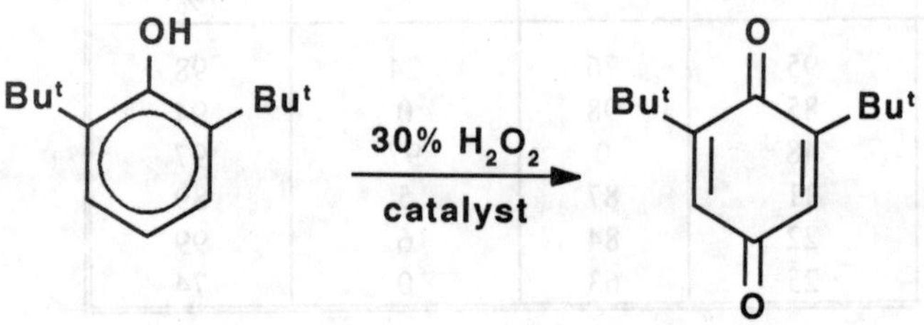


(11)

| Catalyst | Conv. (%) | Sel. (%) |
|---|---|---|
| TS-1 | 6.5 | >95 |
| Ti-MCM-41 | 20 | >98 |
| Ti-MCM | 83 | >95 |

Interestingly, Ti-MCM-41 and Ti-HMS also catalyze the hydroxylation of benzene (reaction 12) with 30% aq. $H_2O_2$, affording phenol in high selectivities (on benzene) at high conversions[29].

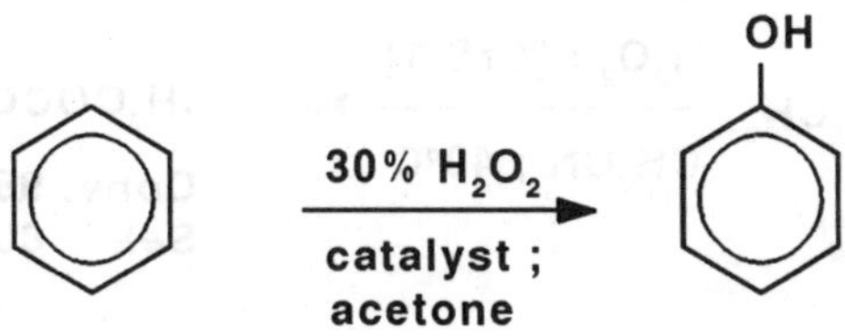


(12)

| Catalyst | Conv. (%) | Sel. (%) |
|---|---|---|
| TS-1 | 31 | >95 |
| Ti-MCM-41 | 68 | >98 |
| Ti-HMS | 37 | >95 |

A large excess of $H_2O_2$ (3 equivs) was used, however, and the selectivity on $H_2O_2$ was not reported.

**9. Other Metal-Substituted Silicalites**

Following the seminal studies of the Enichem group on the TS-1 catalyst, several groups have reported the incorporation of other redox metals into the silicalite framework, e.g. zirconium[30], tin[31] and chromium[32]. ZrS-1 catalyzed the hydroxylation of benzene and phenol with aq. $H_2O_2$ but activities and selectivities were significantly lower than with TS-1[30]. SnS-1, on the other hand, reportedly[31] catalyzes the hydroxylation of phenol with aq. $H_2O_2$ with activities only marginally lower than that of TS-1.

CrS-1 exhibits different types of catalytic activity to the above mentioned catalysts. For example, it catalyzes the selective oxidative cleavage of olefins with aq. $H_2O_2$ in acetonitrile[33]. Methyl acrylate and methyl methacrylate were converted to the methyl esters of glyoxylic acid (reaction 13) and pyruvic acid (reaction 14), respectively.

$$H_2C{=}CCO_2CH_3 \xrightarrow[CH_3CN\ ;\ 40°C]{H_2O_2\ ;\ [CrS\text{-}1]} CHOCO_2CH_3 \qquad (13)$$

Conv. 98%
Sel. 90%

$$H_2C{=}C(CH_3)CO_2CH_3 \xrightarrow[CH_3CN\ ;\ 40°C]{H_2O_2\ ;\ [CrS\text{-}1]} CH_3COCO_2CH_3 \qquad (14)$$

Conv. 95%
Sel. 80%

## 10. Metal-Substituted Aluminophosphates (MeAPOs)

Substitution of framework aluminium in aluminophosphate molecular sieves[34] by redox metal ions may be expected to form novel heterogeneous catalysts for liquid phase oxidations. Indeed, MeAPO catalysts should be complementary to the metal-substituted silicalites. In contrast to silicalite, aluminophosphates possess hydrophilic cavities, making them unsuitable for oxidations in aqueous media, i.e. with aq. $H_2O_2$. However, they are compatible with TBHP or dioxygen and the larger pores of AlPO-5, for example, compared to silicalite, render it amenable to larger substrates.

One catalyst which has been extensively studied[35-38] is CrAPO-5. Substitution of aluminium in AlPO-5 by a trivalent metal ion, such as chromium, maintains the electroneutrality of the framework. As-synthesized CrAPO-5 contains chromium in the trivalent state and is most likely octahedrally coordinated, two molecules of water occupying the fifth and sixth coordination positions (Figure 8). When this catalyst is calcined, to remove the template, the chromium undergoes oxidation to afford dioxochromium(VI) which is still bonded to the micropores of the $AlPO_4$ structure. In order to balance the charges $Cr^{VI}$-APO-5 must contain one acidic P-OH group per chromium and could have a tetrahedral or octahedral configuration (see Figure 8). This was confirmed by measuring the amount of irreversible adsorption of $NH_3$, which corresponded to ca. one molecule of $NH_3$ per Cr. The extra H atom in the calcined catalyst is presumed to be derived from decomposition of the template.

CrAPO-5 is an excellent catalyst for a variety of industrially relevant transformations, such as the decomposition of alkyl hydroperoxides[35], and the oxidation of secondary alcohols[36], alkylaromatics[37,38] and cycloalkanes[37,38] using TBHP or $O_2$ as the primary oxidant.

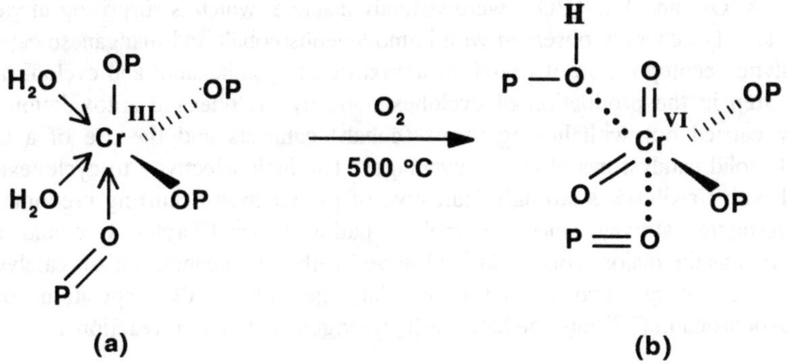


Figure 8. Proposed structures of CrAPO-5: (a) as-synthesized; (b) after calcination.

As shown in Table 7 both CrAPO-5 and CrS-1 are active catalysts for the decomposition of cyclohexyl hydroperoxide (CHHP) in cyclohexane at 70 °C, the highest selectivity for cyclohexanone (86%) being observed with CrAPO-5. Other MeAPOs and silicalites gave both lower activities and selectivities.

Table 7. Decomposition of cyclohexyl hydroperoxide (CHHP) over various redox molecular sieves[a].

| Catalyst | CHHP conv. (%) | Selectivity (%) | |
|---|---|---|---|
| | | Cyclohexanone | Cyclohexanol |
| CrAPO-5 | 87 | 86 | 13 |
| CrS-1 | 98 | 64 | 36 |
| VAPO-11 | 76 | 50 | 50 |
| Co-ZSM-5 | 24 | 43 | 50 |
| VAPO-5 | 17 | 51 | 43 |
| CoAPO-5 | 2 | 50 | 50 |
| MnAPO-5 | 2 | 50 | 50 |
| VS-1 | 0 | -- | -- |
| TS-1 | 0 | -- | -- |
| None | 0 | -- | -- |

a. CHHP (2.9 mmol) in cyclohexane (12 ml) stirred with the catalyst (0.029 mmol metal) for 5 h at 70 °C (data taken from ref. 35).

CoAPO-5 and Mn-APO-5 were virtually inactive, which is surprising in view of the high activity generally observed with homogeneous cobalt and manganese catalysts. The catalytic decomposition of CHHP to a mixture of cyclohexanol and cyclohexanone is a key step in the production of cyclohexanone by cyclohexane autoxidation. It is generally carried out with homogeneous (cobalt) catalysts and the use of a stable, recyclable solid catalyst has obvious advantages. The high selectivity to cyclohexanone observed with CrAPO-5 is strongly indicative of the reaction occurring predominantly via a heterolytic pathway, since a homolytic pathway (see Chapter 8) would afford cyclohexanol as the major product, as is observed with homogeneous cobalt catalysts. A plausible mechanism involves intramolecular, heterolytic decomposition of an alkylperoxochromium(VI) intermediate via β-hydrogen elimination (reaction 15).

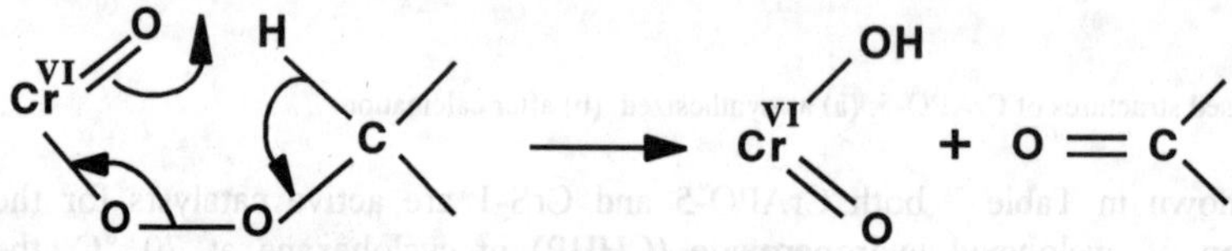


(15)

An intermolecular mechanism via further oxidation of initially produced cyclohexanol with CHHP can only be a minor pathway since reaction of cyclohexanol with cyclohexene-3-hydroperoxide or TBHP afforded only about 6% cyclohexanone under the same conditions.

Evidence for the reaction taking place inside the micropores of CrAPO-5 was provided by the observation[35] that triphenylmethyl hydroperoxide, which is too bulky to be accommodated in the micropores of CrAPO-5, was hardly decomposed (Table 8). In contrast, soluble chromium(III) acetylacetonate and the supported $CrO_2Cl_2$/silica-alumina were effective catalysts for the decomposition of this hydroperoxide.

Table 8. Decomposition of triphenylmethyl hydroperoxide over chromium-containing catalysts at 70 °C.

| Catalyst | Decomposition (%) |
|---|---|
| $Cr(acac)_3$ | 75 |
| $CrO_2Cl_2/SiO_2$-$Al_2O_3$ | 72 |
| CrAPO-5 | 1 |

CrAPO-5 catalyzes the selective oxidation of secondary alcohols to the corresponding ketones, using TBHP as the oxidant (see reactions 16-18)[36]. Interestingly, carveol underwent chemoselective oxidation at the alcohol group, to give carvone (reaction 18), without any attack at the double bonds.

$$PhCH(R)OH \xrightarrow[\text{CrAPO-5; PhCl ; 85°C}]{\text{TBHP}} PhCOR \qquad (16)$$

| R | Sel. (%) |
|---|---|
| $CH_3$ | 96 |
| $C_2H_5$ | 100 |

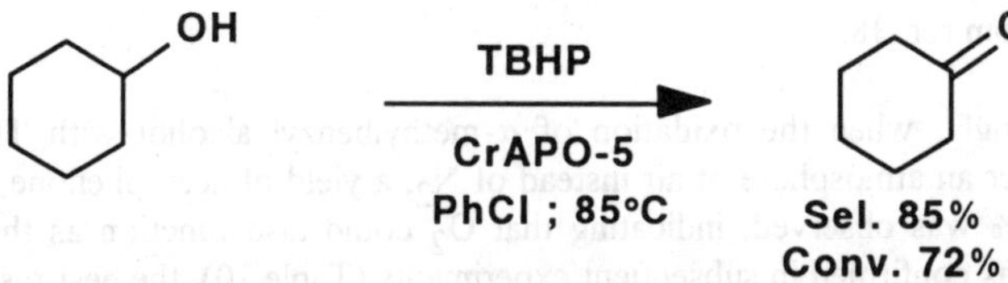


(17)

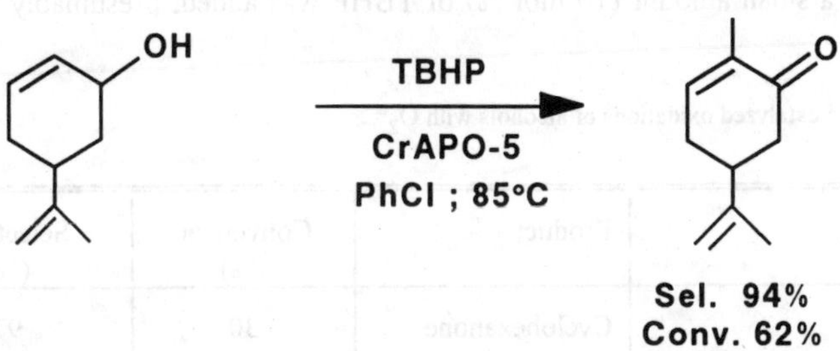


(18)

The CrAPO-5 catalyst was stable and recyclable, the reused catalyst being even more active than the fresh one in the oxidation of α-methylbenzyl alcohol. Moreover, treatment of the catalyst with sodium acetate prior to use led to a catalyst that was both more active and selective (see Table 9)[39].

As in the case of alkyl hydroperoxide decomposition (see earlier) evidence for the reaction taking place inside the micropores was obtained by using the bulky triphenylmethyl hydroperoxide as the oxidant in the oxidation of α-methylbenzyl alcohol[38]. Hardly any reaction was observed with CrAPO-5 while homogeneous $Cr(acac)_3$, in contrast, afforded roughly the same conversion with triphenylmethyl hydroperoxide as with TBHP.

Table 9. Catalyst recycle in the CrAPO-5 catalyzed oxidation of α methylbenzyl alcohol with TBHP in PhCl at 100 °C[a].

| Cycle | Na-CrAPO-5 | | H-CrAPO-5 | |
|---|---|---|---|---|
| | conv. (%) | sel. (%) | conv. (%) | sel. (%) |
| 1 | 85 | 100 | 65 | 95 |
| 2 | 96 | 96 | 79 | 93 |
| 3 | 97 | 89 | 83 | 86 |

a. Data taken from ref. 38.

Interestingly, when the oxidation of α-methylbenzyl alcohol with TBHP was carried out under an atmosphere of air instead of $N_2$, a yield of acetophenone, based on TBHP, of 216% was observed, indicating that $O_2$ could also function as the primary oxidant. This was confirmed in subsequent experiments (Table 10), the best results being obtained when a small amount (10 mol %) of TBHP was added, presumably to initiate the reaction[36].

Table 10. CrAPO-5 catalyzed oxidations of alcohols with $O_2$[a].

| Substrate | Product | Conversion (%) | Selectivity (%) |
|---|---|---|---|
| Cyclohexanol | Cyclohexanone | 30 | 97 |
| α-Methylbenzyl alcohol | Acetophenone | 31 | 96 |
| α-Ethylbenzyl alcohol | Propiophenone | 38 | 90 |
| α-Tetralol | α-Tetralone | 26 | 73 |
| α-Indanol | α-Indanone | 78 | 72 |

a. Data taken from ref. 36.

CrAPO-5 is also an excellent solid catalyst for the oxidation of alkylaromatics to the corresponding ketones using TBHP (reaction 19) as the primary oxidant[37-39].

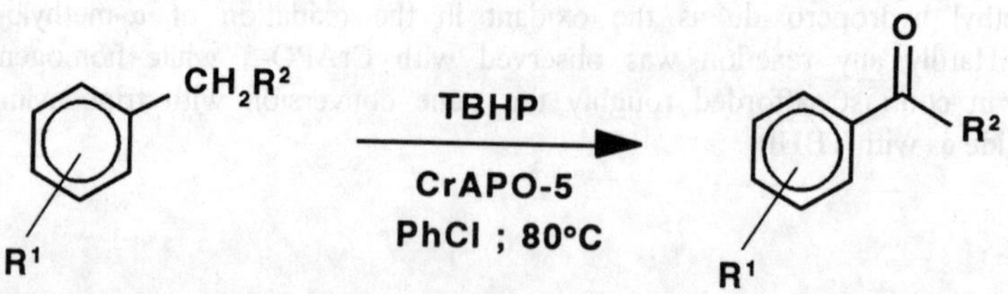


(19)

High selectivities to the corresponding ketones were observed with a variety of alkylaromatic substrates (Table 11). CrAPO-5 was also an effective catalyst for the autoxidation of alkylaromatics to the corresponding ketones when 10 mol % of TBHP was added to initiate the reaction[39]. In some cases, e.g. with ethylbenzene, it was also necessary to neutralize the acid sites in CrAPO-5 in order to circumvent acid-catalyzed decomposition of the intermediate hydroperoxide into phenolic inhibitors. Here again, recycling experiments showed that the CrAPO-5 is stable and recyclable and elemental analysis of mother liquors confirmed that no (below the detection limit) chromium had been leached from the catalyst.

**Table 11.Oxidation of alkylaromatics with TBHP catalyzed by CrAPO-5[a].**

$$ArCH_2R \xrightarrow[PhCl\ /\ 80°C\ /\ 16h]{TBHP\ /\ CrAPO\text{-}5} ArCOR$$

| **Substrate** | **Conv. (%)[b]** | **Sel. (%)[c]** |
|---|---|---|
| Ethylbenzene | 70 | 90 |
| *p*-Ethyltoluene | 68 | 97 |
| n-Propylbenzene | 59 | 93 |
| n-Butylbenzene | 59 | 92 |
| Diphenylmethane | 50 | 94 |
| *p*-Ethylanisole | 13 | 41 |

a. Substrate / TBHP / Cr molar ratio = 1:5:0.03. b. Conversion of substrate. c. To ketone product based on substrate converted.

CrAPO-5 also catalyzes the autoxidation of cyclohexane at 115 °C (reaction 20), giving (at 3% conversion) a mixture of cyclohexanone (64%), cyclohexanol (10%), cyclohexyl hydroperoxide (9%) and dicarboxylic acids (13%).

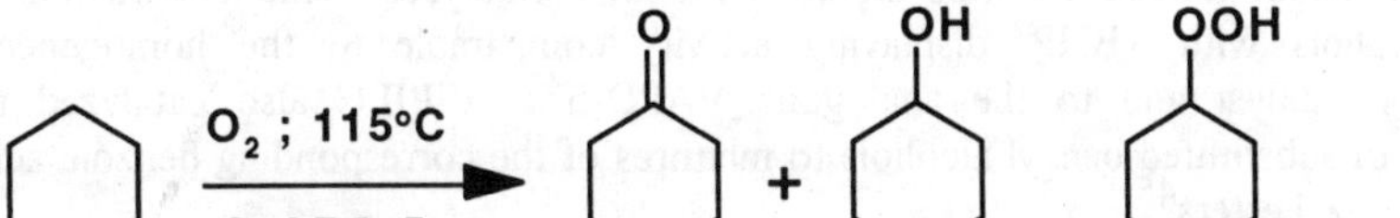


(20)

**Selectivity: 68% 10% 9%**

**at 3% cyclohexane conversion**

In short, CrAPO-5 appears to form the basis for a variety of clean oxidation processes in the liquid phase, whereby the catalyst is readily recovered and recycled.

## 11. Redox Pillared Clays

Another approach to designing heterogeneous oxidation catalysts with novel activities and shape selectivities is to incorporate redox metal ions into the interlamellar space of clay minerals. The intercalation of smectite clays, such as montmorillonite, with redox metal ions can lead to the formation of `redox pillared clays' containing oxometal species propped between the silicate sheets[40]. The catalytic properties of such materials in liquid phase oxidations have been studied by Choudary and coworkers.

Chromium-pillared montmorillonite (Cr-PILC), for example, catalyzes the benzylic oxidation of alkylbenzenes, to the corresponding ketones[41] and the selective oxidation of primary and secondary alcohols to the corresponding aldehydes and ketones[42], using TBHP in dichloromethane at room temperature. Allylic alcohols underwent chemoselective oxidation of the alcohol functionality, e.g.

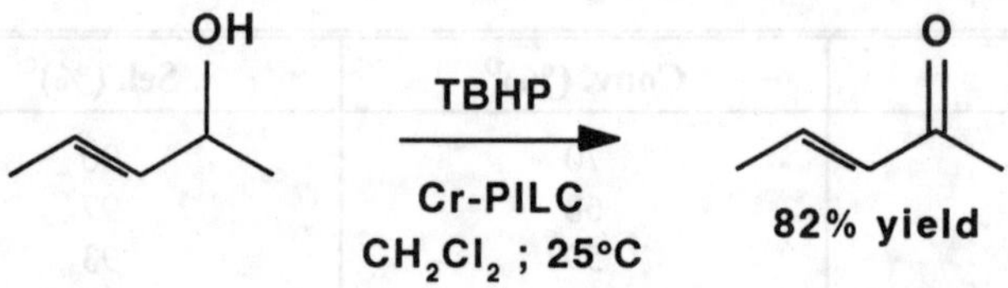


(21)

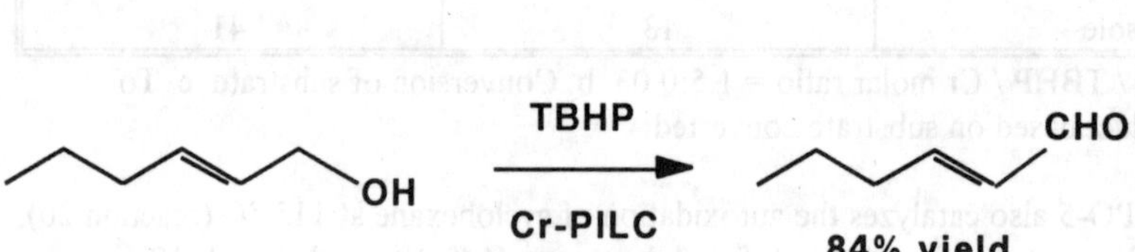


(22)

Vanadium-pillared montmorillonite (V-PILC) catalyzes[43] the epoxidation of allylic alcohols with TBHP, displaying activity comparable to the homogeneous $VO(acac)_2$ catalyst and to the analogous VAPO-5[44]. V-PILC also catalyzed the oxidation of substituted benzyl alcohols to mixtures of the corresponding benzoic acids and their benzyl esters[45].

## 12. Metal Complexes Encapsulated in Molecular Sieves: Ship-in-the-Bottle Complexes

Another approach to creating novel molecular sieves with catalytic redox properties involves the construction of metal complexes, by intrazeolite synthesis, from ligand components that have access to the zeolite cages. The encapsulated metal-ligand complex is too large to allow diffusion out of the zeolite cages. Such so-called ship-in-the-bottle complexes[46,47] combine, in principle, the advantages of homogeneous and heterogeneous catalysts. Moreover, the zeolite cage provides a shape-selective environment analogous to the protein tertiary structure in redox enzymes.

For example, iron phthalocyanine (FePc) encapsulated in zeolite Y is prepared[48] by treating iron(II)-exchanged zeolite Y with dicyanobenzene at 150 °C. Alternatively, the iron can be introduced as ferrocene[49] or as a carbonyl complex[47]. In this way the presence of unchelated metal ion in the zeolite cages is largely avoided.

FePc encapsulated in zeolite Y catalyzed the oxidation of cyclohexane with TBHP with a turnover of 6000 compared to 25 with homogeneous FePc, indicating that occlusion of the catalyst in the zeolite enhances its stability towards oxidative degradation[49].

More recently, Jacobs and coworkers[50] have described the use of zeolite encapsulated manganese(II) bipyridine (bipy) complexes, e.g. cis-$Mn(bipy)^{2+}$-Y, as catalysts for the oxidation of olefins with 30% aq. $H_2O_2$ in acetone at ambient temperature (Table 12). At low conversions the epoxide and diol were the major products. At complete conversion, on the other hand, the dicarboxylic acids, resulting from oxidative cleavage of the double bond, were the only products observed with cyclohexene and cyclododecene.

Table 12. cis-$Mn(bipy)_2^{2+}$-Y catalyzed oxidations of olefins with 30% aq. $H_2O_2$ in acetone at 20 °C[a].

| Olefin | Time (h) | Conversion (%) | Selectivity (%) | | |
|---|---|---|---|---|---|
| | | | epoxide | diol | diacid |
| 1-Hexene | 18 | 20 | 50 | 40 | -- |
| Cyclohexene | 18 | 62 | 6 | 79 | -- |
| Cyclohexene | 40 | 100 | -- | -- | 80 |
| 1-Dodecene | 18 | 20 | 10 | 88 | -- |
| Cyclododecene | 18 | 56 | 4 | 87 | -- |
| Cyclododecene | 40 | 100 | -- | -- | 84 |

a. Data taken from ref. 50.

Up to 1000 turnovers were observed in the oxidation of cyclohexene with 5 equivalents of $H_2O_2$. Moreover, the catalyst could be recycled, after drying at 50 °C, without loss of activity, indicating that encapsulation strongly suppresses the propensity of such metal complexes for oxidative destruction.

Based on the exciting results described above the ship-in-the-bottle approach appears to hold much promise for the development of heterogeneous catalysts capable of emulating redox enzymes. Furthermore, we note that this approach lends itself to the design of chiral heterogeneous catalysts for asymmetric oxidations, a topic whivch is very relevant in the context of fine chemicals manufacture (see Chapter 11).

## References


1. R. A. Sheldon and J. Dakka, *Catalysis Today* **19** (1994) 215-246.
2. T. J. Collins, *Acc. Chem. Res.* **27** (1994) 279.
3. R. A. Sheldon, *Aspects of Homogeneous Catalysis*, Vol. 4, ed. R. Ugo (Reidel, Dordrecht, 1981) 1.
4. R. A. Sheldon, in *The Chemistry of Functional Groups - Peroxides*, ed. S. Patai (Wiley, New York, 1982) 161.
5. R. A. Sheldon, *Stud. Surf. Sci. Catal.* **59** (1991) 33, and references cited therein.
6. M. Besson, P. Gallezot, F. Lahmer, G. Flèche and P. Fuertes, in *Catalysis of Organic Reactions*, eds. J. R. Kosak and T. A. Johnson (Marcel Dekker, New York, 1993) 169.
7. G. Flèche and P. Fuertes, Eur. Pat. 233816 (1987) to Roquette Frères.
8. H. C. Shen and H. S. Weng, *Ind. Eng. Chem. Res.* **27** (1988) 2246 and 2254; see also F. Waller, *Catal. Rev. Sci. Eng.* **28** (1986) 1.
9. H. Turk and W. T. Ford, *J. Org. Chem.* **53** (1988) 460.
10. R. A. Sheldon and J. K. Kochi, *Metal-Catalyzed Oxidations of Organic Compounds* (Academic Press, New York, 1981).
11. H. C. Shen and H. S. Weng, *Ind. Eng. Chem.* **29** (1990) 713; see also C. S. Yao and H. S. Weng, *Chem. Eng. Sci.* **47** (1992) 2745.
12. Y. Ito, M. Inoue and S. Enomoto, *Chem. Pharm. Bull.* **33** (1985) 3583.
13. T. R. Felthouse, *J. Am. Chem. Soc.* **109** (1987) 7566.
14. R. A. Sheldon, *Topics Curr. Chem.* **164** (1993) 21.
15. R. A. Sheldon, *CHEMTECH*, (1991) 566.
16. J. C. Jansen and S. T. Wilson, in *Introduction to Zeolite Science and Practice*, eds. H. van Bekkum, E. M. Flanigen and J. C. Jansen (Elsevier, Amsterdam, 1991) 77.
17. B. Notari, *Catalysis Today* **18** (1993) 163.
18. U. Romano, A. Esposito, F. Maspero, C. Neri and M. G. Clerici, in *New Developments in Selective Oxidation*, eds. G. Centi and F. Trifiro (Elsevier, Amsterdam, 1990) 33.
19. U. Romano, A. Esposito, F. Maspero, C. Neri and M. G. Clerici, *Chim. Ind. (Milan)* **72** (1990) 610.
20. B. Notari, *Stud. Surf. Sci. Catal. 37 (1988) 413.*
21. *M. G. Clerici and P. Ingallina,* J. Catal. ***140*** *(1993) 71.*
22. *M. J. S. Dewar,* Enzyme *36 (1986) 8.*
23. *F. Toda,* Advan. Supramol. Chem. *2 (1992) 141-191; F. Toda,* SynLett., *(1993) 303-312.*
24. R. A. Sheldon and J. Horsley, unpublished results.
25. A. Corma, M. A. Camblor, P. Esteve, A. Martinez and J. Perez-Pariente, *J. Catal.* **145** (1994) 151.
26. T. Sato, J. Dakka and R. A. Sheldon, *Stud. Surf. Sci. Catal.* **84** (1994) 1853.
27. T. Sato, J. Dakka and R. A. Sheldon, *J. Chem. Soc., Chem. Commun.*, (1994)

1887.
28. A. Corma, M. T. Navarro and J. Perez-Pariente, *J. Chem. Soc., Chem. Commun.*, (1994) 147.
29. P. T. Tanev, M. Chibwe and T. J. Pinnavaia, *Nature*, **368**(1994)321
30. M. K. Dongare, P. Singh, P. P. Moghe and P. Ratnasamy, *Zeolites* **11** (1991) 690.
31. N. Kishor Mal, V. Ramaswamy, S. Ganapathy and A.V. Ramaswamy, *J. Chem. Soc., Chem. Commun.*, (1994) 1933.
32. J. S. T. Mambrim, E. J.S. Vichi, H. O. Pastore, C. U. Davanzo, H. Vargas, E. Silva and O. Nakamura, *J. Chem. Soc., Chem. Commun.*, (1991) 922.
33. M. Kawai and T. Kyoura, Jap. Patents 0358,954 and 0356,439 (1991) to Mitsui Toatsu Chemicals, *CA* **115** (1991) 48863d and 48864e.
34. S. T. Wilson, B. M. Lok, C. A. Messina, T. R. Cannan and E. M. Flanigen, *J. Am. Chem. Soc.* **106** (1984) 6092.
35. J. D. Chen, J. Dakka and R. A. Sheldon, *Appl. Catal. A: General* **108** (1994) L1.
36. J. D. Chen, J. Dakka, E. Neeleman and R. A. Sheldon, *J. Chem. Soc., Chem. Commun.*, (1993) 1379.
37. J. D. Chen, M. J. Haanepen, J. H. C. van Hooff and R. A. Sheldon, *Stud. Surf. Sci. Catal.* **84** (1994) 973.
38. R. A. Sheldon, J. D. Chen, J. Dakka and E. Neeleman, in *Zeolites and Microporous Crystals*, eds. T.Hattori and T.Yashima (Elsevier, Amsterdam, 1994)407.
39. J. D. Chen, H. E. B. Lempers and R. A. Sheldon, *Colloids and Surfaces*, submitted for publication.
40. F. Figueras, *Catal. Rev. Sci. Eng.* **30** (1988)457.
41. B. M. Choudary, A. Durga Prasad, V. Bhuma and V. Swapna, *J. Org. Chem.* **57** (1992) 5841.
42. B. M. Choudary, A. Durga Prasad and V. L. K. Valli, *Tetrahedron Lett.* **31** (1990) 5785.
43. B. M. Choudary, V. L. K. Valli and A. Durga Prasad, *J. Chem. Soc., Chem. Commun.*, (1990) 721.
44. M. S. Rigutto and H. van Bekkum, J.Mol.Catal., **81**(1993)77
45. B. M. Choudary, V. L. K. Valli and A. Durga Prasad, *J. Chem. Soc., Chem. Commun.*, (1991) 1115.
46. R. Parton, D. De Vos and P. A. Jacobs, in *Zeolite Microporous Solids: Synthesis, Structure and Reactivity*, ed. E. G. Derouane (Kluwer, Amsterdam, 1992) 552.
47. D. E. De Vos, F. Thibault-Starzyk, P. P. Knops-Gerrits, R. F. Parton and P. A. Jacobs, *Macromol. Symp.* **80** (1994) 157.
48. N. Herron, G. D. Stucky and C. A. Tolman, *J. Chem. Soc., Chem. Commun.*, (1986) 1521.
49. R. F. Parton, L. Uytterhoeven and P. A. Jacobs, *Stud. Surf. Sci. Catal.* **59** (1991) 395.
50. P. P. Knops-Gerrits, D. De Vos, F. Thibault-Starzik and P. A. Jacobs, *Nature* **369** (1994) 543.

**METAL COMPLEX CATALYSIS OF OXIDATION REACTIONS. CATALYSIS WITH PALLADIUM COMPLEXES**

ILYA I. MOISEEV

*N.S.Kurnakov Institute of General & Inorganic Chemistry, Russian Academy of Sciences, Leninsky Prosp. 31, 117907 Moscow GSP-1, Russia*

ABSTRACT

Oxidation of alkenes catalyzed with palladium compounds represents a typical example of organometallic catalysis involving transition metal organyl formation. Ethylene oxidations to produce acetaldehyde and vinyl acetate are being widely used on large scale. In this paper the experimental data relevant to the mechanistic aspects of redox reactions between Pd(II) salts and alkenes are discussed. The latter are compared with those related to the oxidation reactions catalyzed with Pd-561 clusters modeling the active sites of supported Pd metal catalysts. By using solubility measurements and kinetic methods, all stages of the alkene oxidation reactions with Pd(II) salts have been characterized quantitatively. The mechanism of π-complexes and σ-organyls transformation was elucidated by the study of oxidation reactions in nonaqueous hydroxyl-containing solvents. The reactions of β-hydroxyethylpalladium halogenides, their esters and ethers generated *in situ* by reacting palladium (II) salt with corresponding organylmercuric halogenides were used to investigate the pathways of π and σ-bonded Pd-organyl reactions. All the data obtained including isotopic solvent and substrate effects are in a good agreement with a mechanism involving the formation of π-complexes and its transformations involving an π-σ-rearrangement to form palladium (II) σ-organyl derivative and its oxidative decomposition giving rise to the observed product of alkene oxidation and the reduced form of palladium.

Giant cationic palladium clusters approximated as $Pd_{561}L_{60}OAc_{180}$ (L=phen, bipy) and $Pd_{561}Phen_{60}O_{60}(PF_6)_{60}$ were synthesized and characterized with high resolution TEM, HREM, STM, SAXS, EXAFS, IR and magnetic susceptibility data. Under mild conditions the giant palladium clusters catalyze oxidative acetoxylation of ethylene into vinyl acetate, propylene into allyl acetate and other oxidations. In contrary to Pd(II) catalysis, cluster catalysis is not sensitive to the presence of water. A mechanism for the oxidations is proposed involving oxidative addition of alkene to a Pd-Pd fragment and rearrangement of the coordinated alkene to form vinyl or allyl fragment. Oxidation is supposed to take place with the participation of an coordinated oxidant.

## 1. Catalysis with palladium (II) complexes

### 1.1. Introduction.

Liquid-phase oxidation catalyzed with metal complexes constitutes an extensive field of chemical reactions embracing reactions occurring both in living nature and in industry. Industrially important reactions of this type include liquid-phase oxidation of paraffins and alkylarenes, the oxidation of alkens into carbonyl compounds, vinyl and allyl esters, numerous alkene epoxidation processes, *etc.*[1]

The study of oxidation reactions has played an important role in the establishment of metal complex catalysis and in the formulation of its principles. Thus the oxidation of alkenes in the presence of palladium (II) salts has become the first example in metal complex catalysis where it has been possible to characterize reliably all the stages and to clear up the role of the catalytic intermediates in this important multi-step reaction. Despite the fact that the individual details remain controversial, this process is probably the most thoroughly investigated also nowadays.

More than 35 years have passed since elaborate study of the oxidation of olefins by palladium salts was begun, a reaction which was first observed by Phillips in 1894[2] and which 60 years later, owing to Smidt, Hafner, Sedlmeier, Jira et al.[3,4] served as the basis of the most convenient method for the production of carbonyl compounds from olefins.

$$C_2H_4 + PdCl_4^{2-} + H_2O \longrightarrow CH_3CHO + Pd + 2H^+ + 4Cl^-$$
$$Pd + 4Cl^- + 2CuCl_2 \longrightarrow PdCl_4^{2-} + 2CuCl^{2-}$$
$$2CuCl^{2-} + 2H^+ + 1/2\,O_2 \longrightarrow 2CuCl_2 + H_2O$$

$$C_2H_4 + 1/2\,O_2 \xrightarrow{PdCl_4^{2-},\ CuCl_2} CH_3CHO + 52.26\ \text{kcal/mol} \qquad (1)$$

The reaction can be run as a single stage process (Eq.1) or as a two-stage process, in which catalyst solution is treated with ethylene (Eq.2) and solution containing Cu(I) is transferred into another apparatus to be reoxidized by air (Eq.(3)):

$$C_2H_4 + 2CuCl_2 + H_2O \xrightarrow{PdCl_4^{2-}} CH_3CHO + 2CuCl_2^- + 2H^+ - 13.7\ \text{kcal/mol} \qquad (2)$$

$$2CuCl_2^- + 2H^+ + 1/2\,O_2 \longrightarrow 2CuCl_2 + H_2O + 65.43\ \text{kcal/mol} \qquad (3)$$

In two-apparatus system, the oxidation of ethylene is an heat consuming process. The conditions for the oxidation are very mild (100-120°C, 10 atm) and the substrate under oxidation does not react with dioxygen. Because of that the process is a very selective one.

In acetic acid solution, the reduction of Pd(II) salts gives rise to alkenyl esters formation[5]. Vinyl acetate is formed besides ethyledene diacetate from ethylene:

$$C_2H_4 + PdCl_2 + 2NaOAc \longrightarrow CH_2{=}CHOAc + 2NaCl + Pd + AcOH$$
$$C_2H_4 + PdCl_2 + 2NaOAc \longrightarrow CH_3CH(OAc)_2 + 2NaCl + Pd$$

In alcohol media acetals were found to be the products of alkene oxidation[5]:

$$C_2H_4 + PdCl_2 + ROH \longrightarrow CH_3CH(OR)_2 + 2HCl$$

The transformations of high symmetrical ethylene molecule into the acetaldehyde and relative compounds seemed to be intriguing and challenging reactions from the outset. The experience gained in the studying of the mechanism of these reactions is still of importance consisting a body of modern homogeneous catalysis.

### 1.2. Equilibrium of formation of $\pi$-complexes in aqueous solution

At the earliest stages of investigation of the reaction of palladium salts with olefins it was proposed that this reaction proceeds via formation and decomposition of $\pi$-complexes[3,6]. The formation of $\pi$-complexes was confirmed already in the first experiments on interaction of gaseous olefins with solutions of palladium salts. At sufficiently high concentrations of halogen and hydrogen ions, when the oxidation of olefins is hindered, they are rapidly and practically reversibly absorbed in amounts, exceeding their solubility in the absence of palladium salts. The excess solubility of olefins increases with increasing concentration of palladium salts and diminishes with increasing concentration of halogen ions. Obviously this increase in solubility is due to coordination of the olefins with palladium salts. The difference in olefin solubilities between solutions with and without palladium salts ($D[C_nH_{2n}]$) is thereby equal, to the amount of complexed olefin $D[C_nH_{2n}]=[\pi]_\Sigma$. From a small but noticeable slope of the "horizontal" part of the absorption curve, it followed that under the conditions of the solubility measurements, olefin oxidation by the palladium salts took place, albeit slowly. However, even a simple comparison of the absorption rates, which correspond to the period of accumulation of the $\pi$-complexes and to the "horizontal" portion of the absorption curve (where the rate of olefin absorption is equal to the rate of $\pi$-complex consumption) showed that accumulation of the $\pi$-complexes proceeds faster than their oxidative decomposition. Hence the reactions of complex formation were close to equilibrium.

#### *1.2.1. Composition of the $\pi$-Complexes*

On the basis of the known properties of platinum-olefin $\pi$-complexes[7,8] in aqueous solution the reaction of $PdCl_4^{2-}$ with olefin in aqueous solution at sufficiently high hydrogen and chloride ion concentrations can be expected to give $\pi$-complexes $C_nH_{2n}PdCl_3^-$, $C_nH_{2n}PdCl_2OH_2$, $C_nH_{2n}PdCl_2OH^-$ and $(C_nH_{2n})_2PdCl_2$. The olefin solubility $[C_nH_{2n}]_\Sigma$ should be correlated with the concentrations of free olefin and $\pi$-complexes by the following equation of material balance:

$$[C_nH_{2n}]_\Sigma = [C_nH_{2n}] + [C_nH_{2n}PdCl_3^-] + [C_nH_{2n}PdCl_2OH_2] + [C_nH_{2n}PdCl_2OH^-] + 2\,[(C_nH_{2n})_2PdCl_2] \quad (4)$$

where $[C_nH_{2n}]$ is the concentration of non-complexed olefin, which is equal to its solubility in aqueous solution free of palladium salt.

However the treatment of the experimental data[9,10] showed the concentrations of complexes $C_nH_{2n}PdCl_2OH^-$ and $(C_nH_{2n})_2PdCl_2$ are negligibly small relative to the concentrations of the complexes $C_nH_{2n}PdCl_3^-$ and $C_nH_{2n}PdCl_2OH_2$. Therefore, in this case $[C_nH_{2n}]_\Sigma = [C_nH_{2n}] +$ $+ [C_nH_{2n}PdCl_3^-] + [C_nH_{2n}PdCl_2OH_2]$ and the excess of the alkene solubility owes its origin to the equilibrium reactions:

$$C_nH_{2n} + PdCl_4^{2-} \overset{K_1}{\rightleftharpoons} C_nH_{2n}PdCl_3^- + Cl^- \qquad (5)$$

$$C_nH_{2n} + PdCl_4^{2-} + H_2O \overset{K_2}{\rightleftharpoons} C_nH_{2n}PdCl_2OH_2 + 2Cl^- \qquad (6)$$

In a series of experiments devoted to determination of $K_1$ and $K_2$, the concentration of chloride ions introduced as lithium chloride or hydrochloric acid was varied while the ionic strength was maintained constant with the aid of lithium perchlorate or perchloric acid*. In each series both the cation concentration and composition were kept constant. Simultaneously, $Cl^-$ and $ClO_4^-$ concentrations varied in such a way that their total amount was kept constant. The date for ethylene are collected in the Table 1.

Table 1. Equilibrium constants of reactions 5 and 6 for ethylene (see[10] and the references therein).

| Temp., °C | μ, M | $K_1$ | $K_2$, M |
|---|---|---|---|
| 13.4 | 4.5 | 16.3 | 0.4 |
| 13.3 | 3.0 | 15.9 | 0.18 |
| 13.3 | 2.0 | 15.5 | 0.033 |
| 20 | 4.0 | 15.2 | 0.22 |
| 25 | 4.0 | 13.1 | 0.21 |
| 15 | 2.0 | 18.7 | |
| 25 | 2.0 | 17.4 | $10^{-3}$ |
| 35 | 2.0 | 9.4 | |
| 25 | 2.0 | 15.5 | 0.05 |
| 25 | 1.1 | 13.9 | 0.029 |
| 25 | 2.0 | 15.0 | 0.14 |
| 25 | 3.0 | 14.8 | 0.22 |
| 15 | 2.0 | 15.8 | 0.11 |
| 32 | 2.0 | 14.2 | 0.16 |
| 40 | 2.0 | 12.7 | 0.18 |

*The perchlorate anion selected for compensation of the concentrational changes of the chloride ion has apperently the least tendency of entering the inner coordination sphere of palladium (II).

### 1.3. The kinetics of olefin oxidation by $PdCl_4^{2-}$

The study of the olefin oxidation under conditions when the palladium salt is consumed to form Pd metal and there is an increase in the concentration of both hydrogen and chloride ions is connected with a number of inconveniences. For this reason, in order to simplify the observed relationships as much as possible the first papers on the kinetics of this reaction[10-13] were devoted to the investigation of the oxidation of olefins by $PdCl_4^{2-}$ in the presence of p-benzoquinone as oxidizing agent of the reduced form of palladium. In contrast with such oxidants as cupric chloride, the addition of p-benzoquinone to the solution does not cause any changes in the concentration of the hydrogen and chloride ions. The concentration of the olefin at a given moment during the course of the reaction can be followed through the changes in the oxidation-reduction potential of the quinone-hydroquinone system. Under the reaction conditions (lithium chloride, lithium perchlorate and hydrochloric and perchloric acids at 10-40$^o$ C), neither quinone nor hydroquinone participates in any side reactions with palladium chloride, olefins or resulting carbonyl compounds. The change of the quinone-hydroquinone concentration is due only to the following over-all reaction:

$$C_nH_{2n}+H_2O+C_6H_4O_2 \xrightarrow{Pd^{II}} C_nH_{2n}O + C_6H_4(OH)_2$$

The quinone-hydroquinone potential is practically independent of the palladium chloride and olefin concentrations. It shows the absence of the complexation between the oxidant and the reactants.

Kinetics are somewhat less complicated when the $PdCl_4^{2-}$ concentrations do not exceed $2x10^{-2}$ M.

#### *1.3.1. The Reaction Stages at Low Concentration of Palladium Salt*

The reaction was found to strictly obey the first order with respect to the ethylene concentration[10-13]. The observed first order rate constant ($k_1$) is proportional to the concentration of $PdCl_4^{2-}$ and a reciprocal to the hydrogen ion concentration and to the square of the chloride ion concentration. The rate of the reaction decreases by 4.02±0.15 times on replacing water by the deuterium oxide.

On the basis of the aforementioned data the author, Vargaftik and Syrkin[13] proposed in 1963 a reaction mechanism which accounted for all available facts concerning this reaction (Eq.(7.1)-(7.6)):

1. $C_2H_4 + PdCl_4^{2-} \rightleftharpoons C_2H_3PdCl_3^- + Cl^-$

2. $C_2H_4\ PdCl_3^- + H_2O \rightleftharpoons C_2H_4PdCl_2OH_2 + Cl^-$

3. $C_2H_4PdCl_2OH_2 + H_2O \rightleftharpoons C_2H_4PdCl_2OH^- + H_3O^+$ (7)

4. $C_2H_4PdCl_2OH^- + H_2O \longrightarrow H_2O.PdCl_2\text{-}CH_2CH_2OH^-$

5. $H_2O.PdCl_2\text{-}CH_2CH_2OH^- + H_2O \longrightarrow Cl^- + PdCl_{(aq)}^- + H_3O^+ + CH_3CHO$

6. $PdCl_{(aq)}^- + C_6H_4O_2 + 2H_3O^+ + 3Cl^- \longrightarrow PdCl_4^{2-} + C_6H_4(OH)_2$

Stages 1 to 3 are reversible. Stage 4 in principle may be reversible as well. Reaction 5 is practically irreversible. The rate of olefin oxidation is independent of p-benzoquinone concentration. This proves its non-participation either in the rate-determining stage of the reaction or in the preceding stages. In the absence of p-benzoquinone the complexes of lower valence palladium $PdCl_{(aq)}^-$ or $PdCl_{2(aq)}^{2-}$ are decomposed to form Pd metal:

$$PdCl_{(aq)}^- \longrightarrow Pd_{(solid)} + Cl^- \quad (8)$$

Since in p-benzoquinone containing solution the metallic palladium phase is absent and the concentration of the $PdCl_{(aq)}^-$ type complexes is very low, it may be concluded that Reaction (7.6) is very rapid, its rate exceeding that of Reaction (8).

The kinetic equation corresponded to the mechanism under discussion is:

$$w = k_1 \frac{[PdCl_4^{2-}][C_2H_4]}{[H_3O^+][Cl^-]^2} \quad (9)$$

Almost a year after publication[13] of mechanistic scheme (7) and the kinetic data, Henry has shown that this equation, originally obtained for the reaction in the presence of p-benzoquinone, quite satisfactorily describes the kinetics of $PdCl_4^{2-}$ reaction with ethylene in the absence of an oxidant[14]. Despite the fact that under these conditions stage (8) takes place instead of stage (7.6), the same kinetic relations were observed and, as Henry himself noted, even the absolute values of the rate constants obtained in both studies are close to each other. The coincidence of our values with those of Henry lends support to the concept of the role of p-benzoquinone in this reaction.* Henry has also

* We do not discuss the kinetics of olefin oxidation in the presence of both palladium and copper salts[15-19] or Pd and vanadium oxides[20].

found that the oxidation rate of propylene, but-1-ene, cis- and trans-but-2-enes in oxidant-free solutions obeys the same law. Henry's conclusions as to the sequence and nature of the kinetic steps of the reaction of olefins with $PdCl_4^{2-}$ and his concepts on the mechanism of decomposition of the σ–bonded organopalladium compound are very similar to those presented in[13].

However, already at this early stage of our investigation there were indications that Eq. (9) does not quite satisfactorily describe the reaction kinetics. In particular, it was observed that the reaction rate grows faster with increasing concentration of palladium salt than could be expected on the basis of mechanism (7).

The study of this phenomenon led to the conclusion[21] that in the interval of $PdCl_4^{2-}$ concentrations from 0.02 to 0.2 M the kinetic data obey the Eq. (10):

$$w = k_I \frac{[PdCl_4^{2-}][C_nH_{2n}]}{[H_3O^+][Cl^-]^2} + k_{II} \frac{[PdCl_4^{2-}]^2[C_nH_{2n}]}{[H_3O^+][Cl^-]^3} \quad (10)$$

At $PdCl_4^{2-}$ concentration 0.02 M contribution of the first term is 75% and that of the second one is only 25%. For 0.2 M $PdCl_4^{2-}$ it is vice versa: 25 and 75%. Nevertheless in both cases (and also at intermediate concentrations of $PdCl_4^{2-}$) the $k_1$ - $1/[H_3O^+]$ plots pass through the origin. The same is also true for other olefins.

The validity of the Eq. (10) is exemplified by Fig.1

The presence of a second term leads to the conclusion, that reactions of olefins with $PdCl_4^{2-}$ proceed along two parallel routes. We believe that stages (7.1) and (7.2) and possibly (7.3) of scheme (7) are common to both routes. Further transformations of the π-complex $C_nH_{2n}PdCl_2OH^-$ can proceed either without participation of other ions (first route) or with participation of a palladium acido complex (second route). In the latter case the following reactions should proceed alongside with stages (7.4) and (7.5):

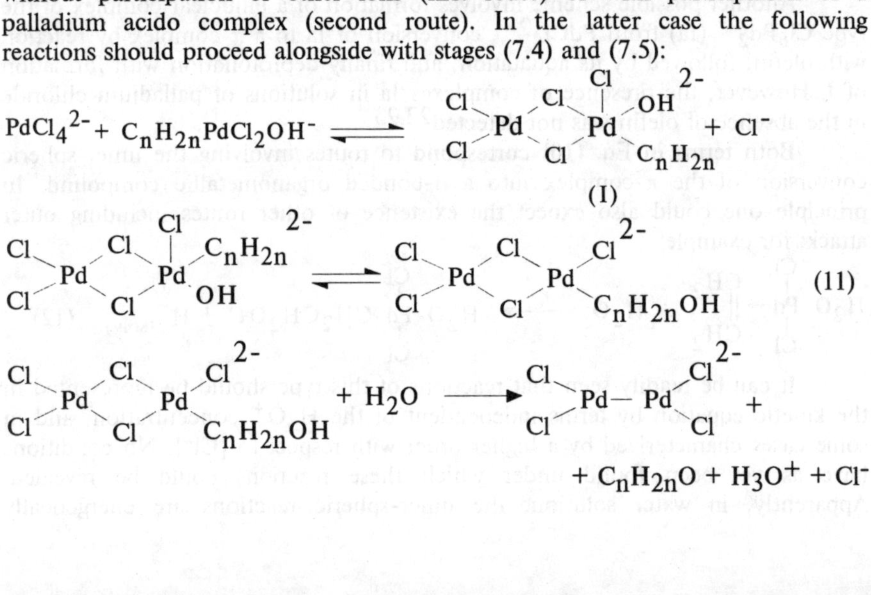

(11)

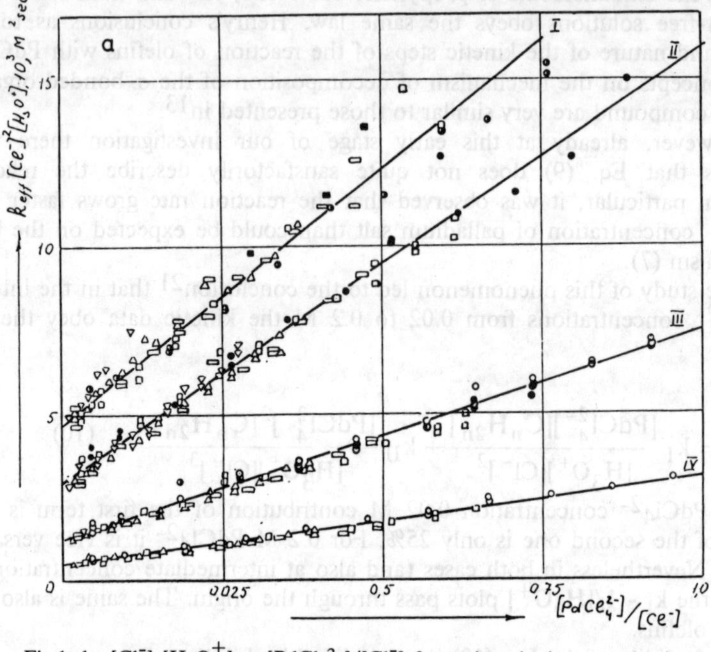


Fig.1. $k_{eff}[Cl^-]_2[H_3O^+]$ *vs* $[PdCl_4^{2-}]/[Cl^-]$ for various ionic strengths (μ=1.1 M(I), 2.0 M(II), 3.0 M (III), 4.0 M(IV)) for ethylene[10]. Concentration (M) of $Cl^-$ ions: o, 0.2; □, 0.3; ▭, 0.4; Δ, 0.5; ∇, 0.7;◊, 1.0; ×, 1.4. All unfilled circles correspond to $[H_3O^+]$=0.2 M; all black circles correspond to $[H_3O^+]$=0.5 M; ◐, $[H_3O^+]$=0.1 M, $[Cl^-]$=0.2 M; ◑, $[H_3O^+]$=0.8 M, $[Cl^-]$=0.2 M; ◓, $[H_3O^+]$=1.0 M, $[Cl^-]$=0.2 M; ◒, $[H_3O^+]$=1.5 M, $[Cl^-]$=0.2 M.

Another possible scheme involves formation of a binuclear complex of the type $Cl_6Pd_2^{2-}$ (Ia) from $PdCl_4^{2-}$, conversion of Ia to a π-complex by reaction with olefin, followed by its aquatation, and finally deprotonation with formation of I. However, the presence of complexes Ia in solutions of palladium chloride in the absence of olefin was not detected[22,23].

Both terms in Eq. (10) correspond to routes involving the inner-spheric conversion of the π-complex into a σ-bonded organometallic compound. In principle one could also expect the existence of other routes including outer attack; for example:

$$H_2O\text{-}\underset{Cl}{\overset{Cl}{Pd}}\text{—}\begin{matrix}CH_2\\ \| \\ CH_2\end{matrix} + H_2O \longrightarrow H_2O\text{-}\underset{Cl}{\overset{Cl}{Pd}}\text{-}CH_2CH_2OH + H^+_{(solv)} \qquad (12)$$

It can be readily seen that reactions of this type should be represented in the kinetic equation by terms independent of the $H_3O^+$ concentration, and in some cases characterized by a higher order with respect to $[Cl^-]$. No conditions have as yet been found under which these reactions could be revealed. Apparently, in water solution, the inner-spheric reactions are energetically somewhat more favorable and occur more easily than the outer-spheric ones. In amine media, a more basic solution, the amination of an coordinated alkene proceeds as trans-addition like the expected one by[24,25] the Eq. (12). If the first and second routes are suppressed by high chloride ion concentrations, in the presence of Cu(II) salts or nitrate ion, glycol esters are formed instead of acetaldehyde. This reaction supposedly involves intermediate complex formation between Pd(II) and the oxidant used, e.g. Cu(II) salt or $NO_3^-$ ion[26]. These reactions have been shown to involve trans-attack of the nucleophile on the coordinated alkene[27-30].

### 1.4. Mechanism of oxidative reductive decomposition of the π-complexes

The reactions of olefins with palladium (II) salts in water and in nonaqueous solvents have a number of common features. Both in water and in the nonaqueous solvents decomposition of the π-complex results in Pd metal and the product of the organic ligand oxidation. An easily detectable genetic relationship exists between the products of the reaction of ethylene in a variety of solvents (vinyl ester and ethylidene diacetate in acetic acid[5,31-33], acetal in alcohol[5,33], acetaldehyde in water[3,4,6]).

These facts lead to the conclusion that the carbonyl compound in water and the olefin oxidation products in nonaqueous media have precursors of similar structure. Of course, both the formation of these intermediates and their further transformations are not completely equivalent in different media. This section is devoted to certain aspects of the mechanisms of formation and decomposition of these hypothetical intermediates.

#### *1.4.1. σ-Bonded Organo-palladium Compounds in the Oxidation of Olefins*

The formation of acetaldehyde by hydrolytic decomposition of the anion of the Zeise's salt was once considered[34] as a support for the ethylidene structure of the organic ligand in this complex:

$$Cl_3Pt^{2-}\text{-}\overset{+}{C}H\text{-}CH_3 \longleftrightarrow Cl_3Pt^{-}{=}CH\text{-}CH_3$$

It is not clear, however, why only the oxocomplex can undergo such reaction and what is the function of the bonded hydroxyl. Moreover, if conversion of the π- to the carbene complex is an equilibrium reaction as it was proposed by Aguilo[15], then palladium (II) catalyzed isomerization of the olefin should take place. Actually the isomerization is catalyzed by complexes of low valence palladium. No isomerization was observed in $PdCl_2$ solutions containing oxidants, where palladium (II) π-complexes including $C_nH_{2n}PdCl_2OH^-$ are formed, and Pd(I) and Pd(O) are practically absent[35-36]. Henry has found $C_2D_4$ and $C_2H_4$ to react with equal rates[14]. Hence, if the above discussed mechanism is considered to be valid, it must be assumed that π-complex to carbene-complex conversion is the slowest stage of the over-all reaction and, moreover, proceeds without an isotopic effect. At present there is no actual evidence confirming this mechanism.

The accumulated experimental data are probably in the best accordance with the proposal[5,13], that the reaction proceeds via the formation and decomposition of a σ-bonded organo-palladium compound (σ-carbocomplex) of $X_3PdCH_2CH_2OH$ type. It was found that the σ-carbocomplexes $XPdCH_2CH_2OR$ formed by exchange of $PdCl_2$ with mercury derivatives X-$HgCH_2CH_2OR$ (R=H, $CH_3$, Ac, X=Cl, Br) *in situ*[10,37] are decomposed to the products identical to those formed in the oxidation of ethylene in the corresponding hydroxyl-containing solvent; for example, in the reaction with β-chloromercuroethanol:

$$PdCl_2 + Cl\text{-}HgCH_2CH_2OH \xrightarrow{-HgCl_2} Cl\text{-}PdCH_2CH_2OH \longrightarrow$$
$$\longrightarrow Pd + CH_3CHO + HCl$$

The hypothesis that reduction of palladium (II) π-complexes involves the intermediate formation of σ-bonded organopalladium compounds, first regarded rather skeptically by many authors, now has been widely accepted[4,38,39,40]. At the same time opinions as to the mechanism of the formation and decomposition of these compounds still differ.

*1.4.2. Rearrangement of π-Complexes to σ-Carbocomplexes*

The kinetic data gave information on the composition of the activated complex in the σ- to π-complex transition (not taking into account the number of solvent molecules). Thus, in water the reaction proceeds mainly *via* the complexes

$C_nH_{2n}PdCl_2OH^-$ and 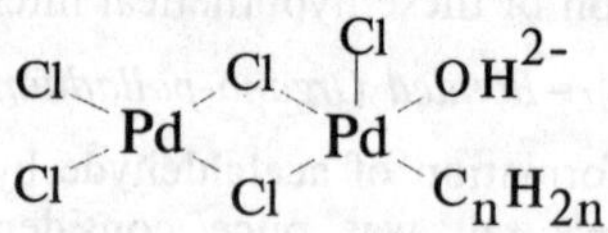

The data on the equilibrium of complex formation and kinetic studies of ethylene oxidation by palladium acetate in acetic acid led to the conclusion that in this medium the reaction proceeds by inner-spheric π-σ-transformation of the complex $C_2H_4Pd(OAc)_2HOAc$. The reaction rate depends only on the concentration of this complex and does not depend on the concentrations of free AcONa.

Oxymercuration, the most completely studied reaction of oxymetallation, with few exceptions, proceeds via trans-addition[41,38]. On the contrary, oxypalladation in the cases when this reaction is inner-spheric should typify cis-addition. In analyzing the reactions for such differences as polarities and strength of the metal-oxygen and metal-carbon bonds in the Pd(II) and Hg(II) complexes, nucleophilic activities of the migrating ligands in these complexes, etc., should be taken into account. The absence of pertinent data precludes any detailed discussion. However, two facts should be mentioned which perhaps could shed some light on this problem.

In contrast with comparatively stable palladium complexes, π-complexes of mercury are unstable because of the lower ability of Hg(II) to back-donation.

These labile compounds evidently very readily undergo transformation to σ-complexes owing to both thermodynamic and kinetic factors. If formation of Hg-C and C-O bonds occurs simultaneously when a mercury ion approaches the olefin molecule, one should expect a predominantly trans-attack of the solvent.

Isomerization of the π- to σ-complex is considered in references[14,42,43] to be a monomolecular process in which only cis-ligands participate:

$$cis\text{-}(C_nH_{2n}PdCl_2OH^-) \longrightarrow (Cl_2Pd\text{-}C_nH_{2n}OH)^- \qquad (13)$$

In reaction (13) the C=C π-bond is replaced by a stronger C-O bond but the Pd atom now has only one ligand instead of two. Owing to this, reaction (13) should have a highly negative heat effect absolute value being probably several times larger than the activation energy of this elementary act*. Obviously in order to compensate the energy loss associated with the loss of one ligand some bonds should arise in the resultant complex which are not shown in the structure of the σ-carbocomplex in Eq. (13); for example, "hydrogen bond" (II) or Pd-O bond (III):

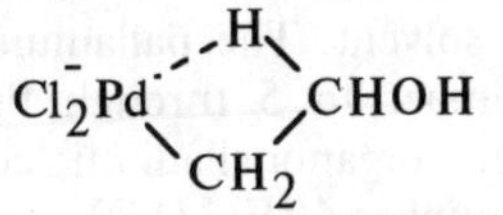

II

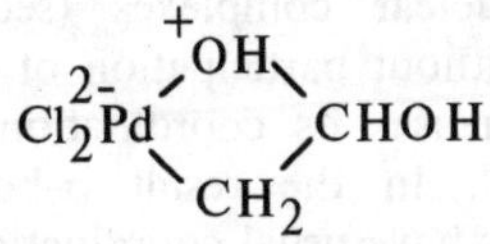

III

At present no data are available either to support or invalidate these assumption.

However, one cannot exclude the possibility that π-σ-isomerization is not monomolecular but proceeds with participation of solvent molecules which enter the inner coordination sphere of palladium and occupy the fourth coordination site of the organopalladium compound. For the reaction in water the following scheme appears to be plausible.

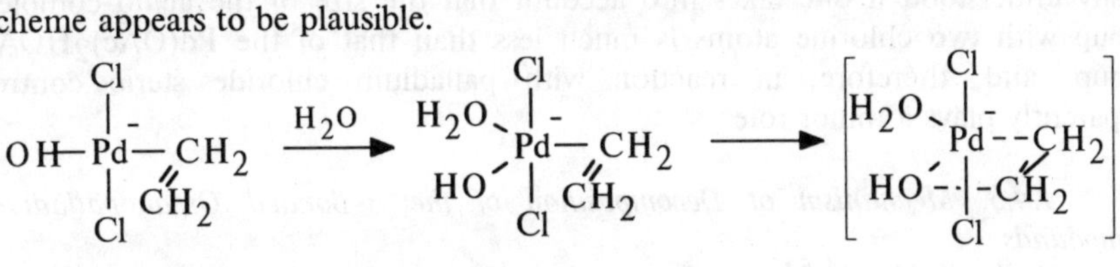

IV IVa

$$\longrightarrow (H_2O)Cl_2Pd^-CH_2CH_2OH$$

---

* The heat effect of reaction (13) was estimated to be about 80-100 kcal/mol. The activation enthalpy of ethylene oxidation in water (for the first rout) is 16-19 kcal/mol[10,13]; in acetic acid 17 kcal/mol[44]. The activation enthalpy of the elementary act under discussion probably does not differ from the observed values by more than 5-10 kcal/mol.

One may assume that formation of weak five-coordinated complexes, even if it is accompanied by certain perhaps not very large energy losses is more advantageous than a process like reaction (13). Even in the case of six-coordinated octahedral complexes, the substitution and insertion reactions often proceed *via* intermediate seven-coordinated complexes.

Ninomija et al.[45] found the rate of palladium chloride reduction by ethylene in acetic acid-p-xylene mixture to be proportional to the square of the acetic acid concentration. This fact supports the idea of solvent participation in the insertion reaction.

The olefin and the trans-ligand are much closer to each other within the five-coordinated complex than in the starting square complex. We believe this to be the reason why ethylene can react with both *cis*- and *trans*-ligands of the starting complex, especially if the olefin in the complex IV can rotate around the axis passing through the metal atom and perpendicular to the C=C bond as in the case of rhodium and platinum complexes[46,47]. There are apparently no grounds for assuming that in the square complexes the inner-spheric reaction can occur only between cis-ligands.

For binuclear complexes (second route) the $\pi$-$\sigma$-isomerization can proceed also without participation of the solvent. The palladium atom bonded with olefin increases its coordination number to 5 through the two bridged chlorine atoms*. In the result $\sigma$-bonded organopalladium compound both palladium atoms have usual coordination number 4 (Eq. (11)).

The greater size of the $H_2OCl_2Pd$ group in comparison with that of the hydroxyl favors Markovnikov's rule addition in the oxydation of $\alpha$-olefins in water[4]. Palladium acetate in acetic acid reacts with olefins also mainly by Markovnikov's rule[48]. Here also the size of the inserting acido-complex group is greater than that of the acetoxy group. However, the olefin oxidation by palladium chloride in acetic acid is considerably less selective[49,50]. This can be easily understood if one takes into account that the size of the acido-complex group with two chlorine atoms is much less than that of the $Pd(OAc)_2HOAc$ group, and, therefore, in reaction with palladium chloride steric control apparently plays a minor role.

### *1.4.3. Mechanism of Decomposition of the $\sigma$-Bonded Organopalladium compounds*

Krekeler found[4,51] that the reaction of Karasch's complex with deuterium oxide gives rise to non-deuterated acetic aldehyde.

$$(C_2H_4PdCl_2)_2 + 2D_2O = 2CH_3CHO + 2Pd + 4DCl$$

---

* The interaction between palladium atoms in the binuclear complex via bridged chlorines is likely to increase the acceptor ability of the $\pi$-complexed palladium atom and, therefore, facititate the $\pi$-$\sigma$-isomerization.

Oxidation of ethylene in $CH_3OD$ results in dimethylacetal practically free of deuterium[31,37]:

$$C_2H_4 + PdCl_2 + 2CH_3OD = CH_3CH(OCH_3)_2 + 2DCl + Pd$$

Oxidation of ethylene in deuteroacetic acid $CH_3COOD$ leads to ethylidene diacetate practically free of deuterium:

$$C_2H_4 + PdCl_2 + 2CH_3COONa \xrightleftharpoons{CH_3COOD} CH_3CH(OCOCH_3)_2 + 2NaCl + Pd$$

These facts seem to indicate that the 1,2-hydride shift occurs simultaneously with departure of the acido-complex group from the organopalladium molecule.

Chatt and Shaw[52,53,54] have shown that even in aryl platinum derivatives such as trans-$(R_3P)_2PtXAr$ the positive end of the Pd-C dipole is on the carbon atom. The more grounds there are to expect that this is also the case for the alkyl derivatives of palladium is discussed. The author, Vargaftik and Syrkin[13,31,37] assumed that the positive charge on the β-carbon of β-palladium substituted ethanol increases in the transition state of decomposition. The "carbonium ion" nature of the β-carbon enhances on removal of the acido-complex group and may be the cause of the 1,2-hydride shift. The carbonium ion is probably not a kinetic entity in the oxidation of olefins in water because the hydroxyl proton can split off simultaneously with elimination of palladium and the hydride shift:

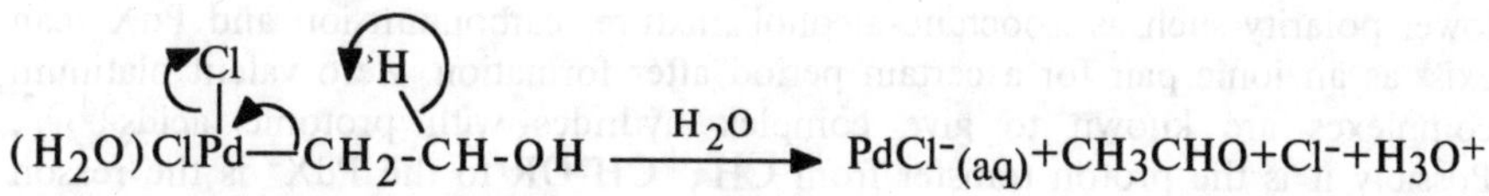


In the light of the above mechanism the absence of glycols among the products of the reaction of olefins and palladium salts in aqueous solution can be regarded as indication that heterolysis of the Pd-C bond and the hydride shift proceed faster than reaction of the organopalladium compound with the solvent molecule, resulting in its alkylation. It is possible that the 1,2-hydride shift in this case occurs not only in the field of two carbon atoms but also under influence of the palladium atom as Henry postulated[14].

In the reaction of olefins in alcohol or acetic acid, elimination of $R^+$, when R is $CH_3$, $C_2H_5$ or Ac in the course of decomposition of the organopalladium compound is hindered because of the lesser tendency of alkyls and acyls than hydrogen to exist in the form of positively charged ions. Perhaps in these solvents one could isolate the kinetic stage of carbonium ion formation:

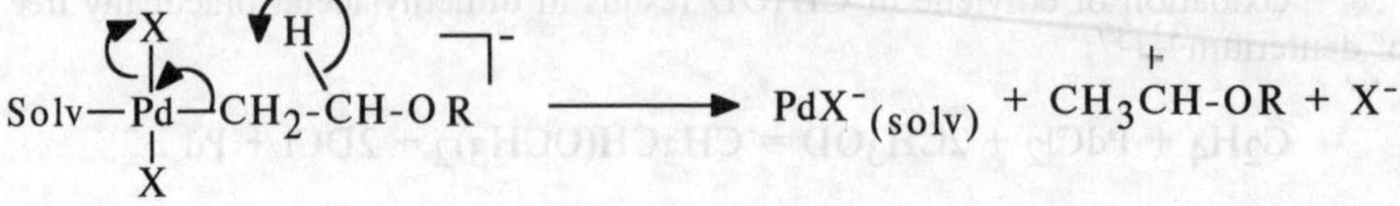


Reactions of carbonium ions with solvent molecules (O-alkylation) should lead to ketals and acetals in alcohols and to alkylidene diacetates in acetic acid. Vinyl and allyl ethers were not formed in alcohol. The yield of vinyl ether did not exceed 0.3% even in the reaction of sodium alcoholate in ethanol with Karasch's complex $(C_2H_4PdCl_2)_2$, although under these conditions vinyl ether is sufficiently stable and is not converted into acetal.

At the same time Stern and Spector[33] found that vinyl ethers are formed in significant amounts if ethylene and alcohol react with $PdCl_2$ in the presence of a diluent (isooctane). The yield of vinyl derivatives is much higher in acetic acid than in alcohol[5]. The amount of alkylidenediacetate was shown by Stern[32,33] to be very small when $PdCl_2$, $C_2H_4$, and $CH_3COO^-$ reacted in saturated hydrocarbon medium. The above said suggests that the yield of vinyl derivatives is among other factors greatly dependent on the solvating power of the medium and grows with decrease in ability of the latter for solvation of ions formed in decomposition of σ-carbo-complex.

It seems to be possible that in alcohol the carbonium ion and the acido-complex of zero valent palladium, $PdX^-$, are solvated and separated immediately on formation, and then react independently of each other. In a solvent with lower polarity such as isooctane-alcohol mixture, carbonium ion and $PdX^-$ can exist as an ionic pair for a certain period after formation. Zero valent platinum complexes are known to give complex hydrides with protonic acids[54,55]. Possibly it is the proton transfer from $CH_3{}^+CH\text{-}OR$ to the $PdX^-$ is the reason for formation of vinyl ether and palladium hydride. A comparatively high energy of the Pd-H bond plays perhaps as essential role in formation of H-PdX.

The reaction of palladium salts with olefins has certain common features with the oxidation of olefins by thallium (III), lead (IV) and mercury (II) salts[56,57,58]. All these reactions proceed via addition of metal atom and lyate ion to the olefin double bond and decomposition of the adduct with formation of the oxidation products and the reduced form of metal. There seem to be little grounds for assumption that hydrides are formed on any stage of thallium (III) and lead (IV) salts reactions. On the other hand, Crigee[57] and Kabbe[58] represented a strong evidence that the reaction of these salts leads to formation of carbonium ions. The stability of mercury and thallium σ-bonded organometallic adducts is considerably higher than that of palladium σ-complexes with the same acido-ligands. Correspondingly changes also may take place in the rates of formation and decomposition of organo metallic compounds. In the case of the Tl(III) and Hg(II) glycol formation becomes an important reaction path[56-58].

## 2. Catalysis with giant clusters

### 2.1. Introduction

Besides the Wacker oxidation, syntheses of vinyl acetate and allyl acetate from ethylene and propylene correspondingly and acetic acid are of industrial importance. The oxidative acetoxylation of olefins was firstly reported as homogeneous oxidations by using Pd(II) salts and sodium acetate as the catalysts[5]. Later on, however, industrial processes with supported Pd-metal catalysts have been developed. The addition of an alkaline acetate in acetic acid solution to the charging feed is a necessary condition for the reaction to proceed[59-62]. Pd black also has been found to be active in this reaction under mild conditions (310-360 K, 1 atm) with the reactants in the liquid phase, provided that sodium acetate was brought into the reaction solution before reducing Pd(II) to Pd black, but not afterwards[63]. Catalytic activity seemed to be provided by palladium atoms in some low oxidation state, rather than by palladium metal or Pd(II)[63,64].

Catalytically active solutions were found to be formed via reduction of $Pd(OAc)_2$, in the presence of the ligands L (L=1,10-phenanthroline, 3,3"-dicarboxy-2,2"-bipyridine, 2,2"-biquinoline, etc.), with various reductants, the most convenient being $H_2$[63,64].Catalytic activity for oxidation of olefins, as a function of total content of a "mild" base L, exhibited an extreme patter.

In order to elucidate the nature of these compounds, the stoichiometry of the reactions leading to their formation, as well as their compositions and structures, were studied.

The primary isolated product of palladium (II) acetate reduction with $H_2$ in AcOH solution (containing 0.5 mol of phen per Pd atom) was an X-ray-amorphous substance with $Pd_4phen(OCOCH_3)_2$ as the simplest compositional formula, as ascertained by elemental analysis[65]. However, as was found by volumetric measurement, $H_2$ consumption was 1.3 ± 0.05 mol per Pd (II) atom, corresponding to the stoicheiometric equation:

$$4Pd_3(OCOCH_3)_6 + 3phen + 15H_2 = 3Pd_4phen(OCOCH_3)_2H_4 + 18CH_3COOH$$

The suggestion that the complex **1** is a hydride is supported by $^1H$ NMR spectra[65,66].

The metal core size for **1** molecules found from TEM and SAXS data led to conclusion that the value of *n* in the idealized $[Pd_4phen(OAc)_2H_4]_n$ formula as approximately 100, supposing the packing density of Pd atoms in **1** core to be nearly the same as that for bulk palladium metal.

### 2.2. Palladium-561 cluster (2)

When reacted with $O_2$, compound **1** lost hydride atoms to form $H_2O$. In this process only a small portion of the palladium atoms was oxidized to Pd(II), while the majority of substance **1** was transformed into a polynuclear compound **2** with $Pd_9phen(OAc)_3$ as the simplest formula, according to the elemental analysis data. The substance **2** is stable in air and soluble in water and polar organic solvents[65,66].

The molecular mass of **2** was estimated as $(1.0 \pm 0.5)x10^5$ from data on sedimentation rates for **2** in aqueous solutions obtained by ultracentrifuging.

More accurate data on the size of the molecules of **2** were obtained by SAXS, TEM and electron diffraction[65,66].

In TEM micrographs metallic skeletons of **2** molecules were observed as nearly spherical particles 26 ± 3.5 A in diameter. In the electron diffractogram of the same sample **2**, there were several diffuse rings with arrangement of maxima close to those for the metallic palladium (see Table 2).

Table 2. Interplanar distances in the metal skeleton of **2** according to ED data

| Substance | Interplanar distances (A) | | | | |
|---|---|---|---|---|---|
| **2** | 2.26 | 1.95 | 1.39 | 1.17 | 0.89 |
| Pd metal | 2.23 | 1.94 | 1.37 | 1.17 | 0.89 |

The nearly identical patterns of the interplanar distances found for **2** and for Pd metal might indicate the coincidence of the main symmetry features for the **2** metal skeleton with those of the lattice for bulky Pd metal. However this suggestion contradicts the EXAFS data, which definitely indicate icosahedral packing of Pd atoms in **2** (see below). In this situation it seems more reasonable to assume that the similarity in the ED patterns of **2** and metallic Pd results from the destruction of **2** molecules to produce particles of metallic Pd under the influence of the electron beam. A destructive influence of the electron beam upon cluster molecules exposed in the electron microscope is indeed known to occur under the conditions of TEM and ED experiments. For example, the loss of cluster ligands, agglomeration and beam damage during TEM studies have been observed for $Au_{55}(PPh_3)_{12}Cl_6$[67,68] and $[Ni_{38}Pt_6(CO)_{48}H](NMe_3CH_2Ph)_5$[69] clusters at high electron beam intensities and long experiment times. In TEM experiments performed at low beam intensities, neither agglomeration nor beam damage of the shape of palladium clusters was found upon variation of the exposure time from 10 s to 10 min[65]. However, total or partial loss of the cluster ligands cannot be excluded, even at the low beam intensities used. Ligand loss may result in a relaxation of the initial icosahedral metal skeleton into a f.c.c. one (*e.g.* a cubooctahedral).

In the absence of the agglomeration of palladium particles, relaxation of the metal skeleton may be expected not to affect noticeably the size of the metal particles under investigation. Under this assumption, the electron diffractograms were used for an independent estimation of the size of the metal core of **2** molecules. From the half-width of diffraction rings, the size of the particles responsible for the diffraction pattern was found to be *ca.* 25 A, in agreement with TEM and SAXS data.

The consistency of the data on metal skeleton size obtained with various techniques (Table 3) seems to confirm the lack of considerable destruction of the metal skeleton or agglomeration of Pd clusters in the TEM and ED experiments[65]. The size distribution of **2** cores was found to be monomodal.

The packing of Pd atoms in **2** was elucidated with EXAFS spectroscopy[65,70]. The four intense peaks corresponding to the four shortest Pd-Pd distances have been found (Table 4). The set of Pd-Pd distances obtained is seen to be consistent with the icosahedral packing of Pd atoms in the metal skeleton of the cluster **2** (the interatomic distance ratios expected for the four nearest neighbor atoms of the icosahedral skeleton are 1:1.2:1.4:1.6) and to deviate notably from the patterns of distances expected for f.c.c. and h.c.p. packing.

Table 3. Metal skeleton size (d) of **2** molecules as evaluated by various techniques[65]

| Method | d (A) |
|---|---|
| TEM | 26 ± 3.5 |
| ED | ~25 |
| SAXS | 20 ± 5 |

TABLE 4. The four shortest Pd-Pd distances in the metal skeleton of **2** found from EXAFS data, compared with those distances expected for different packings of the Pd atoms

| Data type | Pd-Pd distances, A | | | | |
|---|---|---|---|---|---|
| EXAFS | 2.60±0.04 | 3.1±0.1 | 3.66±0.1 | 4.08±0.1 | - |
| packing[a] | | | | | |
| f.c.c. | 2.60 | - | 3.66 | - | - |
| h.c.p. | 2.60 | - | 3.66 | - | 4.50 |
| icosahedron | 2.60 | 3.10 | 3.66 | 4.10 | - |

[a]In calculations, the shortest Pd-Pd distance was taken to be 2.60 A for all packings; f.c.c. = face-centered cubic packing; h.c.p. = hexagonal close packing.

The mean atomic volume of Pd in the icosahedral metal skeleton of **2** calculated from the interatomic distances of Table 4 is *ca.* 16 $A^3$, *i.e.* only slightly exceeding that of Pd metal (14.7 $A^3$). This result excludes the possibility of coordination of the cluster ligands phen and OAc by the inner Pd atoms. Therefore, the phen and OAc groups must be situated at the periphery of the metal core of cluster **2**.

With the known character of packing of Pd atoms and the distances between the nearest neighboring Pd atoms in the **2** core, the total number $N_{\Sigma}$ of palladium atoms in cluster **2** was estimated . As was found for a sphere *ca* 25 A in diameter, the number, $N_{\Sigma}$, is approximately equal to 570. On the basis of this value for $N_{\Sigma}$, and of the chemical composition of **2**, suggested by the elemental analysis, the molecules of **2** was approximated by the formula $Pd_{570 \pm 30}phen_{63 \pm 3}(OAc)_{190 \pm 10}$. The molecular mass corresponding to this formula ($M$ = 83200) agrees with the result of direct determination, by the rates of sedimentation in solution, of $M = (1 \pm 0.5)x10^5$.

The value found, $N_{\Sigma}$=570, matches quite well the idealized 5-layer icosahedron containing, according to the formula[71] $N_{\Sigma} = 1/3(10m^3 + 15m^2 + 11m + 3)$, $m$ (the number of layers) = 5, $N_{\Sigma}$ = 561 metal atoms. Taking into consideration the chemical analysis, the overall $Pd_{561}phen_{60}(OCOCH_3)_{180}$ idealized formula was suggested for the icosahedral cluster **2** (Fig.2).

The idealized formula $Pd_{561}phen_{60}(OCOCH_3)_{180}$ seems to correspond to some average size and composition of this cluster, rather than to a certain fixed size and composition. In other words, the existence of distributions in the size and composition of various particles of **2**, around the average value of d ≈ 25 A and the idealized composition $Pd_{561}phen_{60}(OCOCH_3)_{180}$ should be anticipated.

HREM studies[72] of clusters, approximate formula $Pd_{561}L_{60}(OAc)_{180}$ (L=Dipy, Phen), have confirmed the conclusions[65,66] concerning with monomodal size distribution of the cluster cores. Besides Pd species with f.c.c. metal core, larger Pd particles (8 nm) exhibiting multiple twinning and evidencing for a distorted icosahedral structure were observed.

The observed by TEM metal cores of the particles of **2** are indeed characterized by a distribution over the values of their diameters, rather than by a single value of the diameter. This suggests that, in fact, various particles of the isolated substance **2** contain different numbers of palladium atoms in their cores. The situation here seems to resemble to some extent that for organic polymers, in which various molecules contain different numbers of monomer units. Taking into consideration that the average energy of Pd-Pd bonds is rather small (40-60 kJ $mol^{-1}$), the icosahedral metal clusters with d ≈ 25 A can be expected to have some defects ("caps", "nichas" , *etc.*) which perhaps are not detected by TEM or HREM and other techniques used to characterize the cluster.

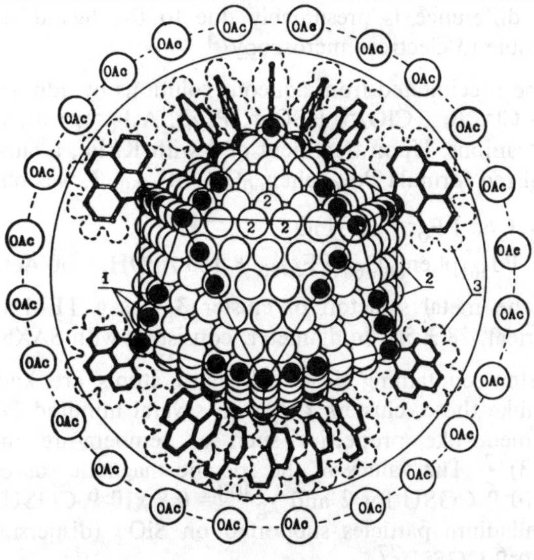


Fig. 2. Idealized model of cluster 2: 1 = Pd atoms coordinated with Phen ligands; 2 = Pd atoms accessible for coordination with $OAc^-$ anions or molecules of substrates or solvent; 3 = van der Waals' shapes of coordinated phen molecules.

Examination of the molecular models shows that bidentately coordinated phen ligands, because of the steric hindrances created by H atoms at their 2 and 2" positions, may be coordinated only at the edges and vertices of the icosahedron. At the outer layer of the idealized icosahedron which contains 252 metal atoms, in fact, *ca* 60 bidentately coordinated phen ligands may be arranged. Note also that the palladium core of **2** with icosahedral packing has a formal total charge of about +180, balanced by some 180 anion ligands.

With this arrangement, almost the whole surface of the metal skeleton is sterically screened by bulky phen ligands. Acetate anions may be located only in the outer sphere of cluster **2**.

Note that a similar outer sphere arrangement of acetate anions was found also by X-ray structural analysis for the tetranuclear cationic cluster $[Pd_4phen_4(CO)_2](OAc)_4$ with a tetrahedral metal skeleton[73].

The conclusion concerning the outer sphere coordination of $CH_3COO$ ligands was confirmed by (I) IR spectra (the frequency difference $\nu_{as}$ (OCO) - $\nu_s$(OCO) = 165 $cm^{-1}$, *i.e.* as in $CH_3COONa$), by (2) data on electroconductivity of aqueous solutions of **2**, and by (3) NMR spectra of **2**, for which the line from the protons of $OAc^-$ groups was observed as the usual narrow singlet with $\delta$ = 2.0 ppm, in contrast to the multiplet signal from the protons of phen that was noticeably broadened; (4) STM observation showed giant palladium-561 clusters to be larger in size than observed by TEM and

HREM studies. The difference is presumably due to the ligand shell of the clusters, which is invisible to electron microscopy[74].

Cluster **2** can be precipitated from aqueous solutions by adding salts such as NaX or KX ($X^-$ = $Cl^-$, $Br^-$, $ClO_4^-$, $HSO_4^-$, $PF_6^-$, $AsF_6^-$), with substitution of $OAc^-$ anions by $X^-$ anions. Upon treatment of **2** with $KPF_6$, a cluster soluble in $CH_3CN$, with idealized formula $Pd_{561}phen_{60}O_{60}(PF_6)_{60}$, **3**, was obtained[75]

$$Pd_{561}phen_{60}(OAc)_{180} + 60\ PF_6^- + 60\ H_2O \longrightarrow$$
$$\longrightarrow Pd_{561}phen_{60}O_{60}(PF_6)_{60} + 120\ AcOH + 60\ AcO^- \quad (14)$$

The shape of the metal skeleton of cluster **3**, as a TEM micrograph shows, is almost spherical, 28 ± 5 A in diameter, consistent with SAXS data.

Molecular clusters containing 4-38 palladium atoms are known to be diamagnetic[73,76]. Unlike these, clusters **1-3**, with several hundred Pd atoms in their cores, reveal metal-like properties such as temperature independent paramagnetism (Fig. 3)[77]. The values of the specific magnetic susceptibility at 300 K ($\chi_g^{300}$= 1.0 x$10^{-6}$ CGSU for **2** and $\chi_g^{300}$= 0.8 x$10^{-6}$ CGSU for **3**) are close to those for palladium particles supported on $SiO_2$ (dispersion 0.2-0.5, $\chi_g^{300}$= (0.8 ± 0.2) x$10^{-6}$ CGSU)[77].

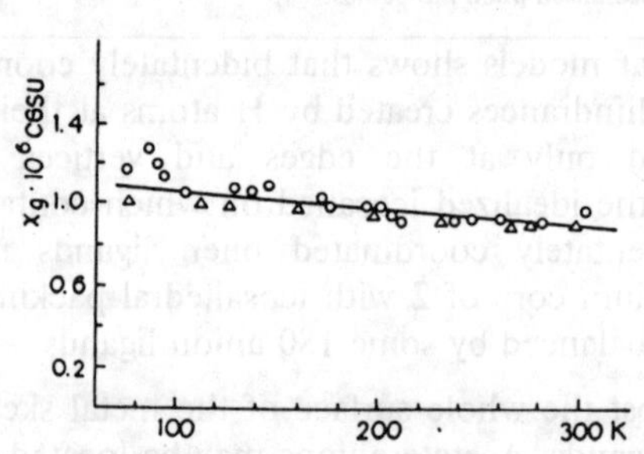


**Fig. 3. Specific magnetic susceptibility, $\chi_g$, of clusters 2 (○) and 3 (△) as a function of temperature.**

An ordinary Fourier transform (FT) of the EXAFS spectrum of **3** resulted in a RDA curve $\rho(R)$ with two maxima for Pd-Pd distances. The ratio of the two Pd-Pd distances, 1.4, tentatively suggests the f.c.c. packing of the Pd atoms in **3**[70].

However, the peak at 2.66 A for the shortest Pd-Pd distance of the $\rho(R)$ curve is essentially broader than two other significant peaks (3.70 A for Pd-Pd and 2.14 A for Pd-light atom distances). This suggests the presence of the some unresolved "fine structure" of the RDA curve. By using a statistical regularization (SR) method[78] for the analysis of the EXAFS spectrum a set of interatomic distances was obtained from the g(R) function for cluster **3**, which was interpreted in terms of a model, assuming that the product of the reaction **2** with $KPF_6$ is similar to the initial cluster **2**. Within this model, the EXAFS data should be treated as evidence of a more complicated arrangement of Pd atoms in **3** than in **2**.

The ligand substitution process (see Eq.14) is assumed to include other anions besides $PF_6^-$. Perhaps the reaction is a hydrolysis resulting in the appearance of the anions $OH^-$ and $O^{2-}$ at the cluster surface, accompanying the lidand substitution in aqueous solutions at pH 5- 6. Unlike large $PF_6^-$ and $OAc^-$ anions, the small $OH^-$ and $O^{2-}$ anions can perhaps be bound directly to Pd atoms, rather than being located in the outer sphere of the cluster.

The lack of an absorption in the 3600 $cm^{-1}$ region (which is characteristic of OH groups in the IR spectrum of **3**) makes the $O^{2-}$ anion a more likely hypothetical additional ligand.

The oxygen ligands may also be inserted between two outer (*i.e.* the 4th and 5th) icosahedral layers of Pd atoms. Both coordination of O atoms at the surface of the skeleton of **3** and their insertion between the outer layers of the skeleton are expected to lengthen the Pd-Pd distances in the outer layer. This can explain the appearance in the g(*R*) function of cluster **3** of additional 4.45 A distance that does not fit the idealized icosahedral structure.

Thus, cluster **3** is inferred to be icosahedral, with the inner layers of its core being packed on the same way as those in cluster **2** (Pd-Pd distances are 2.55, 3.05, 3.60 and 4.05 A), and outer- layer Pd atoms (*ca.* 50% of total Pd) located at perturbed distances.

The giant clusters obtained eventually serve as a bridge between ordinary molecular clusters and colloidal metals. The sizes of the metal skeletons of the giant clusters exceed notably those of large molecular clusters such as $Pd_{38}(CO)_{28}PEt_3)_{12}$ [76] and $Au_{55}(PPh_3)_{12}Cl_6$ [79], being close to lower sizes of colloidal metal particles. Unlike the latter, the giant clusters have a distinct ligand environment with a definite stoichiometry inherent in molecular clusters. However, a set of more or less imperfect metal polyhedra with a certain size distribution, rather than a single perfect polyhedron as in molecular clusters, seems to arise when the number of metal atoms amounts to several hundred. Nevertheless, the idealized formulae based on the data mentioned above are useful as models characterizing the average size and composition of the giant clusters and providing an understanding of their catalytic properties.

## 2.3. Catalytic activity of Pd-561 clusters

### *2.3.1. Acetoxylation reactions*

In solutions of clusters **2** and **3** containing acetic acid, oxidative acetoxylation of olefins and alkylarenes occurs: ethylene is converted into vinyl acetate:

$$C_2H_4 + 1/2\ O_2 + AcOH \longrightarrow CH_2{=}CHOAc + H_2O \qquad (15)$$

propylene into allyl acetate:

$$C_3H_6 + 1/2\ O_2 + AcOH \longrightarrow CH_2{=}CHCH_2OAc + H_2O \qquad (16)$$

and toluene into benzyl acetate:

$$PhCH_3 + 1/2\ O_2 + AcOH \longrightarrow PhCH_2OAc + H_2O \qquad (17)$$

The reactions are not sensitive towards water presence. Even in aqueous (10%) AcOH solution the selectivity of reactions (15) - (17) towards the products of oxidative acetoxylation is 95-98 %. The only side reaction observed with these catalysts is subsequent oxidation of alkenyl and benzyl esters, when they are accumulated, to form ethylidene- and benzylidene diacetates, respectively. In comparison with metallic Pd catalysts, which show their activity at higher temperatures[59-62], clusters **2** and **3** promote the side reactions to a lesser extent.

### *2.3.2. Oxidation of olefins in aqueous solution.*

Ethylene is converted into acetaldehyde by reacting with $O_2$ in water solution containing Pd-561 clusters[65]. Unlike Pd(II) oxidations, no reaction between ethylene and Pd-clusters was observed under anaerobic atmosphere. Moreover, the cluster catalyzed oxidation of propylene gives rise to allylic products (allyl alcohol, acrolein, acrylic acid) mostly instead of acetone[81].

$$CH_2{=}CH\text{-}CH_3 + O_2 \xrightarrow{\text{Pd-561}} \begin{array}{ll} CH_2{=}CH\text{-}CH_2OH & 14\% \\ CH_2{=}CH\text{-}CHO & 2\% \\ CH_2{=}CH\text{-}COOH & 60\% \end{array}$$

The yield of acetone, which is the main product (95% yield[5,51]) of propylene oxidation with Pd(II) salts in aqueous solution, does not exceed 5% under cluster catalysis.

As another distinct from the Pd(II) oxidation, propylene oxidation catalyzed with Pd-clusters was found to be accelerated by addition of 0.3-2 M $H_2SO_4$[81].

Ethylene has been converted into acetic acid in the presence of Pd-561 clusters in aqueous $H_2SO_4$ solution:

$$C_2H_4 + O_2 \longrightarrow CH_3COOH$$

*2.3.3. Oxidation of alcohols*

In the presence of clusters **2** and **3**, normal aliphatic alcohols containing 4-6% $H_2O$ are readily oxidized by dioxygen to form aldehydes, acetals and esters having the same carbon skeleton in both acid and alcohol components as the starting alcohol.[80]

$$RCH_2OH + O_2 \longrightarrow RCOOCH_2R + 2\,H_2O$$
$$RCHO$$
$$RCH(OR)_2$$

$$n\text{-}C_3H_7OH + O_2 \longrightarrow C_2H_5COOC_3H_7 \quad 7\%$$
$$C_2H_5CHO \quad 77\%$$
$$C_2H_5CH(OC_3H_7)_2 \quad 16\%$$

Secondary alcohols are smoothly oxidized into ketones:

$$(CH_3)_2CHOH + 1/2\,O_2 \longrightarrow CH_3COCH_3 + H_2O \quad 100\%$$

Oxidation of ethanol containing no more than 0.2% $H_2O$ in the presence of cluster **2** gives rise to acetic acid and acetic anhydride[82]:

$$C_2H_5OH + O_2 \longrightarrow CH_3COOH$$
$$(CH_3CO)_2O$$

*2.3.4. Acetal formation.*

In the absence of $O_2$, the aldehyde added to the alcohol suspension of cluster **3** is converted into acetal:

$$CH_3CHO + 2\,C_2H_5OH \longrightarrow CH_3CH(OC_2H_5)_2 + H_2O$$

Thus, the giant clusters exhibit a variety of catalytic activities. As the first step in elucidating the nature of these activities, the kinetics of reactions (15) and (16) in solutions of the clusters were studied.

*2.3.5. Kinetics and mechanisms of oxidative acetoxylation of ethylene and propylene in solutions of giant clusters*

Oxidative acetoxylation of alkenes *via* reactions like (3) is known to be carried out homogeneously in solutions of AcOH containing Pd(II) + $OAc^-$ ions[5,49,50], and heterogeneously in the presence of metallic palladium combined with alkali metal acetates[59-62]. Heterogeneous acetoxylation reactions can be performed with both liquid and gas phase reactants.

In solutions of Pd(II), the reaction involves oxidation of olefin by palladium (II):

$$C_nH_{2n} + Pd(OAc)_2 \longrightarrow C_nH_{2n-1}OAc + Pd(O) + AcOH \quad (18)$$

followed by regeneration of Pd(II) from Pd(O) under the action of an oxidant.

$$Pd(O) + 2AcOH + Ox = Pd(OAc)_2 + Red \quad (19)$$
$$Ox = CuCl_2,\ p\text{-}C_6H_4O_2$$

When obtaining vinyl and allyl acetates with solid catalysts under oxygen + acetic acid vapor at 430-470 K[59,60,62], oxidation of metallic palladium to form Pd(II) may, in principle, occur. However, the thermal instability of Pd(II) acetate complexes at temperatures above 373 K, as well as the data on selectivity of the reactions of alkenyl ester formation in systems including Pd(II), contradicts the hypothesis of the heterogeneous process *via* reactions (18) and (19).

In the case of $Pd(OAc)_2$, interaction with propylene in liquid AcOH, a mixture of allyl, isopropenyl and n-propenyl acetates is formed[83]. The allyl acetate yield decreases with increasing temperature up to 380 K and aqueous contents of the solution up to 1-3 %, at the expense of acetone yield increase. Over Pd metal catalysts, the noted side products are not actually formed, and almost no decrease in the yield of allyl acetate is observed upon a temperature increase from 420 to 460 K and addition of water up to 10%[64].

In the case of ethylene, both in heterogeneous and homogeneous systems, the same ester, vinyl acetate, is formed. However, in Pd(II) systems, even small (0.2-1.0 %) quantities of $H_2O$ result in the appearance of acetaldehyde, whereas in the case of Pd metal catalysts, vinyl acetate is the sole product of the oxidation of ethylene even at significantly higher water contents.[59-62,84].

For oxidation of alkyl arenes in the presence of carboxylic acids, substantial differences in selectivity of homogeneous Pd(II) and heterogeneous Pd metal catalysts are also observed. For example, under the action of Pd(II), toluene is converted mainly into bitolyls, which are the products of oxidative coupling of aromatic rings. Over metallic Pd in the presence of AcOH, oxidative carboxylation of the methyl group becomes the main route of toluene oxidation.

$$PhCH_3 + 1/2O_2 + AcOH \longrightarrow PhCH_2OAc + H_2O$$

The above data suggest that vinyl and allyl acetate can be formed by two different routes depending on the Pd-atom oxidation state in the catalyst used to perform the reaction.

Only vinyl and allyl acetates were formed by ethylene and propylene oxidations, respectively, in solutions of clusters **2** and **3** containing up to 10% $H_2O$[65]. The fact that carbonyl compounds are absent from the products seems to exclude the possibility that the reactions occur *via* oxidation of some of the cluster Pd atoms to Pd(II) and subsequent reduction of Pd(II) by olefins, *i.e. via* reactions (18) and (19). Therefore, clusters **2** and **3** may be regarded as good models of solid catalysts for which, in contrast to homogeneous Pd(II) systems, the selectivity of olefin oxidation is rather insensitive to the presence of water.

### *2.3.6. Kinetics of ethylene and propylene oxidation.*

Cluster **2** is soluble in AcOH and in its mixtures with diglyme, while cluster **3** can be dissolved in MeCN-AcOH mixtures. Therefore, the kinetics of reaction (15) were studied in MeCN-AcOH solutions of cluster **3**, and those of reaction (16) both in MeCN-AcOH solutions of cluster **3** and in diglyme-AcOH solutions of cluster **2**. The concentration of AcOH in these mixtures was varied in order to determine the reaction order with respect to AcOH.

In a flow reactor with gaseous olefins and $O_2$ at a constant flow rate of ethylene + oxygen mixture and constant concentration of acetic acid in the solution, the observed rate of vinyl acetate formation, $r_o$ , was found to increase as a linear function of the concentration of cluster **3** in the interval from $2.4 \times 10^{-5}$ to $1.77 \times 10^{-4}$ M.

The observed first-order rate constant, $k_{obs}$ , for ethylene oxidation increased nonlinearly with increasing concentration of ethylene (Fig. 4a). This dependence of $k_{obs}$ on ethylene concentrations can be represented by a linearized form of the Michaelis-Menten equation.

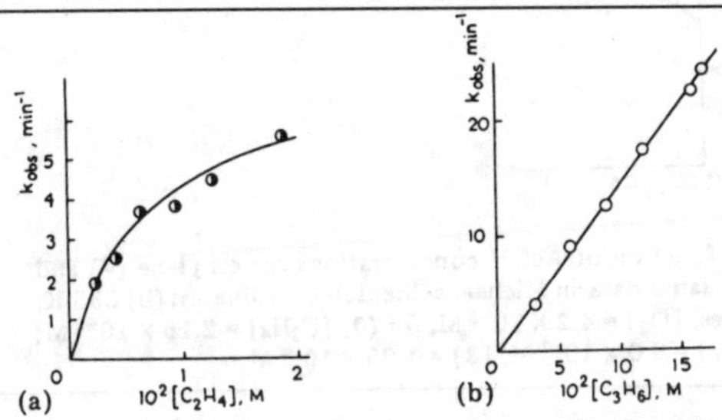

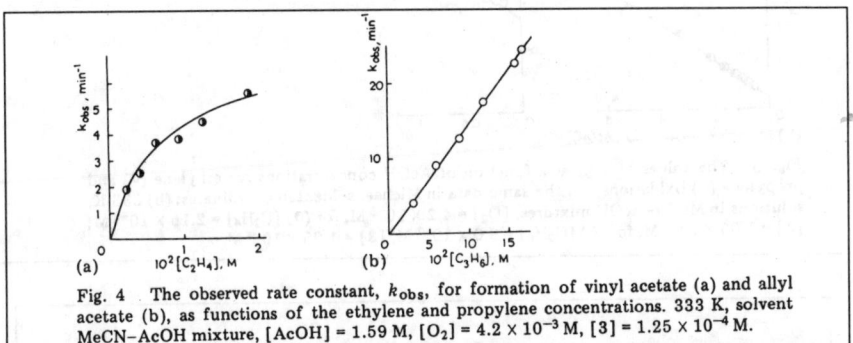

Fig. 4 The observed rate constant, $k_{obs}$, for formation of vinyl acetate (a) and allyl acetate (b), as functions of the ethylene and propylene concentrations. 333 K, solvent MeCN–AcOH mixture, [AcOH] = 1.59 M, $[O_2] = 4.2 \times 10^{-3}$ M, $[3] = 1.25 \times 10^{-4}$ M.

The rates, $r_o$, of allyl acetate formation from propylene in solutions of both cluster **2** and cluster **3** are also proportional to the concentrations of latter. For propylene oxidation, in contrast to ethylene oxidation, the dependence of $r_o$ on the concentration of $C_3H_6$ demonstrates no deviation from linearity (Fig. 4b). The influence of the concentration of acetic acid on $r_o$ for $C_3H_6$ and $C_2H_4$ oxidations is described by Michaelis-Menten kinetics (Fig. 5).

With the oxygen concentration being varied, the variations in the reaction rate for propylene oxidation are more intricate than for ethylene oxidation. Under 0.5 atm of propylene in the $r_o = f(P_{O_2})$ function there appears a maximum (Fig. 6). The decrease in $r_o$ upon increasing $P_{O_2}$ above a certain value, corresponding to the maximum of $r_o$, suggests an inhibition of the reaction by $O_2$ under low pressure of the $C_3H_6$. Under high enough pressures of $C_3H_6$, ($P_{C_3H_6} \geq 0.7$ atm) inhibition of the reaction is imperceptible. For example, under 0.9 atm of $C_3H_6$ partial pressure, this function looks like the curve with saturation and the rate of the propylene oxidation is described by an equation similar to Eq. (20)[86].

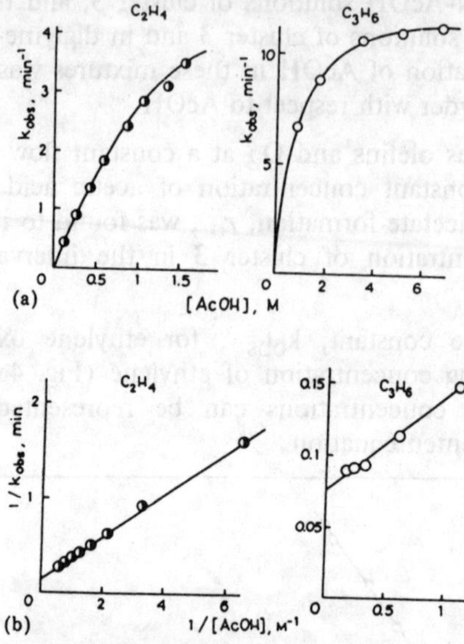


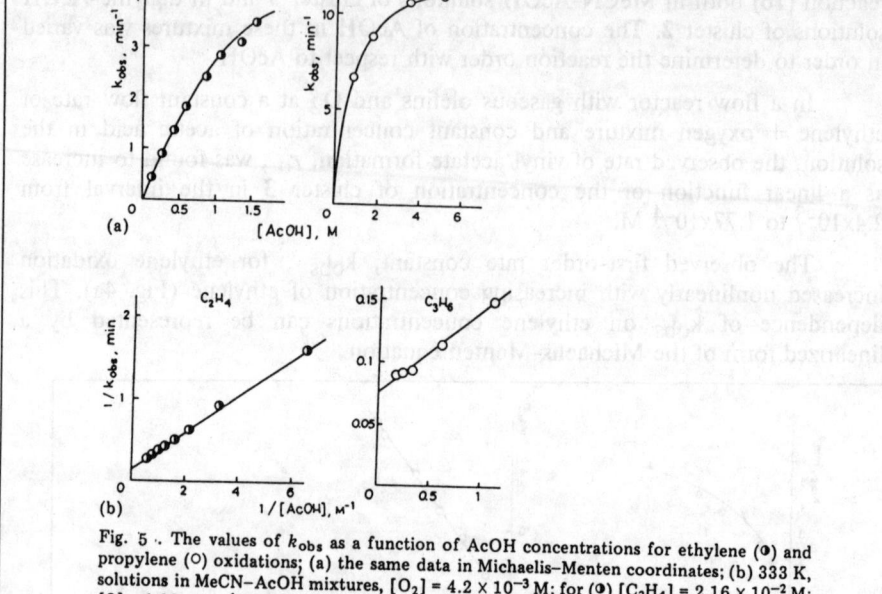

**Fig. 5 . The values of $k_{obs}$ as a function of AcOH concentrations for ethylene (◐) and propylene (○) oxidations; (a) the same data in Michaelis–Menten coordinates; (b) 333 K, solutions in MeCN–AcOH mixtures, $[O_2] = 4.2 \times 10^{-3}$ M; for (◐) $[C_2H_4] = 2.16 \times 10^{-2}$ M; [3] = $1.09 \times 10^{-4}$ M; for (○) $[C_3H_6] = 9.0 \times 10^{-2}$ M, [3] = $1.25 \times 10^{-4}$ M.**

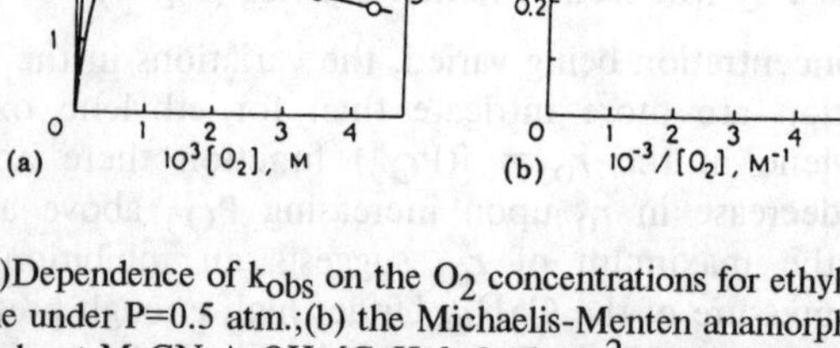

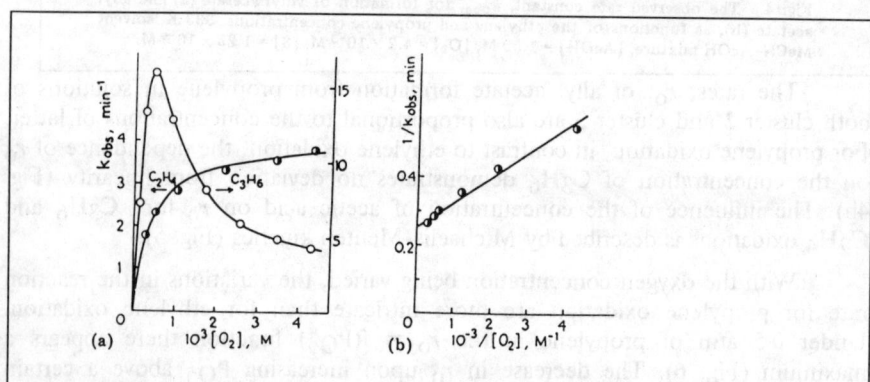


Fig.6. (a)Dependence of $k_{obs}$ on the $O_2$ concentrations for ethylene under P=0.2 atm and propylene under P=0.5 atm.;(b) the Michaelis-Menten anamorphosis for ethylene oxidation. 333 K, solvent MeCN-AcOH; $[C_2H_4]=2.16\times10^{-2}$M, [AcOH]=1.59 M, [3]=$1.09\times10^{-4}$ M; $[C_3H_6]$=0.162 M, [AcOH]=1.75 M, [3]=$1.25\times10^{-4}$M

Thus, experimental values for the reaction rate may be described by the rate equation (20)[65].

$$r_0 = k[\text{cluster}]\frac{[C_2H_4][O_2][AcOH]}{(K_I + [C_2H_4])(K_{II} + [O_2])(K_{III} + [AcOH])} \quad (20)$$

where [Clust] and [Ol] stand for the concentrations of a giant cluster and an olefin, respectively. Note that the reaction between the coordinated $C_nH_{2n}$, $O_2$ and AcOH species characterized by the effective rate constant *k*, is expected to be in fact a complex one, consisting of several elementary steps.

TABLE 5. The constants of the kinetic equation (20) at 333 K

| Substrate | Cluster | k (min$^{-1}$) | $K_I$x$10^3$ (mol l$^{-1}$) | $K_{II}$x$10^4$ (mol l$^{-1}$) | $K_{III}$ (mol l$^{-1}$) |
|---|---|---|---|---|---|
| ethylene | **3** | 8.2±0.7 | 5.8±0.3 | 3.0±0.2 | 1.3±0.1 |
| propylene | **3** | 3.3±0.3 | ≥30 | 5.2±0.3 | 0.67±0.05 |
| propylene | **2** | 5.6±0.5 | ≥30 | 1.2±0.1 | 4.8±0.5 |

From Table 5 it is seen that the rate constants, k, for ethylene and propylene oxidations in solutions of both clusters differ less than three-fold. The Michaelis constants $K_I$, $K_{II}$ and $K_{III}$ for both olefins are also of the same order of magnitude, the largest variation upon the change of the olefin being observed for the constant $K_I$. The Michaelis-Menten character of the kinetics of reactions (15) and (16) suggests the product formation to be preceded by stages of reversible coordination of olefin, $O_2$ and AcOH molecules with the cluster. A comparison of the values of $K_I$ (Table 5) shows propylene to be coordinated notably more weakly than ethylene and $O_2$.

In homogeneous acetoxylation of olefins with Pd(II), the products of the oxidation appear, in the absence of dioxygen, in stoicheiometric quantities with respect to that of the Pd(II) reacted. In contrast, in the case of clusters **2** and **3**, the oxidation of olefins in the absence of dioxygen is not observed. This fact further supports the conclusion that reactions of ethylene and propylene in the presence of the giant clusters cannot be explained by alternating oxidation-reduction reactions of the cluster with substrate and oxidant molecules. The kinetic data can be interpreted in the framework of a scheme including both AcOH and alkene and oxidant adsorption (coordination) at the cluster active sites.

An analysis of Eq. (20) suggests that the stability constants for the complexes formed by cluster **2** or **3** with each substrate (Ol, $O_2$, and AcOH) do not depend on the presence or absence of other coordinated substrates. The assumption of the absence of such influence seems quite reasonable for a big coordination centre such as a giant cluster with several hundred atoms of palladium.

A smaller value of the stability constant for the π-complex of the giant cluster with propylene as compared to ethylene may be explained by the size of the propylene molecule exceeding that of ethylene. The surface of the cluster core is expected to be substantially screened by the bulky molecules of the phenanthroline ligand. Examination of the idealized structure of cluster **2** (Fig. 2) using molecular models shows that only *ca.* 20 palladium atoms on the surface of the metal skeleton of the cluster are sterically accessible for coordination of the olefin molecules. It may happen that because of steric hindrance the more bulky $C_3H_6$ molecules will be coordinated at these sites less strongly that the smaller $C_2H_4$.

For real molecules of giant clusters, with the structure probably deviating somewhat from the perfect polyhedron, the number of accessible coordination sites may deviate from that for the idealized model. The number of these sites can be determined experimentally using poisoning techniques, *i.e.* by carrying out the oxidation reaction in the presence of some ligands that can be strongly bound to those palladium atoms on the surface of the metal skeleton which are active in catalysis (Fig.7).

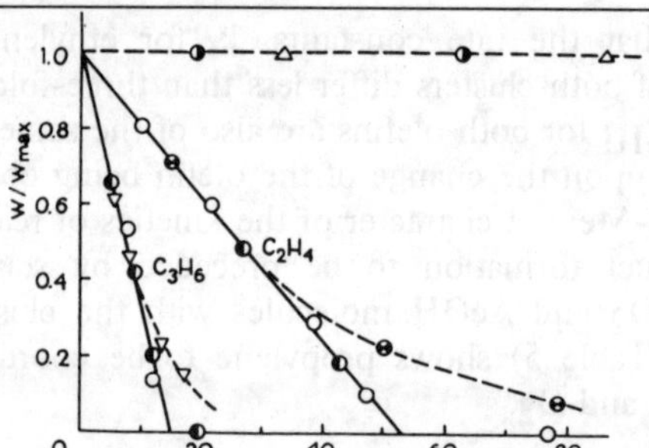

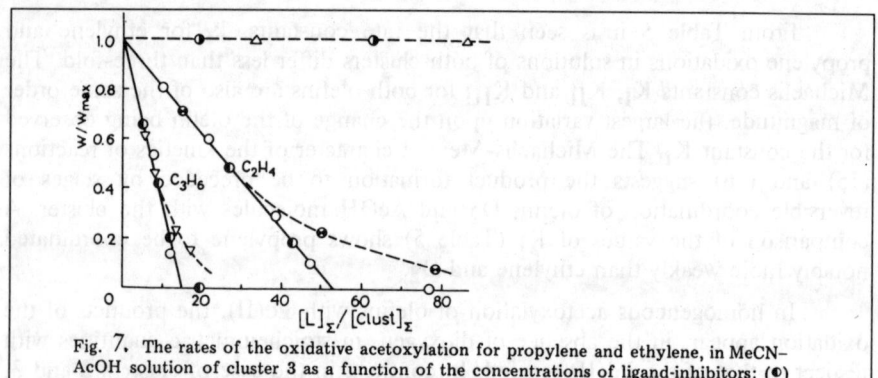

**Fig. 7. The rates of the oxidative acetoxylation for propylene and ethylene, in MeCN–AcOH solution of cluster 3 as a function of the concentrations of ligand-inhibitors: (●) $C_2H_5SH$; (▽) $I_2$; (○) $Et_2NCS_2Na$; (⊗) KSCN; (△) phen; (◑) $PPh_3$.**

The data on the influence of the poisons on the rates of $C_2H_4$ and $C_3H_6$ oxidative acetoxylation in solutions of cluster **3** showed that the bulky ligands that coordinate to Pd(II) atoms, *e.g.* $PPh_3$ and phen, actually have no effect on the rate of the olefin oxidation, apparently because they cannot be coordinated on the sites suitable for olefin, $O_2$ and AcOH binding in the course of the reaction. The smaller ligands, *e.g.* $C_2H_5SH$ and the thiocyanide anion, efficiently suppress the catalytic activity of the cluster. For complete inhibition of $C_2H_4$ oxidation, it is necessary to introduce *ca.* 50 ligand molecules to the surface of one cluster molecule, while for $C_3H_6$ oxidation, it is sufficient to introduce only *ca.* 15 ligand molecules[86].

These data agree with the above supposition about the importance of steric requirements for the ability of a substrate molecule to form a complex with the giant cluster. Indeed about a three-fold decrease in the rate constant, *k* in Eq. (20) when passing from ethylene to propylene (Table 5), is shown by the poisoning experiments to result from a three-fold decrease in the number of the cluster surface sites available for coordination of the olefin molecule.

For further insight into the reaction mechanism, the kinetic isotope effects for $C_2H_4$ and $C_3H_6$ oxidative acetoxylations were studied (Table 6).

TABLE 6. Kinetic isotope effects at 333 K for reactions (15) and (16).

| Substrate | Cluster | $k_{C_nH_{2n}}/k_{C_nD_{2n}}$ | $k_{CH_3COOH}/k_{CD_3COOD}$ |
|---|---|---|---|
| ethylene | **3** | 1.1±0.1 | 1.1±0.1 |
| propylene | **2** | 2.2±0.2 | 1.0±0.05 |
| propylene | **3** | 3.6±0.2 | 1.0±0.05 |
| propylene | Pd black | 1.0±0.1 | 2.0±0.2 |

Besides giant Pd clusters, two other types of species could have also been suspected as responsible for catalytic activity in the oxidation of olefins by $O_2$: (1) mononuclear Pd(II) complexes presumably arising from a giant cluster upon its dissociation; (2) particles of Pd metal, which could be formed as small impurities in reaction solutions upon the coalescence of clusters.

Assumption (1) can be, however, rejected on the basis of the above mentioned differences in the reactivity of Pd clusters and mononuclear Pd(II) complexes.

Assumption (2) can also be rejected, since it is in contradiction with kinetic data. In particular, the difference in the kinetic isotope effects for propylene oxidation catalyzed by giant clusters and Pd black should be noted (Table 6).

Moreover, the oxidation of an equimolar mixture of ethylene and propylene in the presence of clusters **2** or **3** resulted in simultaneous formation of vinyl and allyl acetates in comparable quantities, while in the presence of Pd black, allyl acetate was formed as almost the sole product of the reaction.

*2.3.7. Mechanism of the oxidation of olefins.*

Besides the kinetic equations, the following three experimental facts are important for revealing the reaction mechanism:

(i) both ethylene and propylene are oxidized with nearly equal rate constants *k* in the presence of the same cluster **3**, notwithstanding the pronounced difference between the C-H bond strengths in ethylene (vinyl-H, 445 ± 8 kJ) and in propylene (allyl-H, 362 ± 8 kJ) molecules[87].

(ii) The reaction of the oxidative acetoxylation of propylene in the presence of **2** and **3** yields only allyl acetate, no products of vinyl oxidation being observed at both low and high concentrations of all reacting species.

(iii) For ethylene oxidation, the kinetic isotope effects (KIE), within the limits of experimental error, are equal to unity both for ethylene and for acetic acid molecules. For propylene oxidation, KIE is equal to unity only for acetic acid, while for propylene it considerably exceeds this value (Table 6).

Different KIE observed upon deuteration of $C_2H_4$ and $C_3H_6$ (fact (iii)) suggest different rate-determining steps for the oxidative acetoxylation of these two olefins within the reaction mechanism. In the $C_2H_4$ molecule, the π-bond between two carbon atoms is the weakest one (*ca.* 250 kJ). The data on KIE indicate (Table 6) that no transfer of H atoms from the coordinated $C_2H_4$ and $CH_3COOH$ molecules occurs in the rate-determining stage. In this situation, we suggest that the rate-determining step might be an oxidative addition (with an opening the π-bond) of a π-coordinated $C_2H_4$ molecule to a Pd-Pd group of the cluster forming the σ,σ-coordinated ~Pd-$CH_2$-$CH_2$-Pd~ group. Subsequent splitting of the C-H bond in this group is assumed to be fast and facilitated owing to the formation of the Pd=C multiple bond in the intermediate V[88,89]:

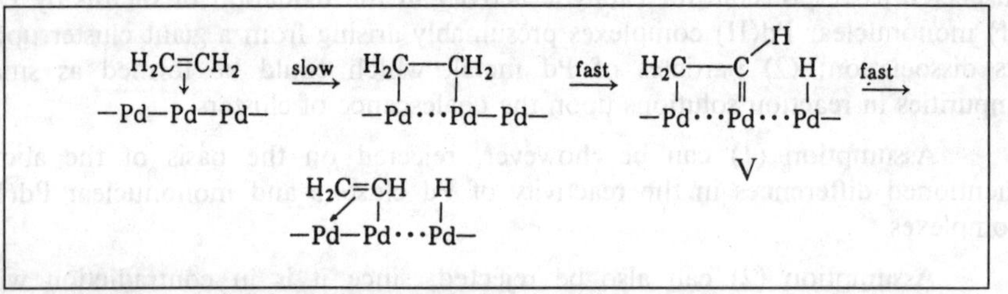

Similar species have been postulated as intermediates for hydrogenation, dehydrogenation, and H-D exchange of ethylene at the surfaces of noble metal catalysts[90,91].

In the case of ethylene, the energy of carbon-carbon π-bond, $E_{\pi C=C}$, is markedly less than that of vinyl-H bond. In the case propylene, allyl-H bond is assumed to be splitted in the slow step. The bond energies of Pd-π-allyl and Pd-H surface species may be expected to compensate the difference between $E_{\pi C=C}$ and $E_{allyl\text{-}H}$ contributing to the activation energy of that step. Thus, the rate-determining step of the oxidative addition of a propylene molecule to a Pd-Pd group may include the splitting of the allyl-H bond, leading to the formation of the π-allyl complex and surface hydride:

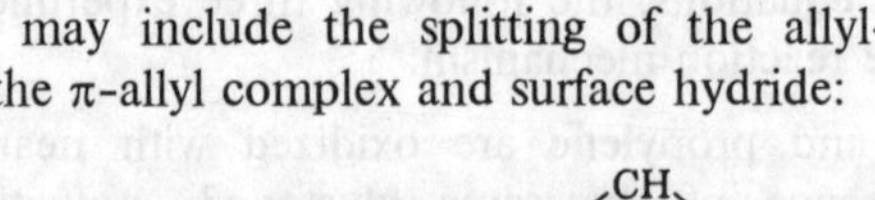

The splitting of the allyl-H bond in the rate-determining step explains a large KIE for oxidative acetoxylation of propylene (Table 6).

Formation of the surface π-allyl group in the reaction intermediate appears to favor the "allyl" direction of the reaction as compared to the "vinyl" one. Further reactions of vinyl or allyl groups and H atoms coordinated at the surface of the cluster metal skeleton are assumed to proceed rapidly and have no influence on the reaction rate.

The proposed mechanisms of olefin oxidative acetoxylation via reaction (15) and (16) indeed assume the rate-determining steps of this reaction to be different for ethylene and propylene.

The data on inhibition of reactions (15) and (16) with poisoning ligands[86] showed that only *ca.* 20% of the surface palladium atoms are available for reagent molecules in the oxidation of ethylene, and *ca.* 6% of these atoms in propylene oxidation. At the sterically screened surface, all three molecules (olefin, $O_2$ and AcOH) are barely coordinated to the neighbour palladium atoms at the same part of the cluster. It is more probable that a $C_2H_4$ or $C_3H_6$ molecule is initially bound at one site of the cluster surface, and an $O_2$ molecule at another site, not necessarily the neighbour to the site where the olefin molecule is located. In this situation, electron transfer from the Pd-alkenyl or Pd-H fragments to the coordinated $O_2$ molecule can perhaps occur through the metal skeleton, the latter acting as an "electron mediator". Thus, the general Scheme A can be assumed for the reaction of olefin acetoxylation over the giant Pd clusters, where Ox = $O_2$, $(PhCOO)_2$ or Pd(II); Red = $H_2O$, PhCOOH or Pd(O), and reaction products $C_nH_{2n-1}OCOCH_3$ and Red formed, respectively, upon recombination of coordinated ($-C_nH_{2n-1}$), ($-OCOCH_3$) and (-Ox), (-H) species that collide during their migration over the surface of the cluster.

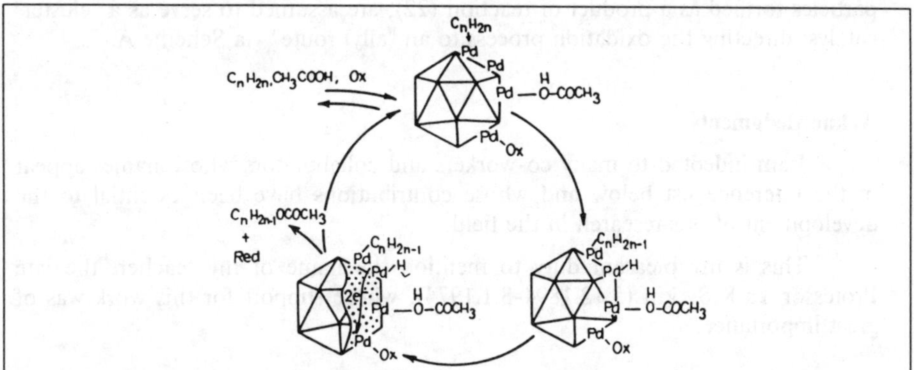


Scheme A. Ox=$O_2$, ROOR, Pd(II)

As is indicated in Scheme A, other oxidants besides $O_2$, *e.g.*, peroxides and Pd(II), can also serve as electron acceptors, in the cluster catalysis. For example, benzoyl peroxide can be used as an electron acceptor in the presence of clusters **2** and **3** [92]. The reaction

$$C_3H_6 + (C_6H_5COO)_2 + CH_3COOH \longrightarrow CH_2{=}CHCH_2OCOCH_3 + 2C_6H_5COOH \quad (21)$$

proceeds in AcOH solution containing giant cluster **2** or **3**, with 98- 100% selectivity for allyl acetate, at 293 K and $10^{-3}$-$10^{-4}$ M stationary concentration of benzoyl peroxide. Benzoyl peroxide must be introduced into the reaction solution in small portions to prevent the oxidation of the cluster. At higher concentrations of benzoyl peroxide, *e.g.* $10^{-1}$-$10^{-2}$ M, clusters **2** and **3** are oxidized to form Pd(II) complexes. If this occurs, ordinary homogeneous oxidation with Pd(II) proceeds, yielding a mixture of alkenyl acetates.

The reaction between $C_3H_6$ and benzoyl peroxides, producing only allyl acetate in high yield, can be also carried out in the presence of palladium black[92]. Under the same conditions, propylene was not oxidized by benzoyl peroxide in the absence of cluster **2** and **3** or Pd black. Therefore, reaction (21) may be regarded as a further example of the cluster-catalyzed oxidative acetoxylation of olefins, proceeding presumably via Scheme A.

The approach based on Scheme A for catalysis with giant clusters may be applied also to the homogeneous oxidative acetoxylation of olefins under the action of Pd(II) in AcOH solutions, giving usually isopropenyl, n-propenyl and allyl acetates with comparable yields[83,93]:

$$C_3H_6 + Pd(OAc)_2 \longrightarrow C_3H_5OAc + Pd(O) + AcOH \quad (22)$$

In the absence of additional oxidants, *e.g.* $O_2$, dispersed metallic palladium particles formed as a product of reaction (22), are assumed to serve as a "cluster" catalyst directing the oxidation process to an "allyl route" via Scheme A.

**Acknowledgments**

I am indebted to many co-workers and collaborators whose names appear in the reference list below and whose contributions have been essential to the development of our research in the field.

This is my pleasant duty to mention the name of my teacher, the late Professor Ya.K.Syrkin (5.12.1894-8.1.1974), whose support for this work was of great importance.

**References**


1. R.A.Sheldon and J.K.Kochi, *Metal-Catalyzed Oxidations of organic Compounds,* Acad. Press, New Jork (1981), 423.
2. F.C.Phillips, *Z. Anorg. Chem.,* **6** (1894), 213.
3. J.Smidt, W.Hafner, R.Jira, J.Sedimeier, R.Sieber, R.Ruttinger, and H.Kojer, *Angew. Chem.,* **71** (1959), 176.
4. R.Jira, and W.Freiesleben, in *Oranometallic Reactions*, Ed. by E.Becker and M. Tsutsui, Wiley&Sons, **3** (1972), 5.
5. I.I.Moiseev, M.N.Vargaftik, and Ya.K.Syrkin, *Doklady Acad. Nauk SSSR*, **133** (1960), 377 (in Russ.).
6. I.I.Moiseev, M.N.Vargaftik, and Ya.K.Syrkin, *Doklady Acad. Nauk SSSR*, **130** (1960), 820 (in Russ.).
7. J.Chatt, *Nature*, **165** (1950), 859.
8. J.Chatt, and R.G.Wilkins, *J. Chem. Soc.* (1952), 2622.
9. I.I.Moiseev, M.N.Vargaftik, and Ya.K.Syrkin, *Doklady Acad. Nauk SSSR*, **152** (1963), 147 (in Russ.).
10. I.I.Moiseev, *Amer.Chem.Soc., Div.Petrol.Chem.*, **14** (1969), B49.
11. M.N.Vargaftik, I.I.Moiseev, and Ya.K.Syrkin, *Doklady Acad. Nauk SSSR*, **139** (1961), 1396 (in Russ.).
12. M.N.Vargaftik, I.I.Moiseev, and Ya.K.Syrkin, *Doklady Acad. Nauk SSSR*, **147** (1962), 399 (in Russ.).
13. I.I.Moiseev, M.N.Vargaftik, and Ya.K.Syrkin, *Doklady Acad. Nauk SSSR*, **153** (1963), 140 (in Russ.).
14. (a) P.M.Henry, *J.Amer.Chem.Soc.*, **86** (1964), 3246; (b) P.M.Henry *Palladium Catalyzed Oxidation of Hydrocarbons*, Reidel, Dordrecht (1980).
15. A.Aguilo, *Adv. Orgonometal. Chem.*, **5 (**1967**)**, 321.
16. T.Dozono, and T.Shiba, *Bull. Japan. Petr. Inst.*, **5** (1965), 8.
17. K.I.Matveev, I.F.Bukhtoyarov, N.N.Schul"tz, and O.A.Emel"yanova, *Kinetika i katalyz,* **5** (1964), 649 (in Russ.).
18. K.I.Matveev, A.M.Osypov, V.F.Odyakov, Y.V.Suzdalnitskaya, I.F.Bukhtoyarov, and O.A.Emel"yanova, *Kinetika i katalyz,* **3** (1962), 661 (in Russ.).
19. K.Teramoto, T.Oga, S.Kikuchi, and M.Ito, *Yuki Gosei Kagaku Kyokai Shi* **21** (1963), 298.
20. E.van der Heide *Ph.D.Thesis,* Delft University (1990).
21. I.I.Moiseev, O.G.Levanda, M.N.Vargaftik, *J.Amer.Chem. Soc,* **96** (1974), 1003 (in Russ.).
22. O.G.Levanda, I.I.Moiseev, and M.N.Vargaftik, *Izvestia Akad. Nauk SSSR, Ser. Khim.* (1968), 2368 (in Russ.).
23. E.D.Weed, see in[15], 328.
24. R.Palumbo, A. De Renzi, A.Panunzi, and G.Paiaro, *J.Amer.Chem.Soc.*, **91**, (1969), 3874.
25. A.Panunzi, A. De Renzi, R.Palumbo and G.Paiaro, *J.Amer.Chem.Soc.*, **91**, (1969), 3879.

26. V.A.Likholobov, in *Perspectives in Catalysis*, Ed. by J.M.Thomas and K.I.Zamaraev, Blackwell Sci. Publ. (1992), 67.
27. J.K.Stille and R.Divakaruni, *J.Amer.Chem.Soc.*, **100** (1978), 239
28. J.T.Backvall, *Acc.Chem.Res.,* **16** (1983), 325.
29. J.E.Backvall, B.Akermark and S.O.Ljunggren, *J.Chem.Soc.*, *Chem. Commun.*, (1977), p.264
30. J.E.Backvall, B.Akermark and S.O.Ljunggren, *J.Amer.Chem.Soc.*, **101** (1979) p.264
31. I.I.Moiseev, and M.N.Vargaftik, *Izvestia Akad. Nauk SSSR, Ser. Khim.,* (1965), 759 (in Russ.).
32. E.W.Stern, *Proc. Chem.Soc., (*1963), 111.
33. E.W.Stern, and M.L.Spector, *Proc. Chem.Soc.* (1961), 370.
34. J.Chatt, *in "Cationic Polymerization and Related Complexes",* Cambridge, England (1953), 57.
35. I.I.Moiseev, and S.V.Pestrikov, *Doklady Akad. Nauk SSSR,* **171** (1966), 151 (in Russ.).
36. S.V.Pestrikov, I.I.Moiseev, and L.M.Sverzh, *Kinetika i Kataliz*, **10** (1969), 74 (in Russ.).
37. I.I.Moiseev, and M.N.Vargaftik, *Doklady Akad. Nauk SSSR,* **166** (1966), 370 (in Russ.).
38. W.Kitching, *Organometallic Chem. Reviews A*, **3** (1968), 61.
39. P.M. Maitlis, *The Organic Chemistry of Palladium*, Vol. **1** and **2**, Academic Press, New York (1971).
40. I.I.Moiseev, *π-Complexes in Liquid-Phase Oxidation of Alkenes* (in Russ.), Nauka, Moscow (1970).
41. J.Chatt, *Chem. Rev.*, **48** (1951), 7.
42. R.Jira, J.Sedlmeier, and J.Smidt, *Ann.* **693** (1966), 99.
43. B.L.Shaw, *Chem. Comm.* (1968), 464.
44. I.I.Moiseev, A.P.Belov, V.A.Igoshin, and Ya.K.Syrkin, *Doklady Akad. Nauk SSSR,* **173** (1967), 863 (in Russ.).
45. R.Ninomija, M.Sato, and T.Shiba, *Bull. Japan Petrol. Inst.* **7** (1965), 31.
46. R.Cramer, *J.Amer.Chem.Soc.*, **86** (1964), 217.
47. B.F.G.Johnson, C.Holloway, G.Hulley, and J.Lewis, *J.Chem.Soc.Chem. Commun.* (1967), 1143.
48. W.Kitching, Z.Rappoport, S.Winstein, and W.G.Young, *J.Amer.Chem.Soc.* **88** (1966), 2054.
49. A.P.Belov, and I.I.Moiseev, *Izvestia Akad. Nauk SSSR, Ser. Khim.* (1966), 139 (in Russ.).
50. A.P.Belov, G.Yu.Pek, and I.I.Moiseev, *Izvestia Akad. Nauk SSSR, Ser. Khim.* (1965), 2204 (in Russ.).
51. J.Smidt, W.Hafner, R.Jira, R.Sieber, J.Sedlmeier, and A.Sabel, *Angew. Chem.* **74** (1962), 93.
52. J.Chatt, and B.L.Shaw, *J. Chem. Soc.* (1959), 705.
53. J.Chatt, and B.L.Shaw, *J. Chem. Soc.* (1959), 4020.

54. R.J.Cross, *Organometal. Chem. Rev.*, **2** (1967), 97,.
55. F.Cariati, R.Ugo, and F.Bonati, *Chim. e Ind.* (1964), 1714.
56. C.B.Anderson, and S.Winstein, *J. Organ. Chem.* **28** (1963), 605,.
57. R.Crigee, *Angew. Chem.,* **70** (1958), 173.
58. H.-J.Kabbe, *Ann.* **656** (1962), 204.
59. T. Shimidzu and M. Tamura, *Shokubai*, **16** (1974), 84.
60. S. Nakamura and T. Yasui, *J. Catal.*, **23** (1971), 315.
61. J.E. Lyons, G. Suld and Chao-Yang Hsu, in Yu. Yermakov and V. Likholobov (eds.), *Proc. Vth Int. Symp. Relations between Homogeneous and Heterogeneous Catalysis,*Novosibirsk, 1986, VNU Science Press. Utrecht, 117.
62. S.F. Polytansky, M.N. Vargaftik, A.M. Shkitov, I.P. Stolarov, I.I. Moiseev and O.M. Nefedov, *Izv. Akad. Nauk SSSR, Ser. Khim.*, (1978), 1913 (in Russ.).
63. M.N. Vargaftik, V.P. Zagorodnikov and I.I. Moiseev, *Kinet. Katal.*, **22** (1981), 951 (in Russ.).
64. I.P. Stolarov,M.N. Vargaftik, O.M. Nefedov and I.I. Moiseev, *Kinet. Katal.* **23** (1982), 376 (in Russ.).
65.(a) M.N. Vargaftik, V.P. Zagorodnikov, I.P. Stolarov, I.I. Moiseev, V.A.Likholobov, D.I. Kotchubey, A.L. Chuvilin, V.I. Zaikovsky, K.I. Zamaraev and G.I. Timofeeva, *J. Chem. Soc. Chem. Commun.*, (1985), 937. (b) M.N. Vargaftik, V.P. Zagorodnikov, I.P. Stolarov, I.I. Moiseev, D.I. Kotchubey, V.A.Likholobov, A.L. Chuvilin, and K.I.Zamaraev, *J.Mol.Catal.*, **53**, 1989, 315.
66. M.N. Vargaftik, V.P. Zagorodnikov, I.P. Stolarov, D.I Kotchubey, V.M. Nekipelov, V.M. Mastikhin, V.D. Chinakov, K.I. Zamaraev and I.I. Moiseev, *Izv. Akad. Nauk SSSR, Ser. Khim.* (1985), 2381 (in Russ.).
67. G. Schmid, *Structure and Bonding*, Vol. **62**, Springer, Berlin-Heidelberg, 1985.
68. L.R. Wallenbert, J.-O. Bovin and G. Schmid, *Surf. Sci.*, **156** (1985) 256.
69. B.T. Heaton, P. Ingallina, R. Devenish, C.J. Humphreys, A.Ceriotti, G. Longoni and M. Marchionna, *J. Chem. Soc. Chem. Commun.* (1987) 765.
70. M.N. Vargaftik, I.I. Moiseev, D.I Kotchubey, K.I. Zamaraev, *Faraday Discuss.*, **92** (1991), 13.
71. B.K. Teo and N.J.A. Sloane, *Inorg. Chem.*, **24** (1985), 4545.
72. V.V.Volkov, G. van Tendeloo, M.N. Vargaftik, I.P. Stolarov, I.I. Moiseev, *Mendeleev Commun.*, (1993), 187.
73. M.N. Vargaftik, T.A. Stromnova,T.S. Khodashova, M.A. Porai- Koshits and I.I. Moiseev, *Koord. Khim.*, **7**, (1981) 132 (in Russ.).
74. J.C.Poulin, H.B.Kagan, M.N. Vargaftik, I.P. Stolarov, I.I. Moiseev, *J.Mol.Catal.* (1994), in press.
75. (a) V.P. Zagorodnikov, M.N. Vargaftik, D.I. Kotchubey, A.L. Chuvilin, S.G.Sakharov, and M.A. Mayfat, *Izv. Akad. Nauk SSSR, Ser. Khim.*, (1986), 253 (in Russ.); (b) V.P. Zagorodnikov, M.N. Vargaftik, D.I. Kotchubey, V.A.Likholobov, V.N. Kolomiychuk, A.N. Naumochkin, A.L.

Chuvilin, V.M.Novotortsev, O.G. Ellert and I.I. Moiseev, *Izv. Akad. Nauk SSSR, Ser. Khim.*, (1989), 849 (in Russ.).
76. E.G. Mednikov, N.K. Eremenko, Yu.L. Slovokhotov and Yu.T. Struchkov, *J.Chem. Soc. Chem. Commun.*, (1987), 218.
77. S. Ladas, B.R.A. Dalla Betta and M. Boudart, *J. Catal.*, **53** (1978), 356.
78. A.N. Naumochkin and D.I. Kotchubey, *Nucl. Instruments and Methods in Physical Research*, **A261** (1987), 163.
79. G. Schmid, R. Pfeil, R. Boese, F. Bandermann, S. Meyer, G.H.M. Calis and J.W.A. van der Velden, *Chem. Ber.*, **114** (1981), 3634.
80. V.P. Zagorodnikov and M.N. Vargaftik, *Izv. Akad. Nauk SSSR, Ser. Khim.*, (1985), 2652 (in Russ.).
81. P.I.Pasichnyk, M.K.Starchevsky, Yu.A.Pazdersky, M.N. Vargaftik, I.I Moiseev, *Mendeleev Commun.* (1994), 1.
82. M.K.Starchevsky, S.L.Gladyi, Yu.A.Pazdersky, M.N. Vargaftik, I.I Moiseev, to be published.
83. S. Winstein, J. McCaskie, H.-B. Lee and P.M. Henry, *J. Am. Chem. Soc.*, *98* (1976), 6913.
84. M. Tamura and T. Yasui, *Shokubai*, **21** (1979), 54.
85. M.K. Starchevsky, M.N. Vargaftik and I.I Moiseev, *Kinet. Katal.*, **20** (1979), 1163 (in Russ.).
86. I.P. Stolarov, M.N. Vargaftik and I.I. Moiseev, *Kinet. Katal.*, **28** (1987), 1359 (in Russ.).
87. V.N. Kondrat'ev, (ed.), *Chemical Bond Energies, Ionisation Potentials and Electron Affinities*, Nauka, Moscow (1974), 39 (in Russ.).
88. I.I. Moiseev, *Advances in Science and Technology*, Ser: Kinetics and Catalysts, VINITI, Moscow, Vol. **13** (1984), 147 (in Russ.).
89. V.P. Zagorodnikov and M.N. Vargaftik, *Kinet. Katal.*, **27** (1986), 851 (in Russ.).
90. A. Farkas and L. Farkas, *J. Am. Chem. Soc.*, **60** (1938), 22.
91. E.M. Stuve and R.J. Madix, *J. Phys. Chem.*, **89** (1985), 105 43. I.P.
92. Stolarov,M.N. Vargaftik, O.M. Nefedov and I.I. Moiseev, *Izv. Akad. Nauk SSSR, Ser. Khim.* (1983), 1455 (in Russ.).
93. I.I. Moiseev, A.P. Belov, V.A. Igoshin and Ya.K. Syrkin, *Dokl. Akad. Nauk SSSR* **173** (1967), 863 (in Russ.).

# CATALYTIC OXIDATION AND FINE CHEMICALS

R.A.SHELDON
*Laboratory for Organic Chemistry and Catalysis,*
*Delft University of Technology, Julianalaan 136,*
*2628 BL Delft, The Netherlands*

## ABSTRACT

The characteristics of fine versus bulk chemicals manufacture and catalytic oxidation versus catalytic oxygen transfer are explained. The applications of a variety of catalytic systems in organic synthesis, including catalytic asymmetric oxidations, are reviewed.

## 1. **Introduction - Why Catalytic Oxidation**?

As a result of increasingly stringent environmental constraints it is becoming prohibitive to perform industrial scale oxidations with traditional stoichiometric oxidants, such as dichromate and permanganate. Consequently, there is a marked trend towards the use of catalytic alternatives that do not generate aqueous effluents containing large quantities of inorganic (heavy metal) salts[1-3].

An illustrative example is the industrial synthesis of hydroquinone (figure 1). Traditionally hydroquinone was manufactured by oxidation of aniline with stoichiometric quantities of $MnO_2$ to give p-benzoquinone, followed by reduction with iron and hydrochloric acid. The aniline was derived from benzene via nitration and reduction. The overall process generates more than 10 kg of inorganic salts ($MnSO_4$, $FeCl_2$, $Na_2SO_4$, NaCl) per kg of hydroquinone. In contrast, the modern route to hydroquinone involves the autoxidation of p-diisopropylbenzene followed by acid-catalyzed rearrangement of the bis-hydroperoxide, analogous to the production of phenol from cumene. This process produces <1 kg of inorganic salts per kg of hydroquinone.

Similarly, resorcinol can be produced in an analogous manner from m-diisopropylbenzene and this process has largely superseded the classical process which involves caustic fusion of benzene-1,3-disulfonic acid. The industrial synthesis of phloroglucinol is also a case in point (figure 2). Traditionally it was manufactured from 2,4,6-trinitrotoluene (TNT) via a dichromate oxidation followed by reduction with iron and hydrochloric acid, a perfect example of nineteenth century chemistry. Alternatively, phloroglucinol can be made via autoxidation of 1,3,5-diisopropylbenzene, affording <1kg of inorganic salts per kg of phloroglucinol, compared to the 37kg generated by the classical process. Nevertheless, to our knowledge the autoxidation process has not been reduced to commercial practice, which may be due to the fact that the process is too complicated for a fine chemical with a volume of a few hundred tons worldwide. Current phloroglucinol manufacturing processes are based on nucleophilic substitutions of trihalobenzenes and there is still a definite need for a low-salt alternative.

## 2. **Characteristics of Fine versus Bulk Chemicals Manufacture**.

Although they share many common features there are several basic differences between fine and bulk chemicals manufacture that can have an important bearing on process selection. Fine chemicals are often complex, multifunctional molecules with low volatility and limited thermal

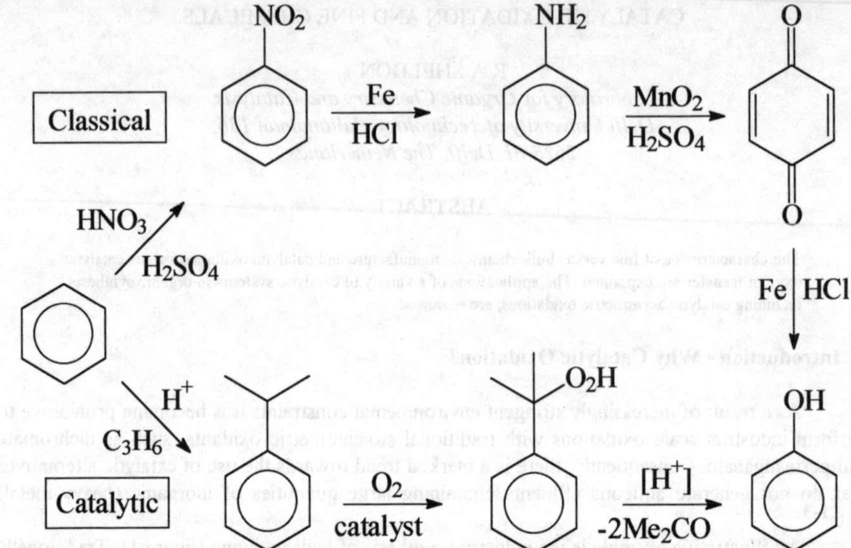


Figure 1. Two routes to hydroquinone.

stability. This necessitates reaction in the liquid phase at moderate temperatures. In addition, many of the desired transformations involve chemo-, regio- or stereoselectivity. Processing tends to be multipurpose and batch-wise, rather than dedicated and continuous as in bulk chemicals production. This means that not only raw materials costs but also simplicity of operation and multipurpose character of the installations are important economic considerations. Furthermore, fine chemicals cannot, generally speaking, bear the costs of the extensive research program characteristic of the development of a proprietory catalyst for a large volume chemical. Consequently, one may have to be content with a catalyst that is perhaps not optimal but is readily available.

## 3. Gas versus Liquid Phase Oxidations.

As noted above the use of gas phase oxidation is often precluded in fine chemicals synthesis. Nevertheless, where feasible it will often be the method of choice and several examples

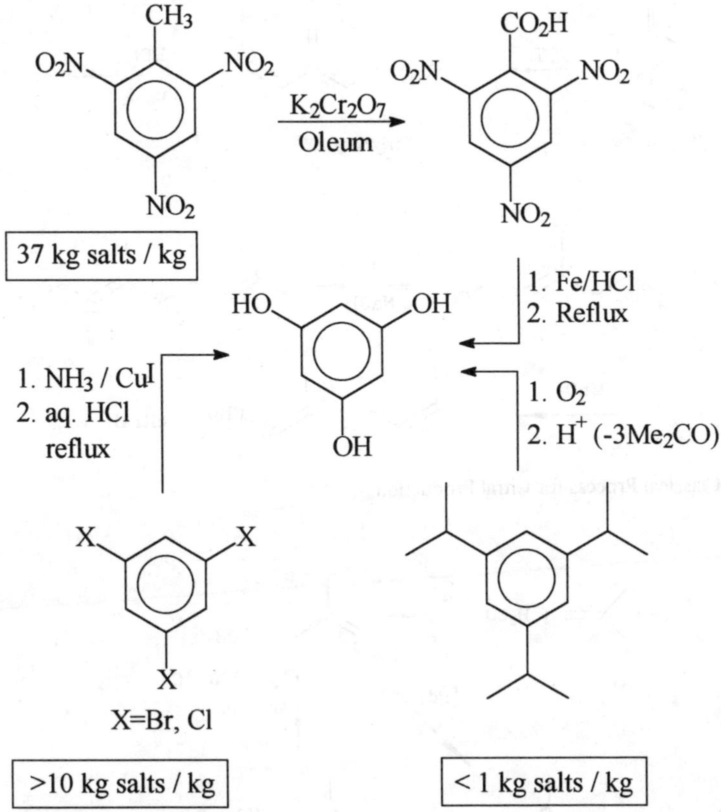


Figure 2. Phoroglucinol production.

of fine chemicals production via gas phase oxidation are known. A prime example is the BASF process[4] for the manufacture of citral, an intermediate in the synthesis of fragrance chemicals and vitamin A. The traditional process involves a five step process, a key step of which is a stoichiometric oxidation with $MnO_2$.

The first step in the BASF route (figure 4) involves acid-catalyzed condensation of the inexpensive raw materials, isobutene and formaldehyde, to give 2-methyl-1-butene-4-ol (isoprenol). A key step in the process is the chemoselective gas phase oxidation of the latter over a supported silver catalyst to yield the corresponding unsaturated aldehyde (figure 4). In order to obtain a high selectivity (95%) the reaction is carried out continuously in a short fixed bed reactor at $500^0$C with very short (0.001 sec.) residence times. The catalyst is the same as that used in the manufacture of formaldehyde by gas phase oxidation of methanol. This process is a perfect example of the commercial production of a fine chemical via gas phase oxidation with dioxygen.

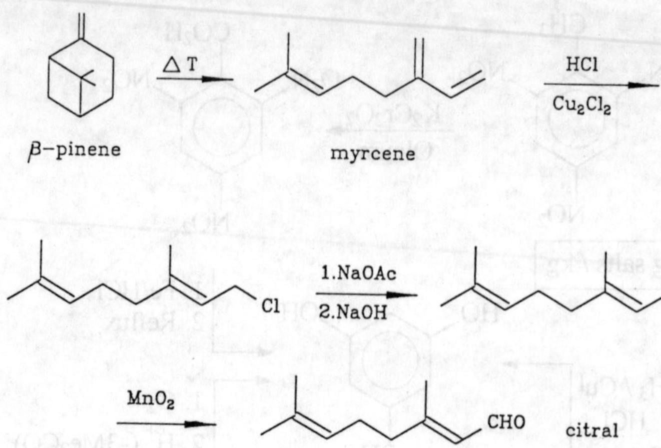


Figure 3. Classical Process for Citral Production.

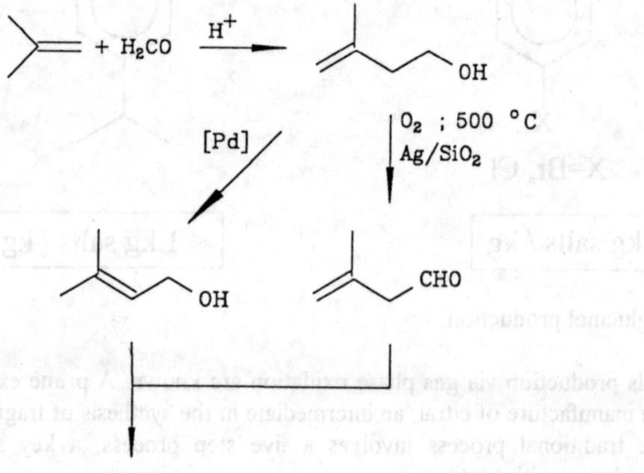


Figure 4. BASF Process for Citral Manufacture.

Another example of the use of gas phase oxidation for fine chemicals production is the Lonza process[5] for nicotinic acid manufacture in which the key step is the ammoxidation of beta-picoline 3-cyanopyridine (reaction 1). Gas phase ammoxidation has also been applied to the production of other (hetero) aromatic nitriles, e.g. 2-cyanopyrazine (reaction 2)[6] and 2,6-dichloro-benzonitrile (reaction 3). The former is the precurser of 2-amidopiperazine, an antituberculostatic, and the latter is an agrochemical intermediate.

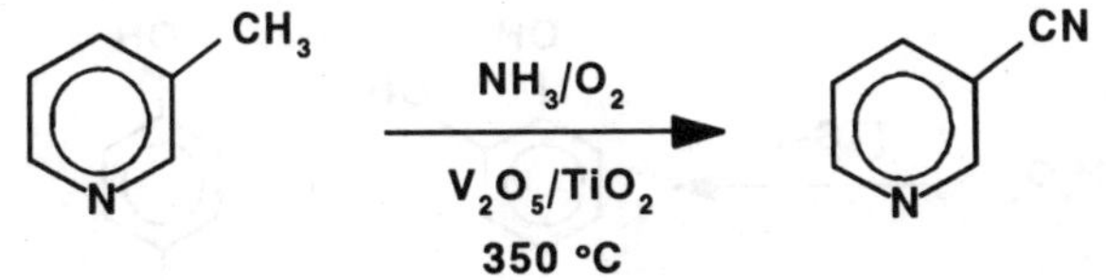


(1)

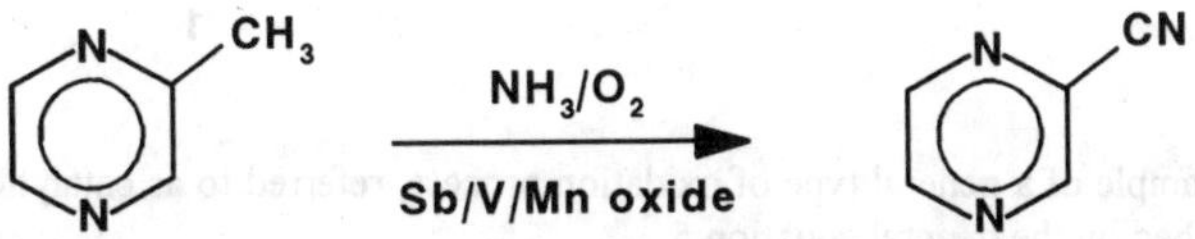


(2)

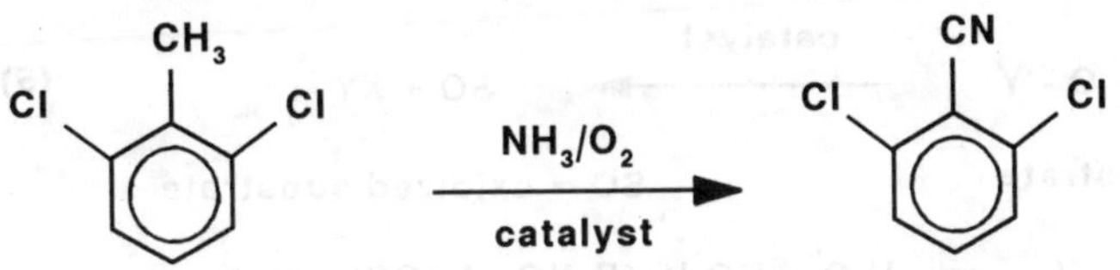


(3)

## 4. Liquid Phase Oxidation-Catalytic Oxidation versus Oxygen Transfer

In bulk chemicals manufacture the choice of oxidant is largely restricted, for economic reasons, to dioxygen. In contrast to catalytic reactions with $H_2$ and CO catalytic oxidations with $O_2$ are complicated by the fact that dioxygen reacts with organic substrates even in the absence of a catalyst. In the liquid phase this blank reaction involves the formation of hydroperoxide intermediates via a free radical chain process (see chapter 8). A major problem associated with autoxidations is that they are largely indiscriminate, i.e. they exhibit poor chemo- and regioselectivities. Moreover, primary oxidation products such as alcohols and aldehydes are generally more susceptible to autoxidation than the hydrocarbon substrates, thus necessitating low conversions and recycling of large quantities of substrate. Consequently, liquid phase autoxidation is synthetically useful only with relatively simple substrates containing one reactive position, e.g. the oxidation of substituted toluenes to the corresponding carboxylic acids (see chapter 8). The economics of fine chemicals manufacture, in contrast. allow for a broader choice of oxidant.

Indeed, even though it is more expensive than dioxygen, hydrogen peroxide is often the oxidant of choice in fine chemicals synthesis because of its simplicity of operation. An example of its commercial use is in the synthesis of a mixture of catechol and hydroquinone by the hydroxylation of phenol (reaction 4) using, for example, the TS-1 catalyst (see chapter 10). In this case, synthesis via the autoxidation of o-diisopropylbenzene is not feasible due to competing intramolecular processes.

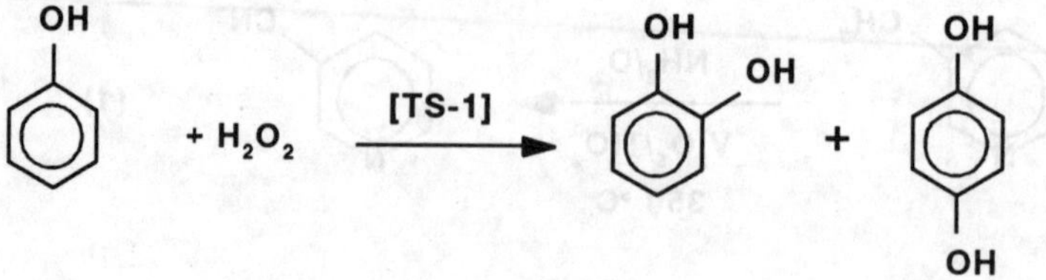


(4)

1 : 1

Reaction 4 is one example of a general type of oxidation process, referred to as catalytic oxygen transfer,[7] that is described by the general equation 5.

$$S + X - O - Y \xrightarrow{\text{catalyst}} SO + XY \qquad (5)$$

S = substrate　　　　SO = oxidized substrate

X - O - Y = $H_2O_2$ , $RO_2H$ , $R_3NO$ , NaOCl , etc.

It is interesting to compare the advantages and limitations of catalytic oxygen transfer with, on the one hand, stoichiometric oxidations and, on the other hand, catalytic oxidations with dioxygen (see Table 1). The latter have the advantage of a cheap oxidant ($O_2$) and few effluent problems but are restricted in scope (see earlier) while the former are broadly applicable but use expensive. environmentally unacceptable reagents. Catalytic oxygen transfer, in contrast, has broad scope and uses relatively inexpensive, environmentally acceptable reagents. It is, therefore, eminently suited for application in fine chemicals production.

It is also instructive to compare the characteristics of catalytic oxidations and reductions (Table 2). Catalytic hydrogenation, in addition to being a high atom utilization, low-salt technology, is relatively simple and inexpensive, has a broad scope in organic synthesis and is a technique well-known to organic chemists. Consequently, there is little incentive to apply catalytic hydrogen transfer techniques in organic synthesis. The situation with regard to catalytic oxidation is completely different (see Table 2). When dioxygen is the oxidant the scope is limited and there is almost always a reaction in the absence of the catalyst. Due to the indiscriminate reactivity of $O_2$ there is also a limited choice of solvents. Hence catalytic oxygen transfer constitutes an

Table 1. Comparison of Oxidation Methods.

| Catalytic oxidation ($O_2$) | Catalytic oxygen transfer | Stoichiometric oxidation |
|---|---|---|
| **Advantages:**<br>- Cheap oxidant ($O_2$)<br>- No effluent problem<br><br>**Disadvantages:**<br>- Limited scope (petrochemicals) | **Advantages:**<br>- Relatively cheap oxidants (e.g. $H_2O_2$, NaOCl)<br>- Environmentally acceptable<br>- Broad scope | **Advantages:**<br>- Broad scope<br><br>**Disadvantages:**<br>- Expensive oxidants ($K_2Cr_2O_7$, $KMnO_4$, $Pb(OAc)_4$, etc. |

attractive, much used alternative. It should be noted, however, that in metal-catalyzed oxidations with peroxidic oxidants one still has to cope with the problem of competing homolytic and heterolytic pathways.

Table 2. Characteristics of Catalytic Oxidation and Catalytic Reduction.

| Catalytic oxidation ($O_2$) | Catalytic hydrogenation($H_2$) |
|---|---|
| - Reaction in absence of catalyst<br>- Limited choice of (inert) solvent<br>- Selectivity problems<br>- Limited scope<br>- Limited application (bulk petrochemicals) | - No reaction without catalyst<br>- Wide choice of solvent<br>- Highly selective<br>- Broad scope in organic synthesis<br>- Widely applied in fine and bulk chemicals |
| **Oxygen transfer** | **Hydrogen transfer** |
| - Attractive alternative with broad scope in fine chemicals | - Only sporadically applied because of broad scope of hydrogenation with $H_2$ |

## 5. Choice of Metal and Oxidant

Virtually all of the transition metals and several main group elements are known to catalyze oxygen transfer processes.[1-3,7] A variety of single oxygen donors can be used (Table 3). Next to price and ease of handling two important considerations which influence the choice of oxidant are the nature of the coproduct and the percentage weight of available oxygen. The former is obviously important in the context of environmental accepability and the latter bears directly on the productivity (kg product per unit reactor volume per unit time).

With these criteria in mind it is readily apparent that hydrogen peroxide is a choice oxidant, its coproduct being water. In principle, it contains 47% active oxygen but in practice it is generally used as a 30-35% aqueous solution which translates to 15% active oxygen.

Table 3. Oxygen Donors.

| Donor | % Active Oxygen | Coproduct |
|---|---|---|
| $H_2O_2$ | 47.0 (14.1)[a] | $H_2O$ |
| $N_2O$ | 36.4 | $N_2$ |
| $O_3$ | 33.3 | $O_2$ |
| $CH_3CO_3H$ | 21.1 | $CH_3CO_2H$ |
| $t\text{-}BuO_2H$ | 17.8 | t-BuOH |
| $HNO_3$ | 25.4 | $NO_x$ |
| NaOCl | 21.6 | NaCl |
| $NaOCl_2$ | 35.6 | NaCl |
| NaOBr | 13.4 | NaBr |
| $C_5H_{11}NO_2$ [b] | 13.7 | $C_5H_{11}NO$ |
| $KHSO_5$ | 10.5 | $KHSO_4$ |
| $NaIO_4$ | 7.5 | $NaIO_3$ |
| PhIO | 7.3 | PhI |

a. Figure in parentheses refers to 30% $H_2)O_2$ (b) N-methylmorpholine-N-oxide

The coproduct from organic oxidants, such as TBHP and amine oxides, is readily recycled via reaction with hydrogen peroxide. The overall process produces water as the coproduct, but requires one extra chemical step compared to the corresponding reaction with hydrogen peroxide. With inorganic oxygen donors environmental considerations are relative. Sodium chloride and potassium bisulfate are obviously preferable to heavy metal (Pb, Cr, Mn, etc) salts. Generally speaking, inorganic oxidants are more difficult to recycle, in an economic manner, than organic ones. Indeed, the ease of recycling may govern the choice of oxidant, e.g. NaOBr may be preferred to NaOCl because NaBr can, in principle, be reoxidized with $H_2O_2$.

## 6. **Mechanism of Oxygen Transfer**

Heterolytic oxygen transfer processes can be divided into two categories based on the nature of the active oxidant: an oxometal or a peroxometal species (figure 5). Generally speaking, catalysis by early transition metals (Mo, W, V, Ti, etc) involves high-valent peroxometal complexes whereas later transition metals (Mn, Fe, Ru), particularly first row elements (e.g. Cr), mediate oxygen transfer via oxometal species. Some elements (e.g. vanadium) can , depending on the substrate, operate via either mechanism. Although the pathways outlined in figure 5 pertain to peroxidic reagents analogous schemes, involving M=O or MOX (X=ClO, $IO_4$, $R_3N$, etc) species, can be envisaged for other oxygen donors.

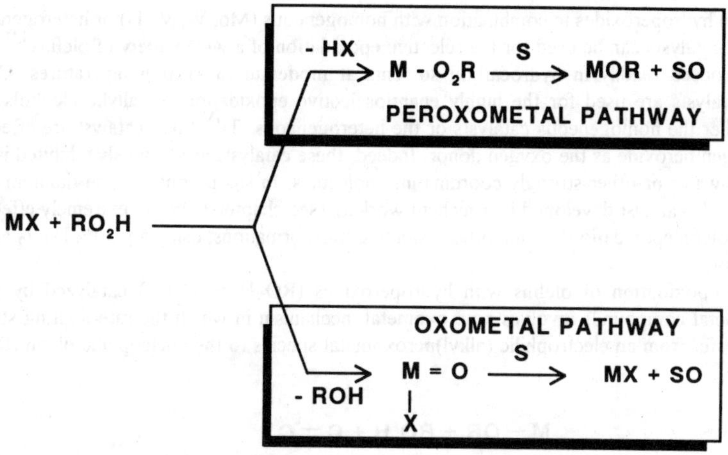


Figure 5. Mechanisms of Oxygen Transfer

## 7. **Olefin Oxidations**

The most important example of a catalytic oxygen transfer process is undoubtedly the metal-catalyzed epoxidation of olefins with alkyl hydroperoxides.[8-11] The epoxidation of propylene with TBHP or ethylbenzene hydroperoxide (EBHP), for example, accounts for more than one million tons of annual, worldwide production of propylene oxide (reaction 6).

$$CH_3CH{=}CH_2 + RO_2H \xrightarrow{\text{catalyst}} CH_3CH\overset{O}{-}CH_2 + ROH \qquad (6)$$

$R = (CH_3)_3C\text{-}$ or $PhCH(CH_3)\text{-}$

**Catalyst : Homogeneous : $Mo^{VI}$ (Arco)**
**Heterogeneous : $Ti^{IV}/SiO_2$ (Shell)**

Reaction (6) is catalyzed by compounds of high-valent, early transition metals such as $Mo^{VI}$, $W^{VI}$, $V^{V}$ and $Ti^{IV}$. Molybdenum compounds are particularly effective homogeneous catalysts[8,11] and are used in the Arco process in combination with either TBHP or EBHP. In the Shell process, on the other hand, a heterogeneous $Ti^{IV}/SiO_2$ catalyst is used with EBHP in a continuous, fixed-bed operation.

Alkyl hydroperoxides in combination with homogeneous (Mo, W, V, Ti) or heterogeneous ($Ti^{IV}/SiO_2$) catalysts can be used for the selective epoxidation of a wide vatiety of olefins[8,9]. The reactions proceed readily in hydrocarbon solvents at moderate (80-120°)temperatures. Chiral titanium catalysts are used for the highly enantioselective epoxidations of allylic alcohols (see later). Neither the homogeneous catalysts or the heterogeneous $Ti^{IV}/SiO_2$ catalyst are effective with hydrogen peroxide as the oxygen donor. Indeed, these catalysts are seriously inhibited in the presence of water or other strongly coordinating molecules. In sharp contrast, the titanium (IV) silicalite (TS-1) catalyst developed by Enichem workers (see chapter 10) is an extremely effective catalyst for olefin epoxidation[12], and other oxidative transformations, using aq. 30% $H_2O_2$ as the oxygen donor.

The epoxidation of olefins with hydroperoxides ($RO_2H$ or $H_2O_2$) catalyzed by early transition metal compounds involves a peroxometal mechanism in which the rate-limiting step is oxygen transfer from an electrophilic (alkyl)peroxometal species to the nucleophilic olefin (figure 6).

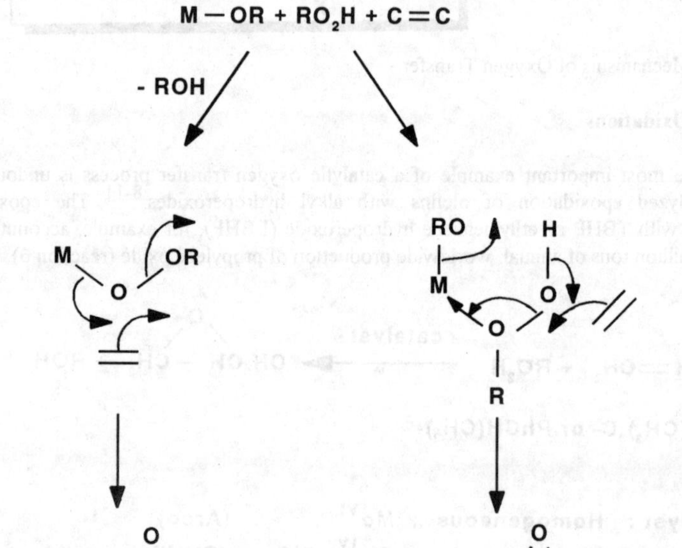


Figure 6. Mechanism of Oxygen Transfer from a Peroxometal Complex to an Olefin.

In addition to epoxidation metal catalyst-oxygen donor reagents effect a variety of synthetically useful oxidative transformations [7-9] of olefins (figure 7).

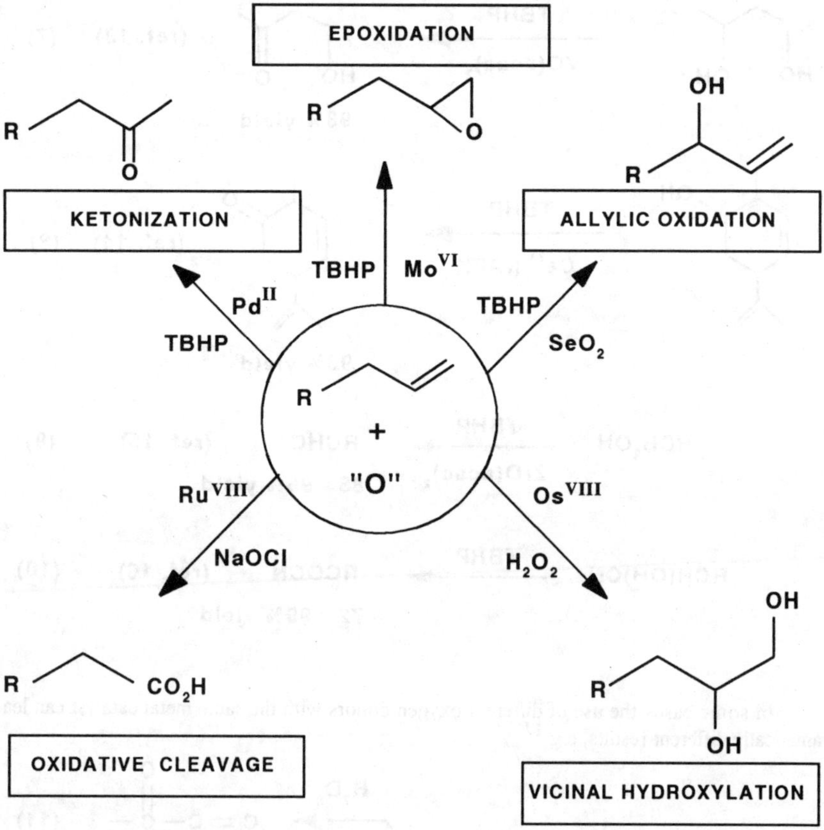


Figure 7. Oxidative Transformations of Olefins.

## 8. **Alcohol Oxidations**

The metal catalyst-hydroperoxide ($RO_2H$ or $H_2O_2$)reagents are also extremely effective for the chemoselective oxidation of alcohols and the regioselective oxidation of diols[2,8,9]. As such they constitute environmentally attractive alternatives to classical procedures e.g. oxidation with stoichiometric quantities of chromium(VI) reagents. A few selected examples are shown below (reactions 7-10).

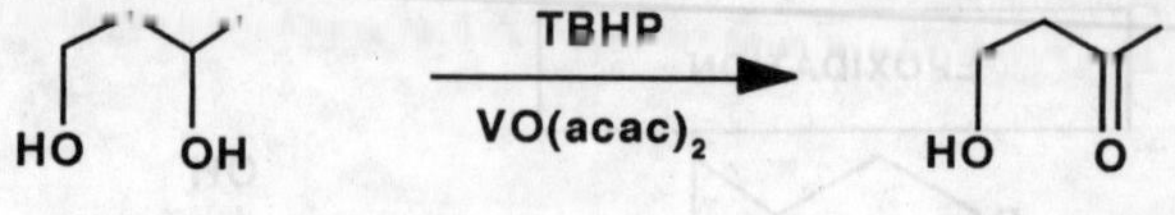


(ref. 10) (7)

98% yield

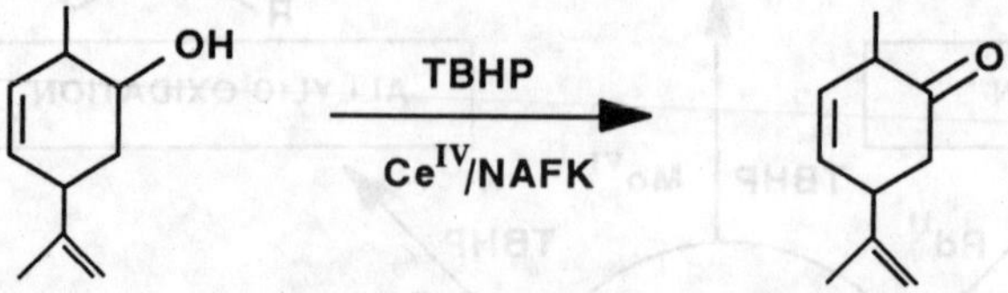


(ref. 14) (8)

98% yield

$$RCH_2OH \xrightarrow[ZrO(acac)_2]{TBHP} RCHO \quad \text{(ref. 15)} \quad (9)$$

85 - 95% yield

$$RCH(OH)CN \xrightarrow{TBHP} RCOCN \quad \text{(ref. 16)} \quad (10)$$

72 - 99% yield

In some cases the use of different oxygen donors with the same metal catalyst can lead to dramatically different results, e.g.[17]

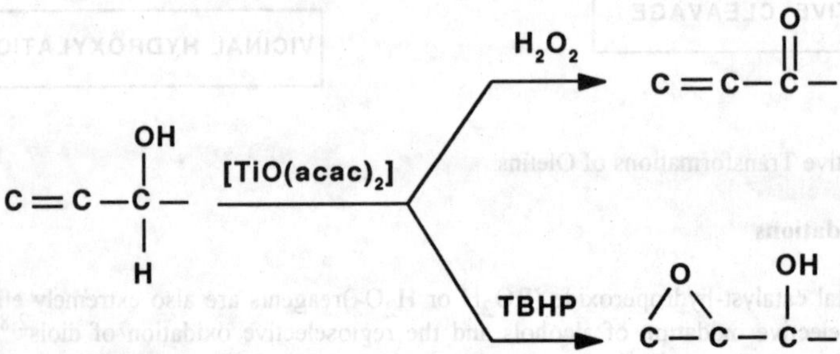


(11)

(12)

A possible explanation is that in the presence of water (i.e. with $H_2O_2$) epoxidation of the double bond is seriously hampered.

The alcohol oxidations outlined above can involve peroxometal or oxometal mechanisms. Which of the two is operating can quite easily be ascertained by carrying out the reaction, with a stoichiometric amount of the catalyst, in the absence of the oxygen donor. Oxidation is observed under these conditions only with catalysts operating via an oxometal mechanism, e.g. vanadium:

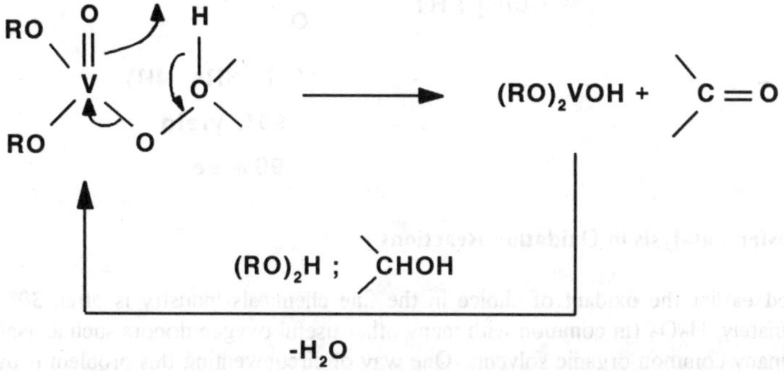


(13)

At this point it is also worth noting that TBHP has several advantages compared to other peroxidic reagents. It has high thermal stability and is safer to handle than $H_2O_2$ or $CH_3CO_3H$. It is non-corrosive and unreactive to most functional groups in the absence of catalysts. It is readily soluble in nonpolar solvents, e.g. hydrocarbons, and reactions are carried out under neutral conditions. Furthermore, the coproduct tert-butanol is readily removed by distillation. Consequently, metal-catalyzed oxidations with TBHP have found wide applications in organic synthesis[1-3,8,9].

## 9. **Ruthenium-Catalyzed Oxidation of Lactams**

The ruthenium-catalyzed acetoxylation of beta-lactams with peracetic acid is a perfect illustration of the application of catalytic oxygen transfer, to a difficult and delicate oxidative transformation (reactions 14 and 15). It constitutes a key step in the Takasago process for the commercial synthesis of a carbapenem intermediate (reaction 15) whereby both high chemoselectivity (99% yield) and diastereoselectivity (>99% de) is observed[18].

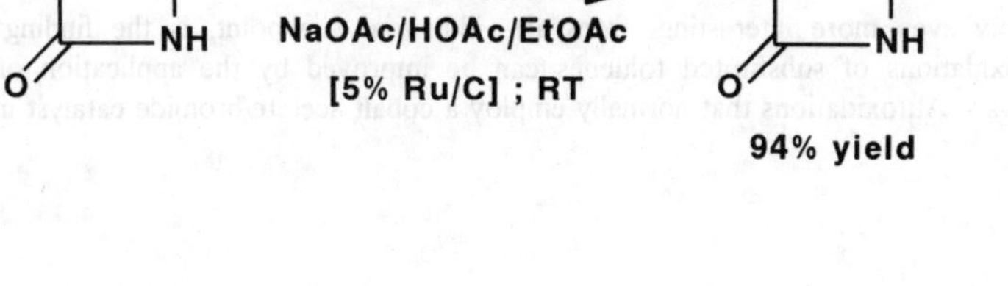


(14)

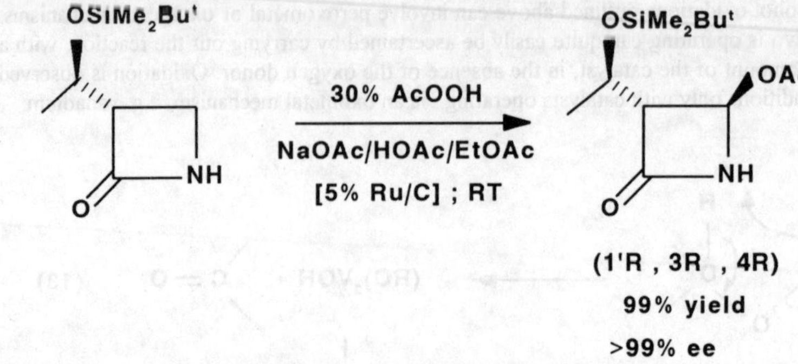


(15)

## 10. **Phase Transfer Catalysis in Oxidation Reactions**

As noted earlier the oxidant of choice in the fine chemicals industry is often 30% aq. $H_2O_2$. Unfortunately, $H_2O_2$ (in common with many other useful oxygen donors such as NaOCl) is insoluble in many common organic solvents. One way of circumventing this problem is by the application of phase transfer catalysis. This usually involves the transfer of the primary oxidant (e.g. $ClO^-$, $S_2O_8^{2-}$) or the catalyst (e.g. $RuO_4^-$, $HMoO_6^-$), as an anion of a quaternary ammonium salt, to the organic phase. It has been used, for example, in the ruthenium-catalyzed cleavage of olefins (reaction 16)[19] and the oxidation of substituted toluenes to the corresponding carboxylic acids (reaction 17)[20] with NaOCl asthe primary oxidant.

$$CH_3(CH_2)_{12}CH{=}CH_2 \xrightarrow[\substack{[RuO_2/Bu_4NBr] \\ CH_2Cl_2/H_2O}]{NaOCl\text{ , }NaOH} CH_3(CH_2)_{12}CO_2Na \quad (16)$$

100% yield

$$ArCH_3 \xrightarrow[\substack{[RuCl_3/Bu_4NBr] \\ ClCH_2CH_2Cl/H_2O \\ 25^\circ C\text{ , }pH = 9}]{NaOCl\text{ , }NaOH} ArCO_2Na \quad (17)$$

98% yield

Probably even more interesting, from an industrial viewpoint, is the finding[21] that catalytic autoxidations of substituted toluenes can be improved by the application of phase transfer catalysis. Autoxidations that normally employ a cobalt acetate/bromide catalyst in acetic acid can be carried out with the neat hydrocarbon by employing a tetraalkylammonium (phosphonium) bromide in combination with cobalt chloride. High selectivities were observed at high substrate conversions.

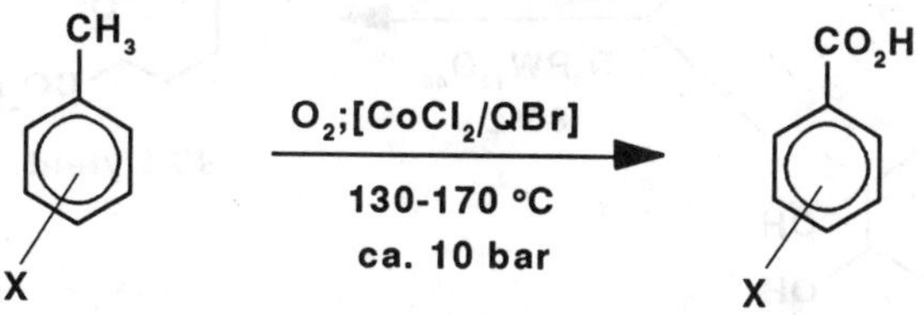


(18)

X = H , o - Me , m - Me , p - Me , p - Br , p - $NO_2$ , p - Ph , o - Ph , p - MeO

Q = $(C_{10}H_{21})_2(CH_3)_2N$ ; $(C_6H_{13})_4P$ , etc.

The first example of a successful catalytic epoxidation of olefins with $H_2O_2$ under phase transfer conditions was reported by Venturello and coworkers[22]:

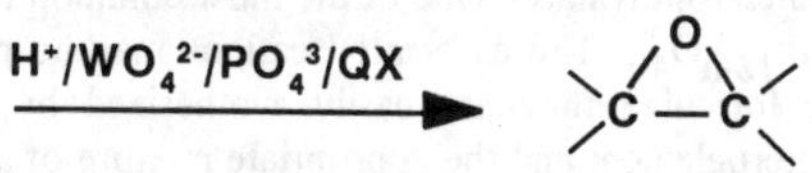


(19)

Q = $(C_8H_{17})_3NCH_3$ (aliquat 336)

Subsequently, tungsten and molybdenum-based catalysts have been widely employed in the epoxidation of olefins, the oxidation of alcohols and the oxidative cleavage of vicinal diols in aqueous/organic biphasic systems[1,2]. Both simple molybdate and tungstate as well as Mo and W heteropolyanions[23] have been employed as catalysts. A typical example of the latter is the $H_3PM_{12}O_{40}$ (M=Mo, W)/cetyl-pyridinium chloride system[23,24] that catalyzes the efficient epoxidation of olefins and allylic alcohols under biphasic conditions. Analogous oxidations of secondary alcohols to ketones and the oxidative cleavage of olefins and vicinal diols, on the other hand, are best performed in a one-phase system with tert-butanol as solvent.

More recently, a detailed physicochemical study[25] of these systems revealed that the same active oxidant is involved irrespective of whether the Venturello system or the heteropolyacid, $H_3PW_{12}O_{40}$ is used. The active species is the less condensed heteropolyoxoperoxo anion, $PO_4[WO(O_2)_2]_4^{3-}$, containing eight peroxo ligands.

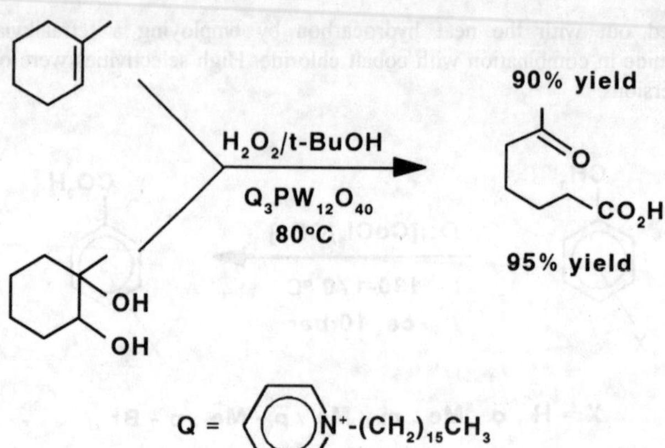


(20)

11. **Heteropolyacids as Oxidation Catalysts**

Heteropolyacids (HPAs)[26-28] and their salts are polyoxo compounds incorporating anions (heteropolyanions) having metal-oxygen octahedra ($MO_6$) as the basic structural unit. They contain one or more heteroatoms, such as Si, Ge, P or As, that are usually located at the centre of the anion. The $MO_6$ octahedra are linked together to form a thermally stable and compact structure for the heteropolyanion. One of the most common types comprises the so-called Keggin anions, $XM_n{}^1M^2{}_{12-n}O_{40}$. The associated cations may be protons or metal ions. Despite their rather awesome formulae they are easily synthesized by acidification of aqueous solutions containing the heteroelement and the appropriate mixture of alkali metal molybdate, tungstate and vanadate.

HPAs are both strong Bronsted acids and multielectron oxidants, i.e. they are potential bifunctional catalysts. They are soluble in oxygen-containing organic solvents which means they can be regarded as `soluble oxides'. Furthermore, the heteropolyanion constitutes a multielectron ligand that can stabilize reactive high-valent oxometal species, i.e. transition metal-substituted HPAs (polyoxometallates) can be regarded as oxidatively resistant analogues of metalloporphyrins.

In the past, HPAs have been widely applied in heterogeneous gas phase reactions[29] where oxidations proceed via Mars-van Krevelen type mechanisms (see chapter 1 ). More recently they have been increasingly applied to liquid phase oxidations[26-28]. As noted in the preceding section Mo-and W-based HPAs are used in combination with $H_2O_2$ as the primary oxidant, under phase transfer conditions, for a variety of oxidative transformations. However, a word of caution is in order: in the presence of $H_2O_2$ heteropolyanions may undergo degradation to less condensed structures (see section 10).

Some HPAs, such as $H_3PMo^{VI}{}_{12-n}V^V{}_nO_{40}$ (PMoV-n) are strong oxidants in their own right and catalyze the oxidation of organic substrates with dioxygen as the primary oxidant. For example, 2-methylnaphthalene is oxidized to 2-methyl-1,4-naphthoquinone (menadione) by dioxygen (3-8 bar) in the presence of 0,02-0.2 M PMoV-n in a $HOAc-H_2O-H_2SO_4$ solution at 120-140°C[26] (reaction 21). Tradionally menadione (vitamin $K_3$) was produced by oxidation of 2-methylnaphthalene with stoichiometric amounts of chromium trioxide, a process that produces 18kg of chromium-containing solid waste per kg of product.

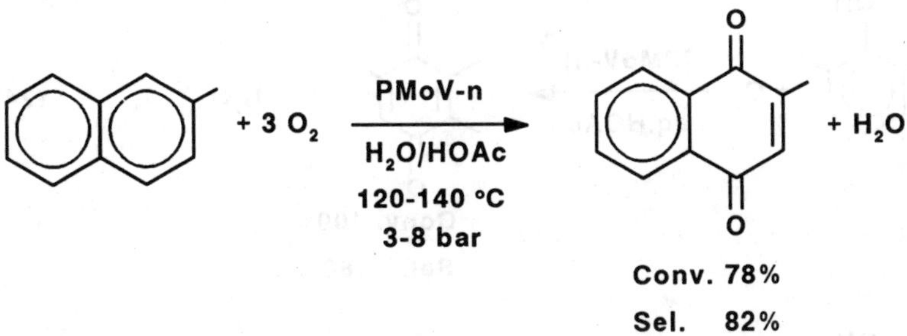


(21)

Interestingly, the same authors[26] have reported an alternative two-step synthesis of menadione from 1-naphthol (reactions 22 and 23). It involves gas phase methylation followed by liquid phase oxidation with dioxygen in the presence of PMoV-n as the catalyst.

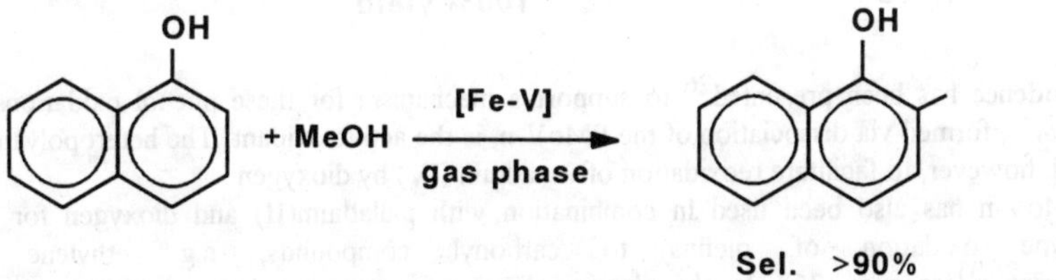


(22)

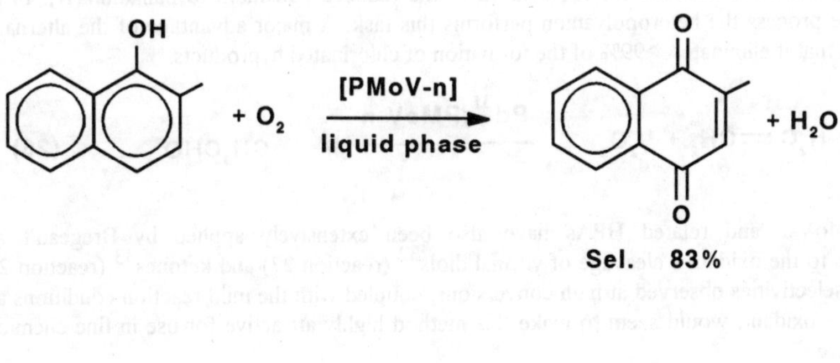


(23)

The same system has been used for the oxidation of 2,3,6-trimethylphenol to the corresponding para-benzoquinone[30] (reaction 24), which is an intermediate in the synthesis of vitamin E, and for the oxidative coupling of 2,6-dialkylphenols to diphenoquinones[26] (reaction 25).

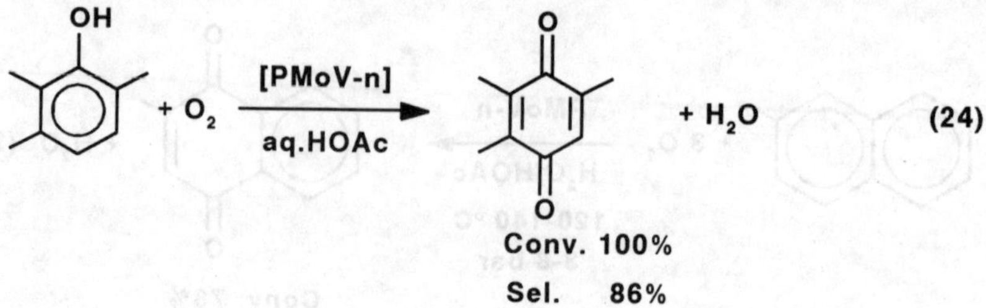


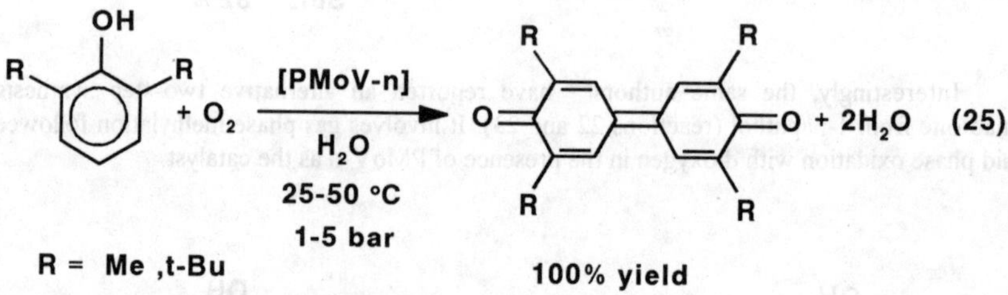


Evidence has been presented[26] to support a mechanism for these phenol oxidations in which $VO_2^+$, formed via dissociation of the PMoV-n, is the active oxidant. The heteropolyanion is required, however, to facilitate reoxidation of vanadium (IV) by dioxygen.

PMoV-n has also been used in combination with palladium(II) and dioxygen for the Wacker-type oxidation of olefins to carbonyl compounds, e.g. ethylene to acetaldehyde[26,31](reaction 26). In the classical Wacker process catalyst a copper chloride cocatalyst is used to mediate the reoxidation of the reduced palladium to palladium(II). In this alternative process the heteropolyanion performs this task. A major advantage of the alternative process is that it eliminates >99% of the formation of chlorinated byproducts.

$$H_2C{=}CH_2 + {}^1/_2O_2 \xrightarrow{Pd^{II}/PMoV\text{-}n} CH_3CHO \qquad (26)$$

PMoV-2 and related HPAs have also been extensively applied by Bregeault and coworkers to the oxidative cleavage of vicanal diols[32] (reaction 27) and ketones[33] (reaction 28). The high selectivities observed at high conversions, coupled with the mild reaction conditions and inexpensive oxidant, would seem to make this method highly attractive for use in fine chemicals manufacture.

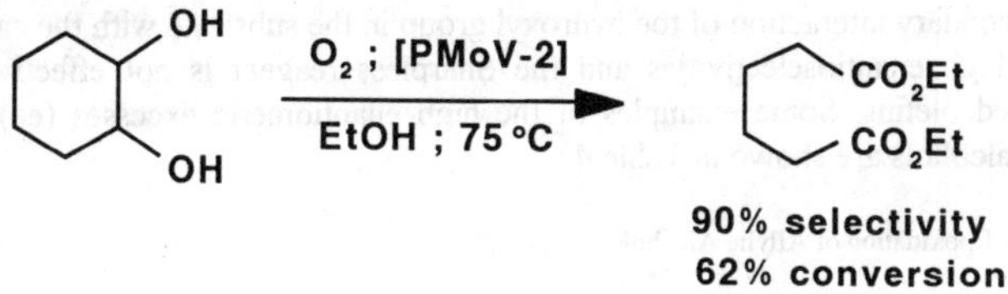


(27)

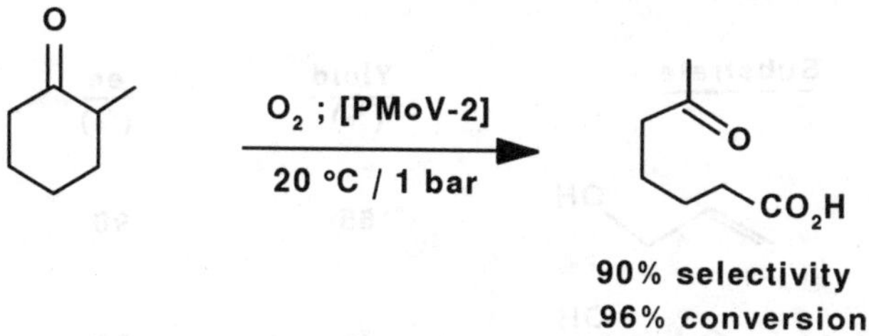


(28)

From the above selected examples it is apparent that HPA-catalyzed oxidations with clean oxidants, such as $O_2$ and $H_2O_2$, have considerable synthetic potential. They may be considered as a bridge between heterogeneous gas-phase oxidations and liquid phase homogeneous oxidations. It should be pointed out, however, that the heterogeneous redox molecular sieve catalysts such as titanium silicalite (TS-1) and related catalysts (see chapter 10) catalyze many of the same reactions. The latter catalysts have the advantage of high stability and ease of recovery and recycling.

## 12. **Catalytic Asymmetric Oxidations**

One of the most challenging goals in catalysis is the design of simple abiological catalysts that can achieve high levels of enantioselectivity. With this goal in mind much effort has been devoted to the development of relatively simple metal catalyst/chiral ligand/oxygen donor combinations capable of mimicking nature's selective and versatile oxidation catalysts, the monooxygenases. Perhaps the most well-known of these is the titanium(IV)/dialkyltartrate/TBHP system developed by Sharpless and coworkers[34]. This system is very effective for the catalytic asymmetric epoxidation of allylic alcohols (reaction 29).

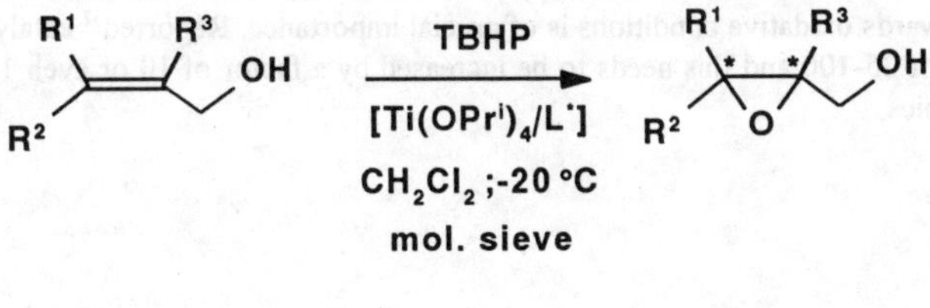


(29)

The secondary interaction of the hydroxyl group in the substrate with the catalyst is crucial for achieving high enantioselectivities and the Sharpless reagent is not effective with simple, unfunctionalized olefins. Some examples of the high enantiomeric excesses (ee) obtained with various allylic alcohols are shown in Table 4.

Table 4. Sharpless Epoxidation of Allylic Alcohols[a].

| Substrate | Yield (%) | ee (%) |
|---|---|---|
| 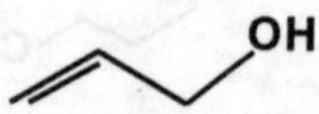 | 65 | 90 |
| 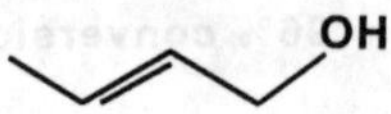 | 70 | 96 |
| 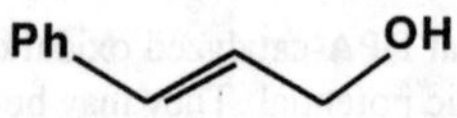 | 89 | >98 |
| 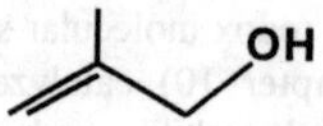 | 50 | >95 |
| 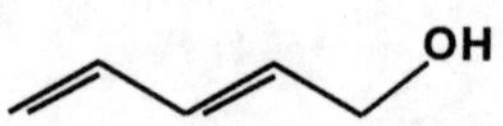 | 40 | 95 |

a. 5 mol % catalyst

More recently, Jacobson[35] and Katsuki[36] have independently developed manganese (III) complexes of chiral Schiff's bases as catalysts for the enantioselective epoxidation of unfunctionalized olefins using NaOCl or PhIO as the oxygen donor. Subsequent fine tuning of the chiral ligand structure afforded highly effective catalysts (e.g. complexes A and B) for the enantioselective epoxidation (reaction 30) of a range of olefins (see Table 5)[37]

From the viewpoint of commercial applications the stability of these relatively expensive chiral ligands towards oxidative conditions is of crucial importance. Reported[37] catalyst turnovers are of the order of 25-100 and this needs to be increased by a factor of 10 or even 100 to obtain attractive economics.

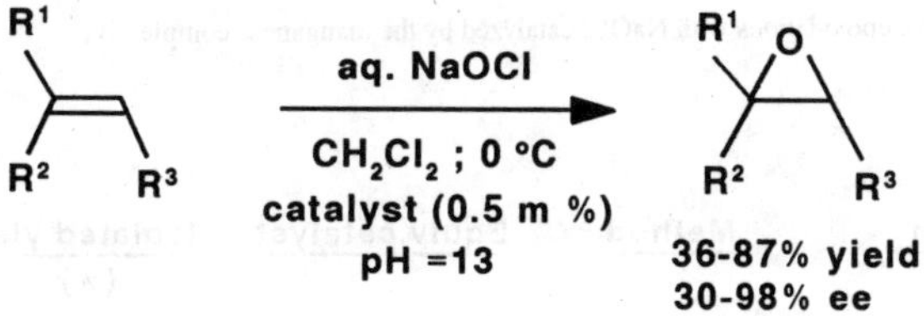


(30)

Catalyst :

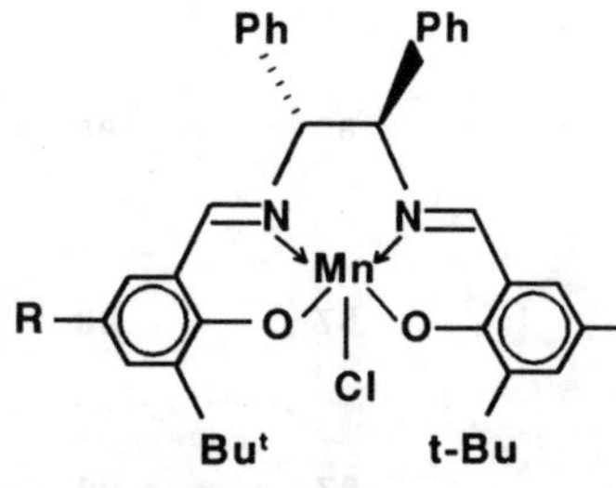


A (R,R)

R = $CH_3$ , t-Bu

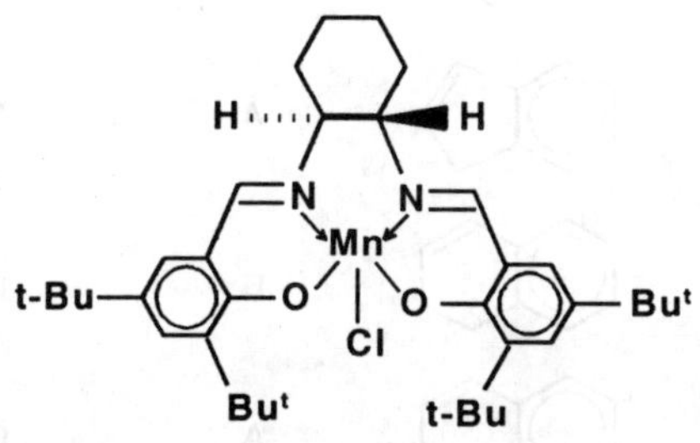


B (S,S)

The Sharpless group[38,39] has also developed a highly effective system for the asymmetric vicinal dihydroxylation of olefins (reaction 31). The system comprises an $OsO_4$ catalyst incombination with dihydroquinine or dihydroquinidine esters or related chiral ligands and N-methylmorpholine-N-oxide (NMO) or potassium ferricyanide, $K_3Fe(CN)_6$, as the primary oxidant.

Table 5. Enantioselective epoxidations with NaOCl catalyzed by the manganese complex B[a].

| Olefin | Method [b] | Equiv.catalyst | Isolated yield (%) | ee (%) |
|---|---|---|---|---|
| 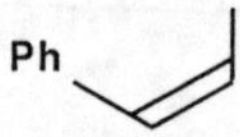 | A | 0.04 | 84 | 92 |
| 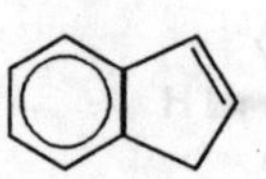 | A | 0.01 | 80 | 88 |
| 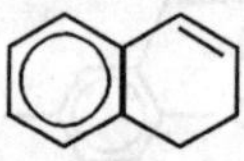 | B | 0.04 | 67 | 88 |
| 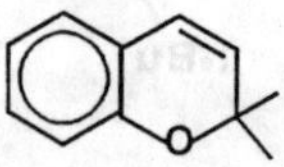 | A | 0.02 | 87 | 98 |
| 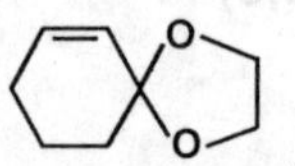 | B | 0.15 | 63 | 94 |
| 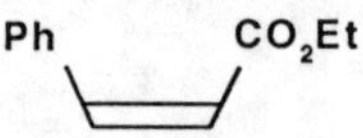 | B | 0.08 | 67 | 97 |

a. Data taken from ref. 37

b. Method A : NaOCl , pH = 11.3 , $CH_2Cl_2$ , 0 °C

Method B : Same as A + 0.2 eq. 4-phenylpyridine-N-oxide

In this case secondary interaction with a coordinating functional group is not essential and the reaction is successful with a broad range of olefin substrates. The efficacy of the Sharpless asymmetric epoxidation (AE) and asymmetric dihydroxylation (AD) methods are compared in Table 6. It is readily apparent that asymmetric dihydroxylation has a much broader scope.

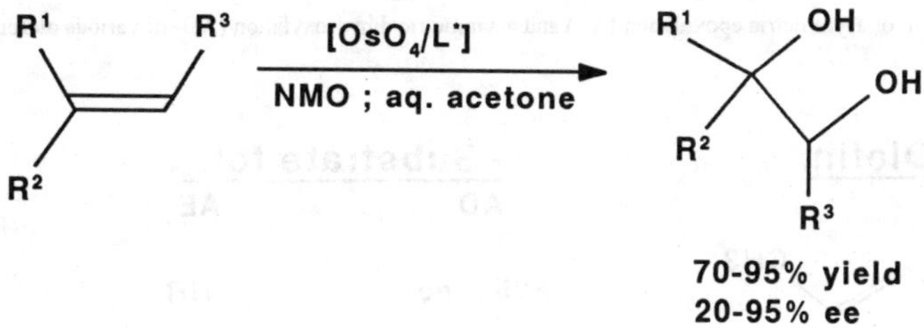


(31)

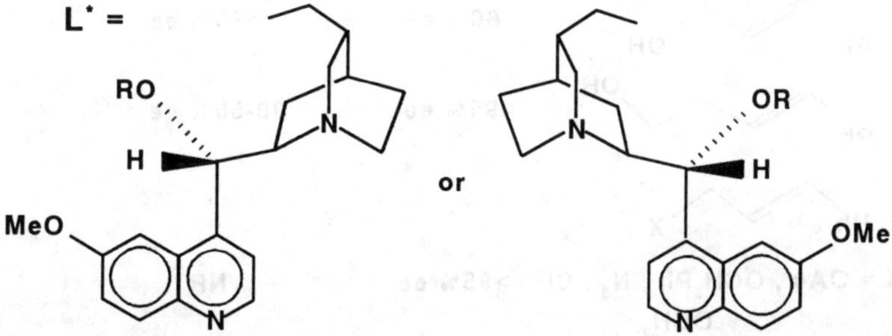


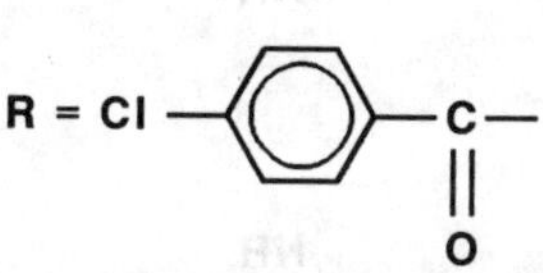


NMO = N-methylmorpholine - N-oxide

If a metal-chiral ligand complex rapidly exchanges its ligands in solution then a prerequisite for high enantioselectivity is that coordination of the chiral ligand leads to a substantial rate accelaration. Sharpless coined the team **ligand-accelerated catalysis** to describe this phenomenon, Thus, if the metal-chiral ligand complex (M-L) rapidly exchanges its ligands in solution, then high enantioselectivities will be observed only when M-L is a much more active catalyst than M :

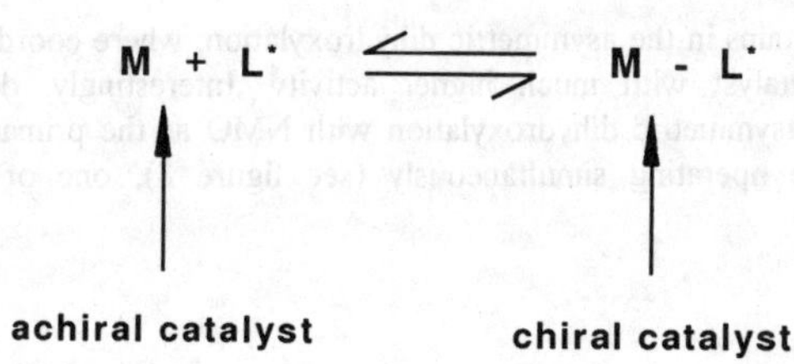


(32)

Table 6. Comparisn of asymmetric epoxidation (AE) and asymmetric dihydroxylation (AD) of various olefins.[a]

| Olefin | Substrate for AD | Substrate for AE |
|---|---|---|
| 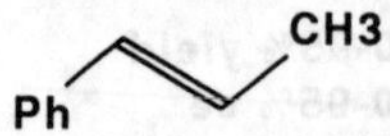 | >95% ee | NR |
|  | 80% ee | >95% ee |
| 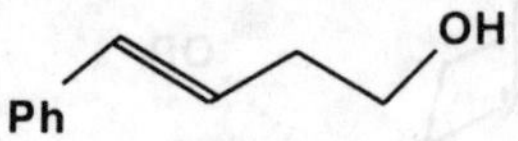 | >95% ee | 30-50% ee |
| 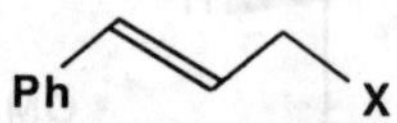 X = OAc , $OCH_2Ph$ , $N_3$ , Cl | >95% ee | NR |
| 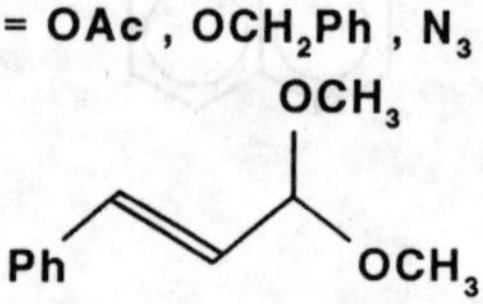 | >95% ee | NR |
| 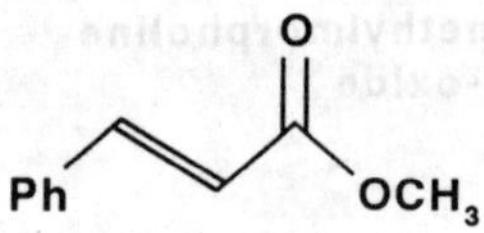 | >95% ee | NR |
| 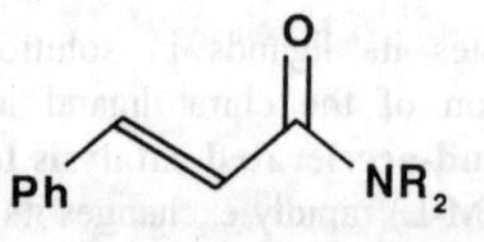 | >95% ee | NR |

a. Data taken from ref. 39.

This situation obtains in the asymmetric dihydroxylation, where coordination of an amine to $OsO_4$ affords a catalyst with much higher activity. Interestingly, detailed mechanistic investigations[39] of the asymmetric dihydroxylation with NMO as the primary oxidant revealed that two pathways are operating simultaneously (see figure 8), one of which has a low enantioselectivity.

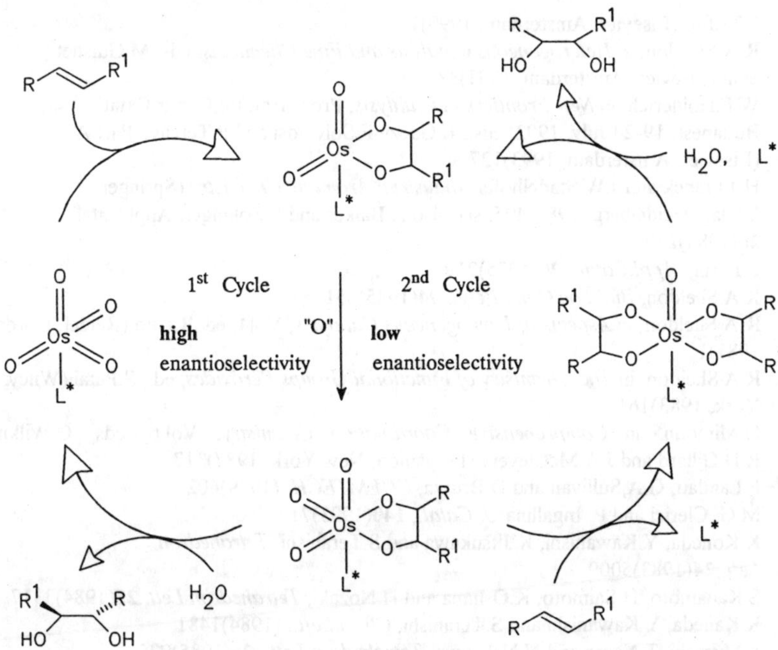


Figure 8. Mechanism of $OsO_4$-catalyzed asymmetric dihydroxylation with N-methylmorpholine-N-oxide (NMO).

This undesirable complication led Sharpless and coworkers to the use of $K_3Fe(CN)_6/K_2CO_3$ as the oxidant. An obvious drawback of the latter is that it can hardly be considered as a high-atom utilization, low-salt system. In order to counter this objection an electrocatalytic version has been developed in which the enantiomerically pure diols are formed from the olefin, water and electricity, with hydrogen gas as the only coproduct[37].

This is an appropriate note on which to end our discussions of the application of clean catalytic technologies to the synthesis of fine chemicals. It is obviously an area eith a golden future.

**References**

1. R.A.Sheldon, CHEMTECH, (1991)566
2. R.A. Sheldon, in *New Development in Selective Oxidation*, eds. G.Centi and

F.Trifiro (Elsevier, Amsterdam, 1990)1
3. R.A.Sheldon, in *Heterogeneous Catalysis and Fine Chemicals,* eds. M.Guisnet et al (Elsevier, Amsterdam, 1991)33
4. W.F.Holderich, in *New Frontiers in Catalysis*, Proc. 10th Int.Congr.Catal., Budapest, 19-24 July, 1992, eds., L.Guczi, F.Solymosi and P.Tetenyi, Part A (Elsevier, Amsterdam, 1993)127
5. H.J.Franck and J.W.Stadelhofer, *Industrial Aromatic Chemistry* (Springer Verlag, Heidelberg, 1988)405; see also A.Baiker and P.Zollinger, Appl.Catal., **20**(1985)219
6. L.Forni, *Appl.Catal.*, **20(**1985)219
7. R.A.Sheldon, *Bull.Soc.Chim.Belg.*, **94**(1985)651
8. R.A.Sheldon, in *Aspects of Homogeneous Catalysis*, Vol4, ed.,R.Ugo (Reidel, Dordrecht, 1981)1
9. R.A.Sheldon, in *The Chemistry of Functional Groups Peroxides*, ed., S.Patai (Wiley, New York, 1983)161
10. H.Mimoun, in *Comprehensive Coordination Chemistry*, Vol.6, eds. G.Wilkinson, R.D.Gillard and J.A.McClevery (Pergamon, New York, 1987)317
11. P.Landau, G.A,Sullivan and D.Brown, *CHEMTECH,* (1979)602
12. M.G. Clerici and P. Ingallina, *J. Catal.*, **140**(1993)71.
13. K.Koneda, Y.Kawanishi, K.Jitsukawa and S.Teranishi, *Tetrahedron Lett.,***24**(1983)5009
14. S.Kanamoto, H.Saimoto, K.Oshima and H.Nozaki, *Tetrahedron Lett.*,**25**(1984)3317
15. K.Kaneda, Y.Kawanishi and S.Teranishi, *Chem.Lett.*, (1984)1481
16. S.Murashi, T.Naota and N.Nakajima, *Tetrahedron Lett.*, **26**(1985)925
17. K.Kaneda and S.Teranishi, in *Activation of Dioxygen and Homogeneous Catalytic Oxidation*, Abstracts of the International Symposium held in Galzignana, Italy, June 24-29, 1984
18. S.Murahashi, T.Naota, T.Kuwabara, T.Saito, H.Kumobayashi and S.Akutagawa, *J.Am.Chem.Soc.*, **112**(1990)7820
19. T.A.Foglia, P.A.Barr and A.J. Malloy, *J.Am.Oil Chem.Soc.*, **54**,(1977) 858A
20. Y.Sasson, G.D.Zappi and R.Neumann, J.Org.Chem., **51**(1986)2880
21. J.Dakka, A.Zoran and Y.Sasson, *Eur.Pat.Appl.*, 0300921 and 0300922(1988) to Gadot Petrochemical Industries
22. C.Venturello, E.Alneri and M.Ricci, *J.Org.Chem.*, **48**(1983)3831
23. Y.Ishii and M.Ogawa, *Rev.Heteroatom Chem.*, **3**(1990)121
24. K.Yamawaki, T.Yoshida, H.Nishihara, Y.Ishii and M.Ogawa, *Synth.Commun.* **16**(1986)537; Y.Ishii, K.Yamawaki, T.Ura, H.Yamada, T.Yoshida and M.Ogawa, *J.Org.Chem.*, **53**(1988)3587
25. C.Aubry, G.Chottard, N.Platzer, J.M.Bregault, R.Thouvenot, F.Chauveau, C.Huet and H.Ledon, *Inorg.Chem.*, **30**(1991)4409; see also C.Venturello, R.D.Alloisio, J.C.J.Bart and M.Ricci *J.Mol.Catal.*,**32**(1985)107
26. I.V.Kozhevnikov, *Russ.Chem.Rev.*, **62**(1993)473
27. R.J.J.Jansen, H.M.van Veldhuizen, M.A.Schwegler and H.van Bekkum, *Recl. Trav. Chim. Pays-Bas*, **113**(1994)115
28. N.Mizuno and M.Misono, *J.Mol.Catal.*, **86**(1994)319

29. M.Misono, *Catal.Rev.Sci.Eng.*,**29**(1987)269

30. O.A.Kholdeeva, A.V.Golovin, R.J.Maksimovskaya and I.V.Kozhevnikov, *J.Mol.Catal.*, **75**(1992)235

31. J.H.Grate, D.R.Hamm and S.Mahajan, in *Catalysis of Organic Reactions*, eds., J.Kosak and T.Johnson (Marcel Dekker, New York 1991)

32. J.M.Bregeault, B.El Ali, J.Mercier, J.Mjartin and C.Martin, *C.R.Acad. Sci. Paris,* **309**(1989)459

33. B.El Ali, J.M.Bregeault, J.Martin and C.Martin, *New J.Chem.*, **13**(1989)173; B.El Ali, J.M.Bregeault, J.Mercier, J.Martin, C.Martin and O.Convert *J.Chem.Soc.Chem. Commun.*, (1989)825

34. R.A.Johnson and K.B.Sharpless, in *Catalytic Asymmetric Synthesis,* ed., I.Ojima (VCH, Berlin, 1993)103; T.Katsuki and K.B.Sharpless, *J.Am.Chem.Soc.*, **102**(1980)5976

35. W.Zhang, J.L.Loebach, S.R.Wilson and E.N.Jacobsen, *J.Am.Chem.Soc.,* **112**(1990)2801

36. R.Irie, K.Noda, Y.Ito and T.Katsuki, *Tet.Lett.,* **32**(1991)1055

37. E.N.Jacobsen in *Catalytic Asymmetric Synthesis,* ed., I.Ojima, (VCH,Berlin,1993)159; see also E.N.Jacobson, W.Zhang, A.R.Muci, J.R.Ecker and L.Deng, *J.Am.Chem. Soc.*,**113**(1991)7063

38. E.N.Jacobsen, I.Marko, W.S.Mungall, G.Schröder and K.B.Sharpless, *J.Am.Chem.Soc.,* **110**,(1988)1968

39. R.A.Johnson and K.B.Sharpless, in *Catalytic Asymmetric Synthesis*, ed., I.Ojima (VCH,Berlin, 1993)227.

# SELECTIVE ELECTROCHEMICAL OXIDATIONS

J.A.R. VAN VEEN

*Shell Research B.V. (Koninklijke/Shell-Laboratorium, Amsterdam), P.O. Box 38000, 1030 BN Amsterdam, The Netherlands*

ABSTRACT

At the present time, electrochemical reactions do not constitute an important part of industrial organic syntheses. Nevertheless, the future of organic electrosyntheses is usually considered to be very bright, especially in the production of fine and specialist chemicals. Reasons for this optimism, which include environmental benignity, are discussed, as are the problematic aspects — expensive separation operations, for example.

The following subjects are addressed: (i) electrochemical regeneration of highly selective redox couples whose stoichiometric application is ruled out for economic and/or environmental reasons — regeneration can be carried out ex situ, in situ, or through fixing the couple onto an electrode — the synthesis of quinones being among the successful applications; (ii) oxygen-transfer anodes, such as $PbO_2$-epoxidation is an important topic here, as it is in (i); (iii) hydrogen abstraction anodes, such as NiOOH, whose action is very similar to the chemical reagent nickel peroxide, an important application of which is the conversion of a sorbose to its corresponding acid — a step in vitamin C synthesis; (iv) Pt anodes, which are o.a. applied in the classical Kolbe synthesis; (v) heterogeneous catalysis of consecutive reactions following the initial charge-transfer step.

Finally, attention will be paid to new developments in electrode materials — polymer-modified electrodes, for example — and to the recent and promising development of solid polymer electrolyte cells.

## 1. Introduction

Up till now, electrochemical reactions have not formed an important part of industrial organic chemistry, despite the fact that these syntheses, e.g. the Kolbe reaction, are among the oldest reactions of organic chemistry. Following the industrial realisation, in the sixties, of the electroorganic syntheses of tetraethyl-lead and adiponitril, the method was expected to become widely established, but this did not happen. The failure has been ascribed to the emphasis on the design of processes for the manufacture of very large tonnage chemicals and it is now generally agreed that the major impact of electrosynthetic processes will be in the production of fine and specialist chemicals. It nevertheless appears that there still are too few reactions which can be carried out with high selectivities or particularly advantageously only by an electrochemical route.

In this chapter we will discuss some examples of electroorganic oxidation reactions, including those which involve the electrochemical regeneration of expensive, highly selective redox systems whose stoichiometric application is ruled out for economic or environmental reasons. First, however, let us summarise the pros and cons of electrochemical synthesis routes in general.

From the point of view of process engineering, electrochemical reactions offer a number of advantages:

- they are usually easily controlled by means of additional parameters such as current density and charge;
- they take place under mild reaction conditions (low temperatures, atmospheric pressure) which often implies that relatively inexpensive equipment can be used;
- they are environmentally benign in that they do not, in general, entail any waste air or waste water problems. This aspect will become even more important in the future. Competition, in this case, will come, of course, from catalytic processes.

Offsetting these advantages are a number of problem areas:

- electroorganic syntheses require, in general, special reactors;
- because such syntheses are phase-boundary reactions, the associated problems may be expected to occur; e.g. electrode deactivation, poisoning and/or corrosion;
- the necessity of using electrolytes frequently results in expensive separation operations during work up of the electrolysis mixtures. It is, in fact, this aspect that renders many electroorganic syntheses uneconomical;
- there is very little possibility of extrapolating the experience gained with one process to the scaling up of new reactions.

It is also to be noted that small processes, such as those producing fine and specialist chemicals, do not usually warrant the development of specific cells, electrode materials, or membranes. Therefore, the process designer will have to select the cell and its components from those already available but probably optimized for other purposes. On the bright side, a number of cells are now commercially available. Firms active in this field include ICI, Electrocell AB, Reilly Tar and Chemical, Steetley Engineering and Electrocatalytic. This reflects the increasing confidence that in the fine-chemicals area, electrochemical processes have an important role to play. Water and methanol are preferred solvents in that they are cheap, give rise to solutions of high conductivity and clean counterelectrode reactions ($H_2$ evolution in the case of electrooxidations). However, they are not always the ideal medium for the organic reaction. Where anodic and cathodic processes interfere with each other, a membrane is used to separate the two electrodes (adding to the resistance of the cell).

As to electrode selection, in the first instance one chooses from a restricted range of readily available materials, e.g. steel, $PbO_2$, Pt (usually as a coating on Ti), diverse forms of C (carbon black, graphite, glassy carbon, etc.) or $RuO_2$-coated $TiO_2$. The choice of the right one is quite empirical as it is thus far very difficult, if not impossible, to predict the success of an electrode material or to define its lifetime without extended studies under realistic process conditions. Accelerated testing is rarely satisfactory except to indicate catastrophic failure.

The present subject can not profitably be discussed from the point of view of electrocatalysis: although adsorbed species are, every now and then, considered to be the key, very little is really known in this area (cf. Section 2.3). The electrical double layer which exists at the electrode/solution interface, on the other hand, would appear to play only a minor role in determining the rates of electron transfer.

What follows is essentially based on several recent review papers [1–8] and aims to give an impression of what is electrochemically possible in the area of selective oxidation. The emphasis is on industrial application and, whenever possible, on the electrocatalytic aspects. We will first discuss some aspects of the use of regenerable redox couples and then focus our attention on oxygen-transfer anodes and on Pt and C anodes. Finally, some recent developments in electrode materials and the emergence of solid-polymer electrolyte cells will be discussed.

## 2. Selected topics

### *2.1. Electrochemically regenerable redox systems: indirect electrosynthesis*

During an indirect electrode reaction, a redox couple is used as a catalyst or an "electron carrier" for the oxidation of another species in the system. That is, the electrode is simply used to continuously reconvert the redox reagent to an oxidation state where it is able to react with an organic compound in a desirable reaction. Indirect electrosynthesis can be carried out in three different ways. The oldest one is the so-called "ex-cell" method. Here, the synthesis reaction and the regeneration are performed in separate vessels, the advantage of which is that the chemical and electrochemical steps can be optimized independently of each other.

The second possibility for electrochemical regeneration of a redox catalyst consists of its continuous retransformation into its active form without isolation, i.e. within the reaction vessel ("in-cell" method). This method is simpler than the previous one but it is, of course, necessary to find conditions under which the organic substrates, reactive intermediates and products do not hinder the electrochemical regeneration of the reagent nor themselves react electrochemically.

A third way of achieving indirect electrosynthesis is by fixing the redox reagent to the electrode surface, so that it can be continuously regenerated there after reacting with the substrate. In this case there is, in principle, the advantage that the separation step is unnecessary.

#### *2.1.1. Homogeneous redox catalysts*

The conversion of anthracene to anthraquinone via electrogenerated chromic acid has been carried out technically for over 50 years. In its latest embodiment (see Fig. 1), the electrolysis is performed in a membrane cell while the chemical step is carried out by allowing the chromic acid to trickle through a column of solid anthracene. The product — anthraquinone — is also insoluble in the aqueous acid so that the organic conversion is effectively solid to solid. The reaction goes to completion provided the particle size of the anthracene falls within a suitable range.

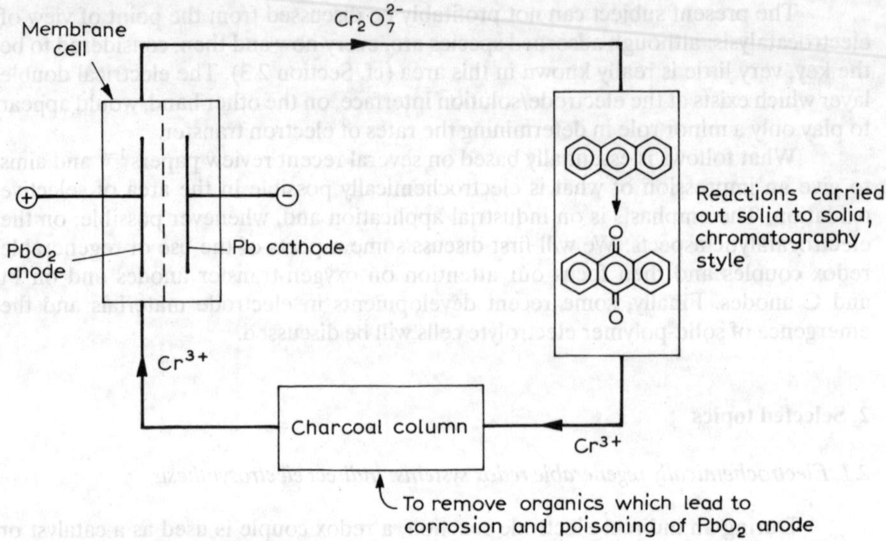


**Fig. 1. Indirect electrosynthetic process for the oxidation of anthracene → anthraquinone.**

The spent redox reagent is then passed through an activated carbon bed to remove traces of organic material which would otherwise lead to loss of current efficiency (increase in the relative rate of oxygen evolution, the principal side reaction) and the $Cr^{3+}$ solution is recycled to the cell. This technique has, however, become less important since the development of catalytic methods ($CrO_3$, liquid phase, 50–150°C; or Iron vanadate, gas phase, 340–390°C [9]).

The oxidation of naphthalenes to yield naphthoquinones (used, for example, as a pulping additive) would be technically interesting. In addition to chromate regeneration, Ce(IV) oxidation and regeneration were, in particular, investigated. The best results (naphthoquinone selectivity: >95%, current efficiency for the regeneration: 95%) were obtained using an ex-cell process and cocatalysts ($Ag^+$, $Co^{2+}$) in the electrochemical regeneration. Because of the poor solubility of the cerium salts, very large reaction volumes are required: it is necessary to apply and regenerate at least 100 l of electrolyte solution per kg of naphthoquinone produced [4]. This has triggered a search for methods to increase the solubility which resulted in the development of a Ce(IV) procedure based on aqueous methanesulfonic acid at WR Grace [16a]. This technology has been scaled up by HydroQuebec to the 100 ton/year scale and it is to be licensed to Taysung Enterprises Co (Taiwan) for the manufacture of a quinone intermediate for a dyestuff [16b].

The oxidation of benzene to benzoquinone, an important target in industrial chemistry, has also been attempted but the results obtained with indirect processes were inferior to those obtained by direct electrochemical oxidation (see Section 2.2).

A different approach to the indirect electrochemical oxidation of aromatic compounds consists of the in-situ generation of Fenton's reagent from cathodically formed hydrogen peroxide and from reductively formed iron(II) ions. On the laboratory scale, nuclear oxidation of benzene and substituted benzenes by this system leads to the selective formation of phenol and substituted phenols. The addition of copper(ll) salts makes the reaction more effective as $Cu^{2+}$ ions are better oxidizing agents than $Fe^{3}+$. The reaction sequence is as follows:

$$\text{Cathode:} \quad Fe^{3+} + e = Fe^{2+}$$

$$\text{Cathode:} \quad O_2 + 2e + 2H^+ = H_2O_2$$

$$Fe^{2+} + H_2O_2 = Fe^{3+} + OH^{\bullet} + OH$$

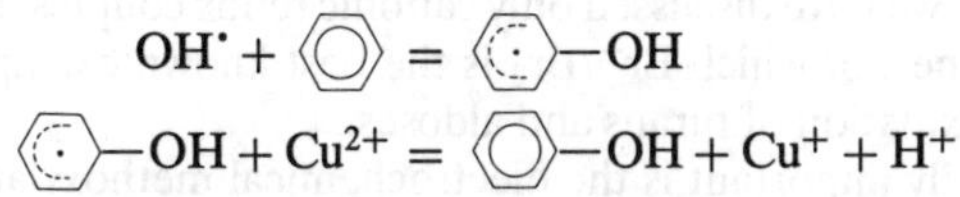

$$Cu^+ + Fe^{3+} = Cu^{2+} + Fe^{2+}$$

So, overall, we have $O_2 + H_2O + 2e + \text{benzene} = \text{phenol} + 2OH^-$. The current yield in phenol reaches 60% while the yield with respect to hydrogen peroxide consumption is 64%. Fluorophenol is generated in 80% yield with respect to the hydrogen peroxide consumption from fluorobenzene in an ortho to para ratio of 85 : 15 [2].

The same idea is being pursued by Otsuka et al. [10] with one difference. They apply a fuel cell system in which the cathode performs the hydroxylation of benzene described above and hydrogen is consumed at the anode, thus cogenerating electricity. Although the current efficiency is still extremely low, i.e. about 2%, the method has already attracted some attention [11]. A similar approach is followed by an Italian group [12] employing a Pd/C hydrophobic thin-layer catalyst and $Fe^{2+}/Fe^{3+}$ redox couple as an "oxygen carrier". They report on the electrooxidation of ethylene (to acetaldehyde) and alkanes, for example, methane to methanol.

Attention has also been paid to the indirect synthesis of aromatic aldehydes via side-chain oxidation of alkyl aromatics by in-cell and ex-cell regeneration of redox systems, such as $Mn^{2+}/Mn^{3+}$ and $Ce^{3+}/Ce^{4+}$. Recent examples for the syntheses of 4-tert-butylbenzaldehyde and anisaldehyde show that the principal disadvantages of this method (large reaction volumes, poor space–time yields, considerable problems with working up and recycling the electrolyte) have not been solved. For example, in the synthesis of anisaldehyde, very respectable yields of about 96% are obtained. However, it is necessary to separate and recycle 17.6 t of cerium ammonium nitrate and 61 t of methanol per one t of anisaldehyde, which militates against the industrial implementation of this process [4]. Interesting is the formation of *o*-nitrobenzaldehyde from the *o*-nitrotoluene (very difficult to oxidize) by electrochemically generated Co(III) under the catalysis of silver ions yielding about 80% of the aldehyde. However, the reaction is rather difficult to carry out [2].

A new sorbic (hexadienoic acid) synthesis based on the addition of carboxymethyl radicals to 1,3-butadiene has reached the pilot-plant stage at Monsanto.

The core of the new synthesis is the in-cell regeneration of the chemical oxidant $Mn(OAc)_3$ in the presence of catalytic amounts of Cu(II) salts. However, although the starting materials butadiene and acetic acid are inexpensive, the cost of the process is still high due to high conductive salt concentrations coupled with low concentrations of the desired product (2–4%) and considerable corrosion problems. Thus, the work-up procedure is rather expensive[4].

Ethylene can be oxidized to acetaldehyde in high yields similar to the Wacker process if electrogenerated palladium(II) is used as catalyst. In this way the copper(II) catalyzed air oxidation of palladium(0) is replaced by its electrooxidation[2].

As an interesting detail, we note here that total oxidation of organic waste — to CO and $CO_2$ — has been recently reported to take place when using $Ag^{2+}/Ag^{+}$ as the in-cell redox reagent[29].

Thus far, we have discussed only cationic redox couples. However, one can also apply anionic ones, of which $Br^-/Br_2$ is the best known example. It is, for example, applied in the oxidation of furans and aldoses.

Technically important is the electrochemical methoxylation of furans to yield the corresponding dimethoxydihydrofurans. It is now carried out industrially by BASF and Otsuka:

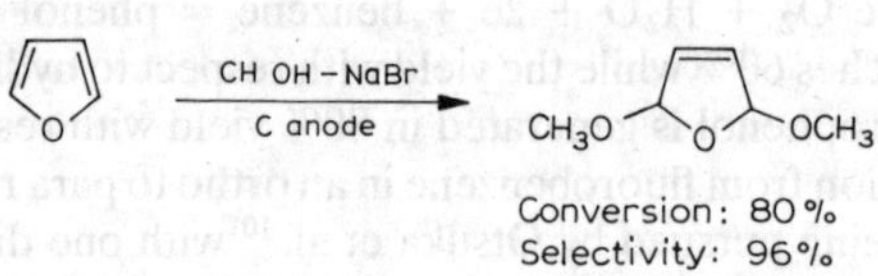


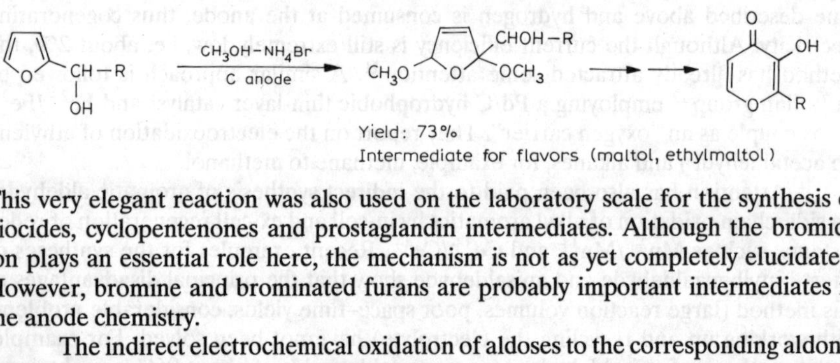


This very elegant reaction was also used on the laboratory scale for the synthesis of biocides, cyclopentenones and prostaglandin intermediates. Although the bromide ion plays an essential role here, the mechanism is not as yet completely elucidated. However, bromine and brominated furans are probably important intermediates in the anode chemistry.

The indirect electrochemical oxidation of aldoses to the corresponding aldonic acids, which was carried out industrially as early as about 1930, is still used today by Sandoz and in India for the production on the tonne scale. Specific examples are the anodic oxidation of lactose to calcium lactobionate and the production of Ca and Na gluconate by electrochemical oxidation of glucose. Most of the gluconic acid and its salts (market volume: about 30,000 tonnes/year), however, are now produced by fermentation of glucose[4].

Although, in general, aldehydes are most conveniently converted into the corresponding carboxylic acids by catalytic air oxidation, sometimes an otherwise difficult-to-achieve selectivity can be obtained electro-chemically. One example, with the $Cl^-/Cl_2$ couple this time, is the indirect, controlled oxidation of glyoxal to glyoxylic acid, CHO—CHO → COOH—CHO (conv. 98%, sel. 82%).

The anodic oxidation of secondary alcohols to the corresponding ketones is also generally inferior to the catalytic dehydrogenation methods. However, electrochemical syntheses can be of interest in special cases. One example of this is the regioselective oxidation of an endo-hydroxyl group in 1,4,3,6-dianhydrohexitols:

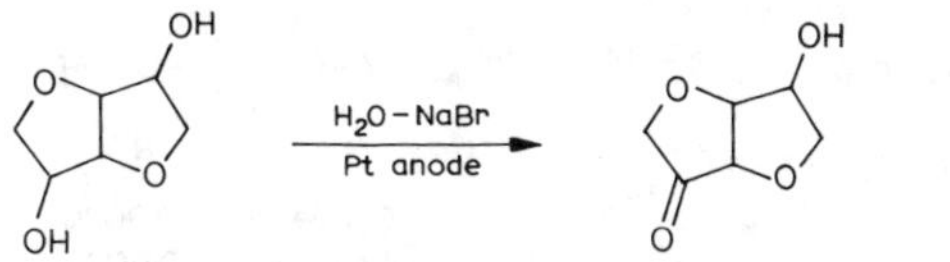


Selectivity: 80%
Current efficiency: 50%

With $I^-/I^+$ mediation, it is possible to obtain $\alpha$-*N*,*N*-dialkylamino ketones from aldehydes and dialkyl amines. The essential step in the proposed reaction mechanism is the coordination of $I^+$ to the double bond of the intermediate alkene, thus facilitating attack by $OH^-$ [2].

A great deal of effort has been put into the search for an electrochemical epoxidation process. In particular, the indirect electrochemical generation of propylene oxide via propylene chloro- or bromohydrin using anodically formed hypochlorite or hypobromite has been studied very intensely. The propylene halohydrins are saponified using the cathodically generated sodium hydroxide:

$$\text{Anode:}\quad 2Br^- = Br_2 + 2e$$

$$\text{Cathode:}\quad 2H_2O + 2e = H_2 + 2OH^-$$

$$\text{Solution:}\quad Br_2 + H_2O = HOBr + HBr$$

$$CH_3{-}CH{=}CH_2 + HOBr = CH_3{-}\underset{OH}{CH}{-}\underset{Br}{CH_2}$$

$$CH_3{-}\underset{OH}{CH}{-}\underset{Br}{CH_2} + OH^- = CH_3{-}\underbrace{CH{-}CH_2}_{O} + Br^- + H_2O$$

Overall:propylene + water = PO + hydrogen.

There is, however, no industrial application for this process. It has proven impossible to suppress the formation of the 1,2-dihalopropane byproduct and, because of the low concentrations of desired product (2–4% of PO in the electrolyte) and the presence of numerous byproducts, the work-up procedure is complicated. The process is thus rendered uneconomical. However, an alternative which combines chloralkali electrolysis with the chlorohydrin process is still being pursued [4]. The electrosynthesis of hypochlorites has been studied in detail by Olin and Steetley Engineering [13].

The indirect electrochemical epoxidation of $C_3$–$C_5$ olefins with the help of anodically generated $[(Py)_2Ag(III){-}O]^-$ species has been studied at Shell[14]. Although high current yields and efficiencies were observed, it is, at present, not industrially applied. The principle of the indirect electrochemical epoxidation can also be applied to complicated structures. For example, ICI utilised this reaction for the preparation of intermediates for the synthesis of fungicides:

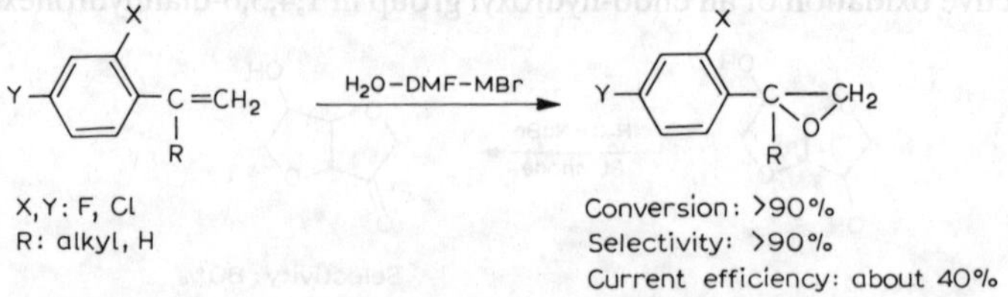


Polyisoprenoids can be epoxidized regioselectively in $\omega$-position to functional groups[2].

In the next section we will discuss an industrially implemented direct epoxidation reaction.

As far as the application of anionic redox couples is concerned, we simply note that it also possible to effect selenation together with N—S and N—P bond formation reactions[2].

It has also proven possible to effect indirect electrochemical oxidations using organic compounds as redox catalysts. A great deal of work has been done on triarylamines where the oxidation potential can be adjusted by the selection of the ortho and para substituents[2]. In this way, selectivities can be obtained in the oxidative removal of protecting groups which would otherwise be accessible, if at all, only with difficulty. The technically interesting indirect electrochemical oxidations of benzylic alcohols, benzaldehyde dimethylacetals, and alkyl aromatic compounds are also possible using triarylamine redox reagents[2,15].

### *2.1.2. Heterogeneous redox catalysts*

Indirect electrolysis where the redox catalyst is dissolved in the electrolyte is now widely practised. It is also widely recognized that it would be advantageous (for example, for product isolation) if the redox catalyst were anchored to the electrode surface. The main problem to be solved here is to achieve sufficient stability while maintaining the redox couple's activity.

The prime example of such systems is the nickel (oxide) hydroxide electrode in aqueous base[3,7]. It can be applied in a wide variety of electrooxidation reactions, such as alcohols to ketones or carboxylic acids, or primary amines to nitriles. The essential steps of the mechanism proposed by Fleischmann and co-workers — more or less generally accepted — are: (a) fast electrochemical conversion of nickel hydroxide to nickel oxide hydroxide; (b) adsorption of the substrate at the nickel oxide hydroxide surface whereby, at least in the case of alcohols, a decreasing adsorption with increasing chain length causes a decrease in the rate of oxidation; (c) abstraction of hydrogen from the carbon atom alpha to the functional group which is usually the

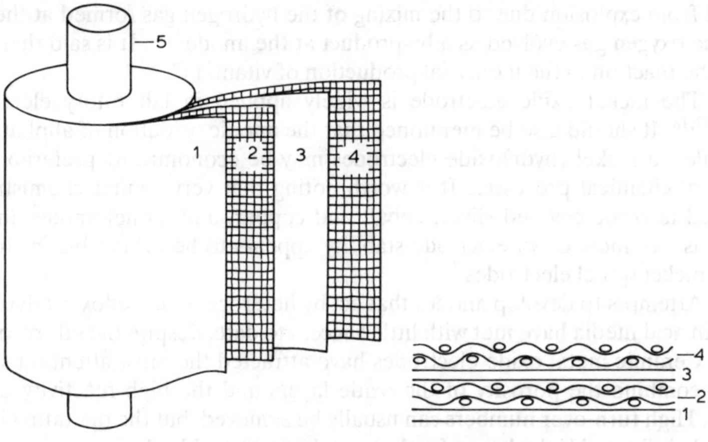


Fig. 2. "Swiss-roll" cell — arrangement of the electrodes: *1* = steel net cathode; *2, 4* = polypropylene net as insulating separator; *3* = nickel net anode; *5* = current feeder.

rate-determining step; (d) further oxidation of the radical formed in step (c):

$$Ni(OH)_2 + OH^- = NiO(OH) + H_2O + e$$

$$NiO(OH) + RCH_2X = Ni(OH)_2 + [RCHX]^{\cdot}$$

$$[RCHX]^{\cdot} + NiO(OH) = Ni(OH)_2 + \text{product}$$

The chemical transformations observed at nickel oxide electrodes in base strongly resemble those reported for the chemical agent nickel peroxide. Indeed, both give the chemistry expected for hydroxyl radicals complexed by the oxide surface. Hence, the products are usually quite different from those obtained with electron-transfer anodes (e.g., Pt, C, cf. Section 2.3). The reactions are characterised by unusually high selectivities although the current densities are uncomfortably low. This has led to various attempts to prepare high-surface-area nickel oxide electrodes — it can be done electrochemically[7]. Sometimes a low concentration of $Ni^{2+}$ in the anolyte is applied to maintain activity. Also, a special cell has been developed to cope with this problem, viz. the so-called "Swiss-roll" cell. This cell (Fig. 2), whose basic design goes back to N.Ibl (ETH Zurich), contains a rolled-up sandwich consisting of an anode and cathode sheet and a separator net. This allows a high electrode area to be applied in a relatively small cell volume.

A reaction that was studied intensely in industry, and for which the Swiss-roll cell was, in fact, developed, is the electrochemical oxidation of diacetone-L-sorbose to diacetone-2-ketogulonic acid (intermediates of the vitamin C synthesis). The process is characterised by a low level of waste water pollution with some of the following complications: the need to add surfactants to improve long-term electrode stability; a special work-up procedure; the necessity to have a system to overcome the safety hazard from explosion due to the mixing of the hydrogen gas formed at the cathode with the oxygen gas evolved as a by-product at the anode[6,14]. It is said that Merck is using the reaction in the industrial production of vitamin C.

The nickel oxide electrode is widely applied in laboratory electrosyntheses[3,4,17,18]. It should also be mentioned that the anodic oxidation of aliphatic amines to nitriles at nickel (hydr)oxide electrodes may be economically preferable to conventional chemical processes. It is worth noting that very similar chemistry can be achieved at oxide-covered silver, cobalt and copper, and monel anodes in alkaline solutions. An increase in electrode stability appears to be achievable in the case of cobalt/nickel spinel electrodes[7].

Attempts to develop anodes that act by heterogeneous redox catalysis and are stable in acid media have met with little success to date, despite the efforts of Beck et al.[8,19]. Ceramic metal oxide electrodes have attracted the most attention since they ideally combine the porosity of the oxide layers and the high reactivity of surface groups. High turn-over numbers can usually be achieved, but the operating life of the electrode is limited by leakage of redox metal into the acid solution.

Recently, reports have appeared on the high surface-area ternary ruthenates, $A_2Ru_2O_7$ (A = Pb, Bi), that are effective in the electrocatalytic oxidation of organic substrates in strongly alkaline media, e.g. sec. alcohols to ketones and alkenes and vicinal diols to carboxylic acids[20], e.g. $R_1{-}CH{=}CH{-}R_2 \rightarrow R_1{-}COOH + R_2{-}COOH$.

It could be shown[21] that activity for the latter reaction is associated with the presence of a Ru(V) surface state, capable of being oxidised to Ru(VI) at higher potentials. Substrates are oxidized by Ru(VI), which is continually regenerated if the potential is kept high enough. This is another case of heterogeneous redox catalysis. As in the case discussed above, operating life is limited by leakage of the active species:

$$\mathrm{Ru(V)} = \mathrm{Ru(VI)} + \mathrm{e}$$

$$\mathrm{Ru(VI)} + \mathrm{S} = \mathrm{Ru(V)} + \text{products}$$

$$\text{side reaction: } \mathrm{Ru(VI)} = (\mathrm{RuO_4^{2-}})_{aq}$$

It appears, that reoxidation of Ru(V) can also be effected by molecular $O_2$, enabling the oxidation reaction to be carried out in a trickle-bed reactor[22].

### 2.2. *Oxygen-transfer anodes*

There are a number of anode reactions that involve the introduction of oxygen into the electroactive species and that are almost exclusive to lead dioxide as an anode material. These include both inorganic and organic reactions. The most extensive use of $PbO_2$ is as an anode for the oxidation of chlorate to perchlorate. It is also used for the conversion of chromium(III) to chromic acid (cf. Section 2.1.1) and of manganese(II) to permanganate where $PbO_2$ generally gives better selectivity than Pt/Ti (although the latter can be improved by using silver(II) ions as cocatalyst[2]).

Some examples of the use of lead dioxide for the oxidation of organic compounds will be given below. Typically, the current efficiencies here are 50–100% (some oxygen is also evolved) but no more than a few percent at electron-transfer electrodes such as platinum.

Although it is not implied that all reactions at $PbO_2$ occur by such mechanisms (nor that they cannot occur elsewhere) it is tempting to propose that it is possible for the lead dioxide surface to transfer oxygen atoms to appropriate acceptors, i.e.[7]

$$PbO_2 + X{-}R = O{=}X{-}R + PbO$$

$$PbO + H_2O = PbO_2 + 2e + 2H^+$$

Traditionally, lead dioxide anodes have been prepared in situ by anodizing lead, usually in sulfuric acid medium. More recently, there has been considerable effort to develop procedures for making high-quality and long-life $PbO_2$ on carbon and titanium. It is still the case, however, that optimization of the electrode requires further testing for every new process.

An electroorganic reaction that is carried out preferably with lead dioxide anodes is the oxidation of benzene rings to quinones (cf. the indirect method discussed above). Considerable efforts were made to develop the electrochemical oxidation of benzene to *p*-benzoquinone to the industrial scale thus forming a basis for a new hydroquinone process. The electrochemical oxidation of benzene in aqueous emulsions containing sulfuric acid using divided cells and $PbO_2$ anodes forms *p*-benzoquinone The product can then be reduced cathodically at a lead electrode to yield hydroquinone in a paired synthesis.

The process has, so far been unable to compete successfully with the $H_20_2$ oxidation of phenol or the Hock Process (air oxidation of *p*-diisopropylbenzene[9]). The reasons for this are the poor benzene conversion, the low quinone concentration in the organic phase (<6%), the complicated work-up procedure (again!) and the unsatisfactory electrode lifetimes. Current densities have to be kept low since, otherwise, quinone will be consumed by further oxidation to maleic acid. Further losses in current yield are then experienced because an increasing percentage of the current goes into oxygen evolution. Some improvements appear to have been made by Dow through the use of a special porous electrode made of $PbO_2$ and PTFE[4,23].

This type of reaction can also be used for the production of substituted quinones and hydroquinones. For example, BASF has developed two lab processes for the synthesis of trimethyl-benzoquinone and trimethylhydroquinone. The latter is required for the synthesis of vitamin E. It is noted, in passing, that the synthesis of (substituted) *p*-benzoquinone can be approached differently (Hoechst[4]). By oxidation in methanol solution at glassy carbon electrodes and in the presence of tetraalkylammonium fluorides as conductive salts, *p*-benzoquinone tetramethyl ketal is formed. This can be converted to *p*-benzoquinone in a simple procedure while methanol can be recycled.

Whereas the indirect process of the production of propylene oxide has not advanced beyond the stage of experimental production, the direct electrosynthesis of hexafluoropropylene oxide at a lead oxide electrode has been implemented industrially by Hoechst:

$$\text{PbO}_2\text{/steel anode: CF2=CF2-CF3} + \text{H}_2\text{O} = \underbrace{\text{CF2-CF2}}_{\text{O}}\text{-CF3} + 2e + 2\text{H}^+$$

$$\text{Steel cathode: } 2e + 2\text{H}^+ = \text{H}_2$$

For the continuous process, a special divided cell (Nafion as cation exchange membrane) based on the principle of a tubular reactor was developed. The final product can be removed in gaseous form, so that the electrolyte ($H_2O/HOAc/HNO_3$) can be recycled in a simple manner [4].

Aromatics side-chain oxidations are not usually carried out at $PbO_2$ anodes. An exception is the electrochemical oxidation of 2-methylpyridine to picolinic acid which is carried out on an industrial scale by Reilly:

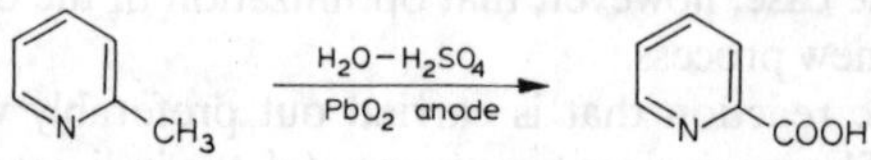


Yield: 80%
Current efficiency: 67%

Very recently, Johnson et al. [26] have described efforts to improve the rate of oxygen-transfer reactions at lead dioxide by including catalytic amounts of group IIIa and Va metals, especially Bi, in the oxide coating. The idea is to incorporate a material with a low oxygen-evolution overpotential in the $PbO_2$ surface in such a way that the catalytic sites are well separated from each other. They will, therefore, generate the active O(H)-radical species with a, nonetheless, relatively low yield of dioxygen. Thus, DMSO can be smoothly oxidized to dimethylsulfone on $PbO_2$ : Bi, but not on $PbO_2$ itself. As is the case with heterogeneous redox electrodes, however, stability of the modified layer can be problematic.

### *2.3. Electron-transfer anodes*

#### *2.3.1. Kolbe electrolysis*

Kolbe electrolysis is a powerful method of generating radicals for synthetic applications. These radicals can combine to form symmetrical dimers, or unsymmetrical coupling products, or can be added to double bonds (Fig. 3). The reaction is performed in the laboratory and on the technical scale. Depending on the reaction conditions (electrode material, cf. below, pH of the electrolyte, current density, additives) and structural parameters of the carboxylates, the intermediate radical can be further oxidised to a carbocation. The cation can rearrange itself, undergo fragmentation and, subsequently, solvolysis or it can eliminate to products. This path is frequently called non-Kolbe electrolysis. In this way radical and carbenium-ion derived products can be obtained from a wide variety of carboxylic acids [24]. Difficulties in the coupling of dicarboxylic acids were overcome when the half esters of the diacids

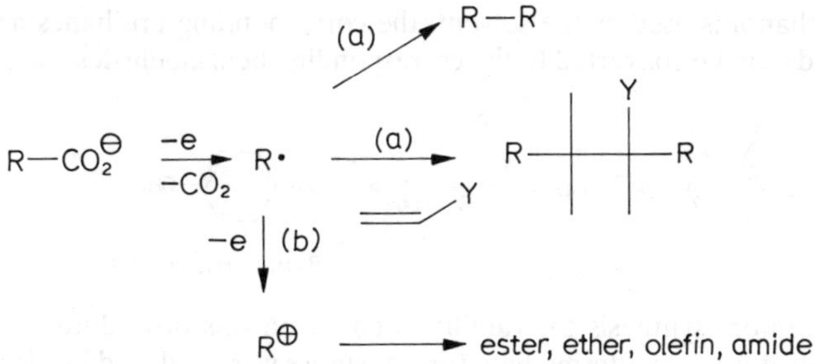


Fig. 3. Scheme of (non-)Kolbe electrolysis.

were electrolysed instead. This Brown-Walker version of the Kolbe reaction is, as yet, the most important one from an industrial point of view.

Classical Kolbe synthesis is carried out at platinum anodes but various forms of carbon can be used as well. Depending on which species is strongly adsorbed, the carboxylate anion (Pt, vitreous carbon) or the alkyl cation (graphite), the Kolbe reaction yields different products: e.g. ethane resp. methylacetate in the electrolysis of potassium acetate, pathways (a) resp. (b) in Fig. 3 [1,7,24,25]. It is worth noting that strong adsorption of either carboxylate or alkyl species is exactly why the Kolbe reaction works at all: otherwise we would only observe the (under the prevailing conditions thermodynamically strongly favoured) evolution of dioxygen.

The anodic oxidation of adipic half esters to the corresponding sebacic acid diesters has been studied in extensive detail:

$$COOH-(CH_2)_4-COOCH_3 \rightarrow COOCH_3-(CH_2)_8-COOCH_3 \quad \text{(Pt/Ti anode)}$$

where the selectivity can be as high as 93%, current efficiency 70% and the energy consumption about 2.6 kWh/kg at current densities of 10–30 A/dm$^2$. Various companies have concerned themselves with this process, but presently the sebacic acid synthesis is carried out industrially only in the USSR (capacity: about 2000 tonnes/year). Fairly large amounts are produced from castor oil, a naturally renewable raw material [4].

The coupling of carboxylic acids has been profitably used in natural product synthesis [24]. Carboxylic acids with certain functional groups can also undergo the Kolbe reaction. The products of the cross Kolbe synthesis are used, for example, as plasticisers and intermediates for musk fragrances [4].

Among the non-Kolbe syntheses we may mention the novel isocyanate synthesis, reported by Shell, which avoids the use of phosgene:

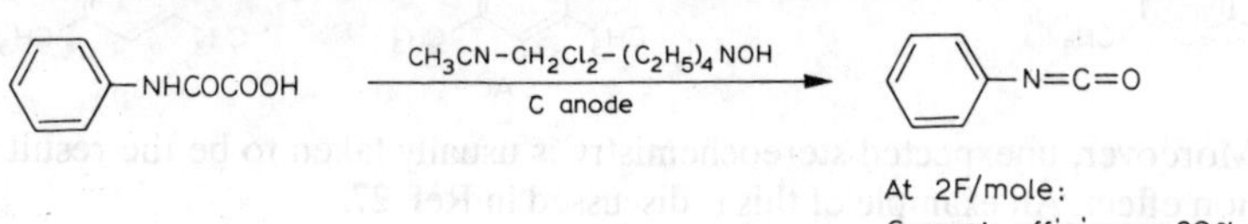

If methanol is used as the solvent, the corresponding urethanes are formed. Mandelic acids can be converted to the corresponding benzaldehydes:

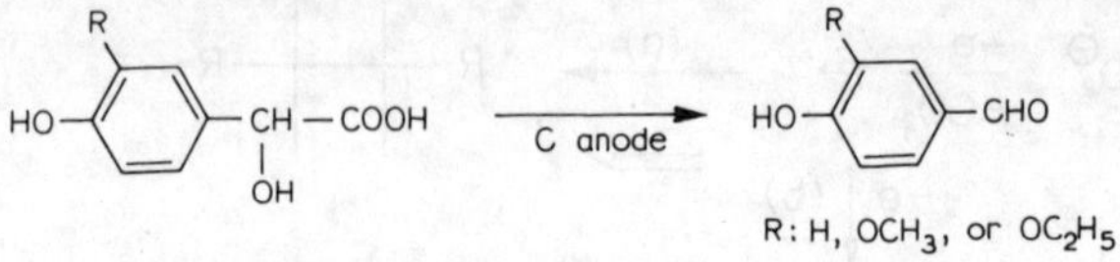


A laboratory synthesis for vanillin is based on this procedure. Lastly, 2-hydroxytetrahydrofuran, an intermediate for cytostatics was produced by electrochemical oxidation from the corresponding carboxylic acid.

*2.3.2. Adsorption effects*

Contrary to the "anticatalytic" Kolbe synthesis at Pt anodes, the anodic coupling of olefin radical cations with olefins appears to be heterogeneously catalysed at carbon anodes. Since, for this reaction, heterogeneous coupling of radical cations with adsorbed olefin competes with solvolysis which eventually yields monomer products, the dimer yield is strongly influenced by the surface concentration of the olefin and, hence, by the adsorbability of olefins at different anode materials [1].

A similar effect has also been noted in the anodic formation of trisarene sulfonium cations by anodic coupling of diarylsulfides to arenes [1]. Since arenes are adsorbed much better on carbon than on Pt anodes, it is possible to avoid undesired self-coupling at C but not at Pt anodes.

There are many other examples of reactions that show a strong dependence on electrode material. This is believed to be caused by the effects of adsorption, but we will limit ourselves to only two of them [7]. The hydroxylation of tetrahydrofuran in aqueous acid

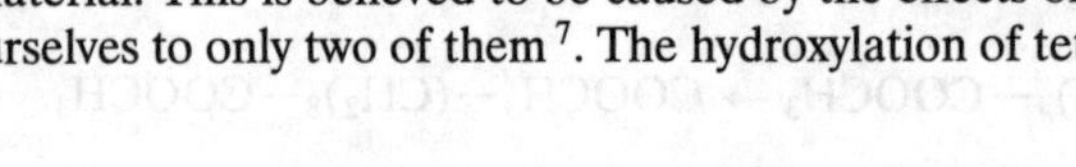


gives yields that vary strongly with the anode. On Pt, for example, the selectivity is 95% and current efficiency is 70% under conditions where it may be shown that THF adsorbs strongly on the platinum oxide surface. On $PbO_2$, on the other hand, the major product is butyrolactone. In the acetoxylation of mesitylene, very different ratios of nuclear to side-chain acetoxylation were observed for different electrodes: from 3.6 at Au to 23 at graphite.

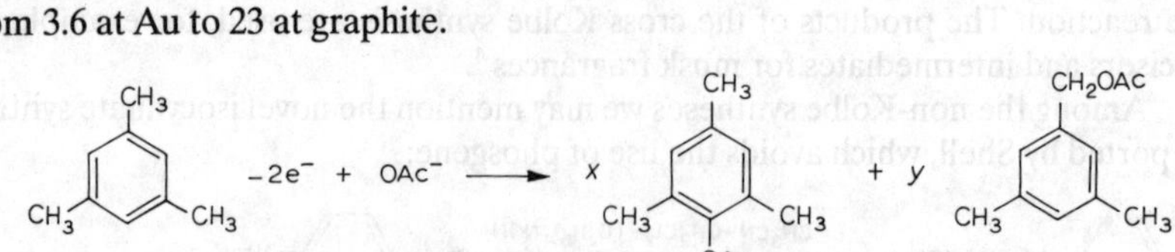


Moreover, unexpected stereochemistry is usually taken to be the result of an adsorption effect. An example of this is discussed in Ref. 27.

It was pointed out above (Section 2.3.1) that the Kolbe electrolysis is successful only because strong reactant adsorption suppresses the evolution of oxygen. A similar effect has been noted in the anodic conversion of cyclohexene in the presence of chloride ions (example cited by Beck[28]). At high potentials, the formation of 3-chloro-cyclohexene, rather than the evolution of $Cl_2$,is observed.

In fuel-cell electrode reactions, of course, adsorption of fuel molecules and the activation of water molecules are of prime importance. However, for a discussion of this topic the reader is referred to the appropriate chapter.

*2.3.3. A miscellany of anodic oxidation reactions*

The anodic oxidation of olefins in the presence of nucleophiles, such as $CH_3OH$ or $CH_3COOH$ is, in principle, a reaction of great industrial interest since it permits allyl oxidation as well as C—C coupling[4]. Nevertheless, it is almost never used in industry today. This is essentially because the selectivities are frequently poor[4]. Over the past few years, the reaction principle has been used in synthesis problems in the area of fine chemicals. For example, the anodic methoxylation of citronellol is a key step in a new rose oxide synthesis by Sumitomo. Also, Kuraray used the addition of anodically generated radicals of 1,3-dicarbonyl compounds for C—C coupling in the preparation of intermediates for $\beta$-blockers.

BASF developed a process for the preparation of 2,5-dimethoxy-2,5-dihydrofuran from butene-1,4-diol, which is an alternative to the anodic methoxylation of furan (see above). The anodic acetoxylation of olefinic terpenes was used for the synthesis of new fragrances and for the intermediates of canthaxanthin. These reactions, with the exception of that by BASF, have not progressed beyond the laboratory stage[4].

Some examples of the anodic functionalization of aromatics have already been presented above. Here we simply note that the electrochemical oxidation of substituted phenols has been used on the laboratory scale for the production of specialties such as antioxidants. Another reaction of interest is the nuclear acyloxylation of aromatics, opening up a new synthetic route to phenols. Much work has been done on the synthesis of naphthyl acetate. The principal problem with this reaction was the large amount of conductive salts which had to be separated and recycled. Another problem was the formation of methylnaphthalenes and their acetoxylation products by Kolbe electrolysis of the solvent acetic acid. The disadvantages have been substantially overcome by using distillable conductive salts and conductive polymers as electrodes:

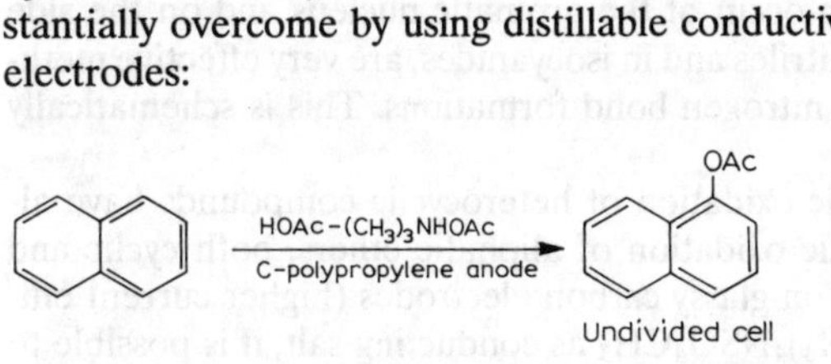


Conversion of naphthalene: 20–50%
Selectivity for naphthyl acetate: 70–85%
Current efficiency: 45–65%
$\beta$-naphthyl acetate fraction: 3–5%

1-Naphthyl acetate can be converted to a-naphthol in a simple manner with recovery of the acetic acid[4].

As far as side-chain substitution is concerned, the electrosyntheses of substituted benzaldehydes are among the few electroorganic reactions which are carried out

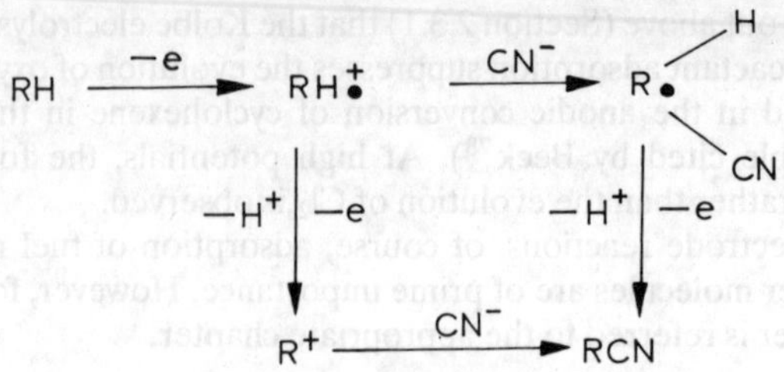


Fig. 4. Schematic representation of cyanation.

industrially on a large scale. Alkyl-substituted aromatic hydrocarbons may lose, upon oxidation, a proton from the alkyl chain, thus affording chain substituted products:

$$ArCH_3 \xrightarrow{-e} ArCH_3^{\overset{+}{\cdot}} \xrightarrow[-H^+]{-e} ArCH_2^+ \xrightarrow{X} ArCH_2X$$

The synthesis of aromatic aldehydes is particularly facile. In general, one has:

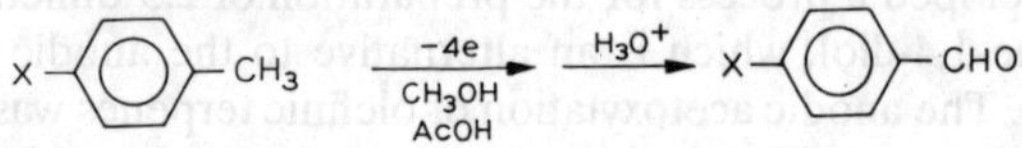


Optimum reaction conditions for the first, acetoxylation, step vary with X; some of the tricks used include: use of quaternary ammonium salts as supporting electrolytes, addition of metal salts (e.g. Co and $Cu(OAc)_2$), application of phase-transfer conditions. If defined amounts of water are added to the electrolyte, the anodic acetoxylation yields the corresponding aldehydes with very good selectivities.

Using methanol instead of acetic acid makes the reaction go through the dimethyl acetal stage. If the phenolic group is protected, it is also possible to obtain *p*-hydroxybenzaldehyde derivatives. Toluene itself, however, cannot be oxidized to benzaldehyde dimethyl acetal under similar conditions: its oxidation to benzaldehyde can, on the other hand, be effected via indirect electrosynthesis using $Ce^{4+}$.

Anodic cyanations which can occur at the aromatic nucleus and on the side chain, and which can result both in nitriles and in isocyanides, are very effective methods for carbon–carbon and carbon–nitrogen bond formations. This is schematically shown in Fig. 4.

Some examples of the anodic oxidation of heterocyclic compounds have already been presented above. Anodic oxidation of aliphatic ethers, both cyclic and non-cyclic, is preferably carried out on glassy carbon electrodes (higher current efficiencies). In methanol and with $(CH_3)_4NSO_4CH_3$ as conducting salt, it is possible to convert THF into 2-methoxy THF and dioxane in methoxydioxane in 80% yields with current efficiencies of 70–75%.

In the anodic oxidation of sulfur compounds, industrial work has been concentrating on the search for alternative processes for the production of tetraalkylthiuram disulfides[4]:

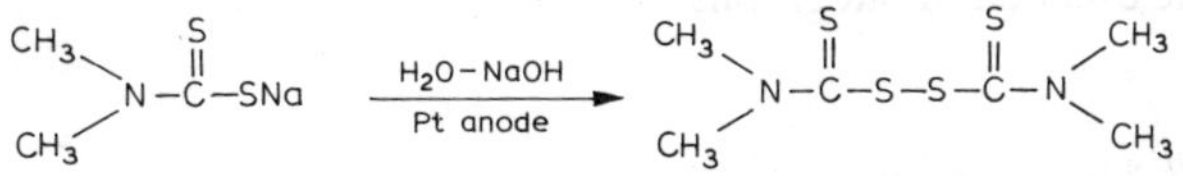

Conversion: 10–25%
Selectivity: 95%
Current efficiency: 88%

Their synthesis can also be carried out as a two-phase electrolysis using ammonium salts. Sulfenamides can be produced by oxidizing tetraalkylthiuram disulfides in the presence of amines:

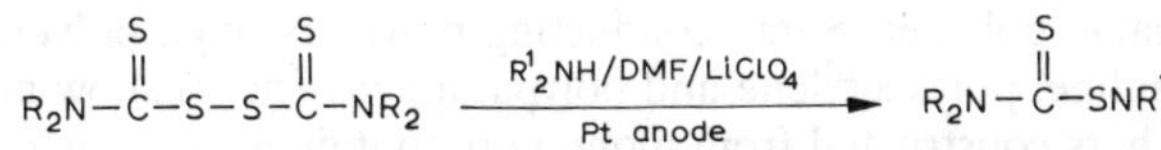

Tetraalkylthiuram disulfides are used as vulcanisation enhancers, fungicides and seed treatment agents. Commercial production is still performed by means of oxidation with $Cl_2$. Although their electrochemical synthesis avoids the production of NaCl, which is inevitable in the other processes, it is currently not being employed in industry.

The above reaction principle can be extended to the synthesis of dibenzothiazyl disulfide and benzothiazolylsulfenamides[4]. The anodic cleavage of disulfides was used on the laboratory scale for the synthesis of other vulcanisation enhancers, the production of phenyl sulfinates and the synthesis of intermediates for penicillins and cephalosporins[4].

Thiolates are often used as protecting groups for carboxylic acids. Deprotection can readily be attained electrooxidatively using bromide salts as electrolytes in $H_2O$—$CH_3CN$ media; for example[30], R—C(O)—S—But → R—COOH (this may, in fact, be an indirect electrooxidation).

Attempts were also made to find an electrosynthesis for sulfoxides. The electrooxidation of dimethyl sulfide to DMSO is reported to be applied industrially by AKZO[31]. Another example is (Rhône-Poulenc)

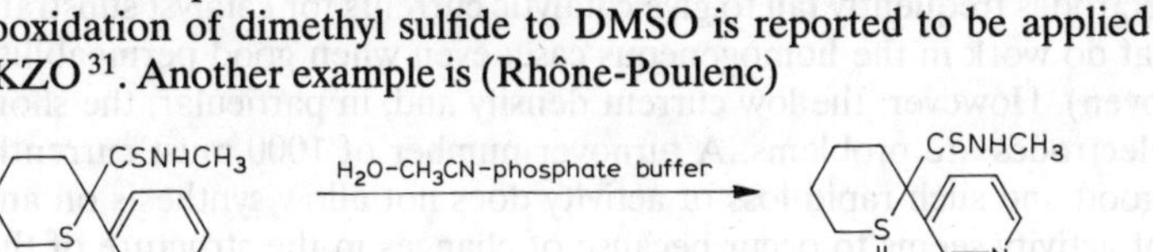

Yield: 73%

The last reaction type of this miscellany concerns indirect organic oxidation involving an electrogenerated superoxide anion radical. Molecular oxygen can be electroreduced in aprotic media to superoxide ion, $O_2^{\overline{\cdot}}$. The superoxide ion can abstract protons from even very weakly acidic substances and, thus, can act as an electrogenerated base. In the presence of $H_2O$ it decomposes rapidly. In a recent application of this method[32], 4-(Di-*n*-propylsulfamyl) toluene is converted in DMF to an important drug, 4-(Di-*n*-propylsulfamyl)benzoic acid. The advantage of the $O_2^{\overline{\cdot}}$ method here is that the sulfamido group remains entirely intact.

It is to be emphasized that the above is only a small selection from the existing literature and some reaction types, such as electrochemical halogenation, although not unimportant industrially, have not been discussed at all. For further information, we direct the reader to some excellent monographs [28,33–35].

### 2.4. Some recent developments

#### 2.4.1. Polymer-modified electrodes

The past 20 years have seen intensive activity aimed at developing chemically modified electrodes, i.e. generally metal or carbon surfaces coated with a layer of a conducting organic polymer. Some conducting polymers (e.g., polypyrrole, polythiophene, polyaniline, polyacetylene and polyparaphenylene [36]) show metallic conductivity while others constructed from monomers containing a redox center where both the oxidised and reduced forms of the couple are stable (e.g., a ferrocene, a nitro aromatic group, a quinone, or a ruthenium complex) conduct by electron hopping between redox centers. Preparation methods include coating with preformed polymers and the formation of polymer coatings from monomers via plasma deposition, thermal curing (sometimes with presilanazation of the electrode surface), or electrochemical coating.

One driving force for the study of such electrodes was the belief that they could be used in synthesis. It is to be expected that the bound redox centers might show the specific chemistry typical of their dissolved counterparts while not needing to be recovered during the product isolation procedure. It was also hoped that the organic polymer might be used to engineer chiral environments for asymmetric synthesis and to act as hosts for redox enzymes that could be driven electrochemically.

While a great deal of progress has certainly been made [5,7,30,36], no modified electrode suitable for large-scale synthesis has yet been described. Many reactions occur with high selectivity and good current efficiency (but it has to be borne in mind that modified electrodes frequently fail to give catalytic currents for catalyst substrate combinations that do work in the homogeneous case, even when good permeability of the film is proven). However, the low current density and, in particular, the short lifetime of the electrodes are problems. A turnover number of 1000 must currently be regarded as good and such rapid loss of activity does not allow synthesis on any scale. The loss of activity seems to occur because of changes in the structure of the polymer layers rather than chemical destruction of the redox centers. Perhaps a way around the stability problem can be found through supporting the organic polymers, a possibility discussed in Ref. 5.

Enantioselective electron transfer reactions are not possible, in principle, because the electron cannot possess chirality. Attempts towards chiral electrochemical synthesis have involved chiral-supporting electrolytes, chiral solvents and chiral adsorbates (mostly alkaloids). However, the observed enantiometric excess values were far from those available with modern organometallic methods. Thus far, however, polymer-coated electrodes have not fared much better and, indeed, it appears that experimental procedures for asymmetric electrochemical reactions at modified electrodes are, as yet, not straightforward enough to allow a general application (reproducibility problems).

The electrochemical nuclear chlorination of substituted aromatics makes it possible, in some cases, to achieve better regioselectivities than the chemical alternatives Thus, in the anodic chlorination of toluene in aprotic solvents, the p/o ratio of the chlorotoluenes can be increased to about 2.2 (chemical alternatives: 0.5–1)[4]:

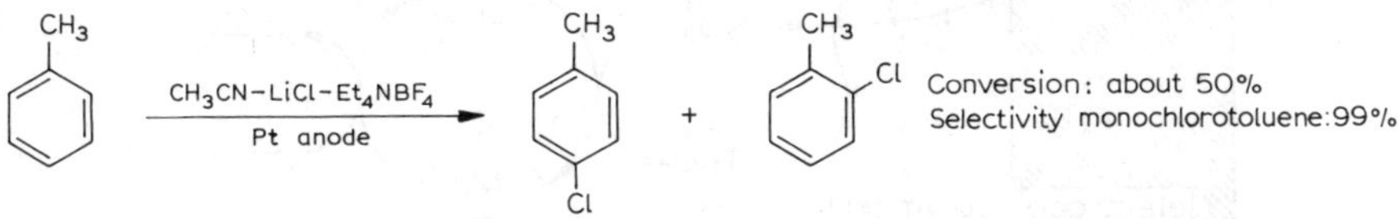


The use of graphite anodes modified with cyclodextrins allows the increase of the p/o ratio to above 4 and, at the same time, the use of aqueous electrolytes (NaCl, $HCl-H_2O$).

*2.4.2. Macrocyclics as (redox) catalysts*

The idea of electrochemistry mediated by transition-metal tetraphenylporphyrins (MeTPP) and related chelates has more frequently been pursued for inorganic electrode reactions, notably the electroreduction of dioxygen which is of eminent importance for fuel cell cathodes[37–40] (see the appropriate chapter in this book).

There are but few examples of electrooxidations carried out with heterogeneous chelate/C catalysts. It is possible to electrochemically oxidize CO to $CO_2$ with Rh and IrTPP/C anodes; in $CO/H_2$ mixtures it is the CO that is selectively oxidized by $H_2O$[41].

Perhaps it will be possible to obtain organic carbonates in a similar way,

$$CO + 2ROH \rightarrow RO-C(O)-OR + 2H^+ + 2e$$

In homogeneous systems, various transition-metal porphyrins have been used to epoxidize olefins or hydroxylate alkanes. In these reactions there is stoichiometric conversion of the added O-donor. Recently, some papers have appeared that describe possible ways to carry out the above reactions electrochemically.

The electrochemical epoxidation of olefins with Mn meso-tetraphenylporphyrin catalyst and $H_2O_2$ generation at polymer-coated electrodes has been reported by Nishihara et al.[42] and is schematically shown in Fig. 5. At high concentration of olefin and porphyrin, the electrocatalytic reaction runs at nearly 100% current efficiency for production of cyclooctene oxide. Porphyrin stability remains a problem, however. Electrocatalytic hydroxylation of alkanes with iron 2,6-difluorotetraphenylporphyrin in dichloromethane, with water from added hydrated fluoride salt providing the oxygen source, is reported in Ref. 44. A very interesting electrocatalytic oxygenation of several hydrocarbons has been accomplished with a manganese porphyrin/periodate system under conditions of phase-transfer catalysis. Epoxides with up to 90% yields were obtained. Unactivated alkanes were hydroxylated, giving yields between 25 and 77%. The periodate was regenerated electrolytically in the aqueous phase[45]. At the time, this system was said to possess potentialities for larger-scale applications.

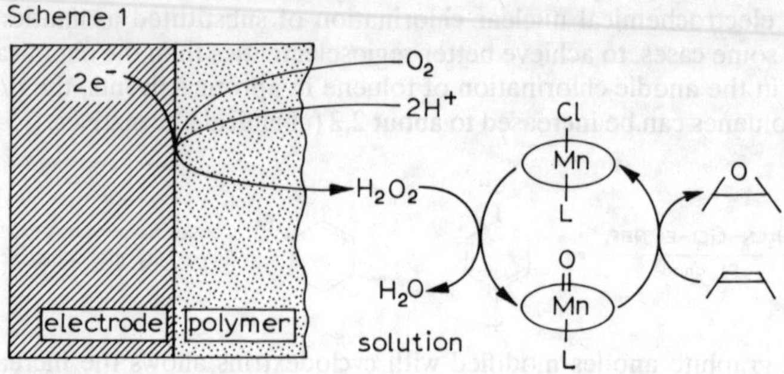


Fig. 5. Schematic representation of the MnTPP-mediated epoxidation of olefins[42].

A really anodic funtionalization of alkanes has recently been reported by Freund et al.[43]. They electrocatalytically hydroxylate *p*-toluenesulfonic acid to the alcohol, *p*-$HO_3SC_6H_4CH_2OH$, with a system consisting of aqueous $PtCl_4^{2-}$ as C—H activation catalyst, phosphomolybdic acid as redox mediator in an electrochemical cell containing a carbon cloth anode. Again, however, the stability of the system will have to be improved.

*2.4.3. Electrodes modified by underpotential deposited metals*

The electrode position of submonolayer amounts of foreign atoms on electrode surfaces (underpotential deposition, UPD) has been actively studied over the last two decades, mostly in connection with fuel-cell research[46–49]

The conversion of larger molecules, including monosaccharides, have been studied as well (for example, in relation to biofuel cells) and here UPD Pb, Tl and Bi have been shown to catalyze electrontransfer reactions and also change products[50]. The powerful in situ spectroscopic techniques developed for the study of fuel-cell electrodes are now also beginning to be used for the study of the electrochemical oxidation of these larger molecules[51].

Electroorganic oxidations for synthetic purposes have hardly been carried out to date. A relatively recent example is the selective electrogenerative (i.e., in a fuel-cell system, thus coproducing D.C. power) oxidation of ethanol to ethyl acetate on sulfur-modified platinum electrodes[52].

The major problem with the use of UPD atom modified electrode surfaces in synthesis would be the long-term control of adatom coverage, although the fuel-cell experience has shown that this problem need not be insurmountable.

A host of other compounds is attracting interest as possible electrode materials, ranging from ceramics (carbides, borides, intrides, substoichiometric oxide)[7,30] to polynuclear compounds, clays and zeolites[53–59] However, they have not, thus far, been applied in electrosynthesis.

*2.4.4. Solid-polymer-electrolyte cells*

Solid-polymer-electrolyte (SPE) cells are another development from other areas of electrochemical technology now being adapted for electrosynthesis. In an SPE cell, the two porous electrodes are fabricated onto the opposite sides of an ion-conducting polymer film, e.g., a Nafion membrane. The chief advantage of such cells is their compactness and, because it is the polymer that provides ionic conduction, it is possible to have reactant feeds and, therefore, product streams that are free of electrolyte. This greatly simplifies product extraction — often a serious bottle-neck as we have seen. SPE cells were first developed for water electrolysis[7] and are now being applied, with extremely encouraging results, in $H_2/O_2$ fuel cells[60]. Ogumi et al. (cited in Refs. 7, 30) have carried out a number of organic syntheses by the SPE method. The promise for such syntheses can be seen from the study of the oxidation of monomethyl adipate to dimethyl sebacate (Kolbe reaction, cf. Section 2.3.1.). Using a 60% solution of the adipate ester in methanol and an SPE cell based on Nafion and a porous Pt anode (formed by hydrazine reduction of chloroplatinic acid), it was possible to obtain the desired product with a current efficiency of 55% and a selectivity >80%. The product extraction is very straightforward. Another example is the indirectly catalysed oxidation of cyclohexanol to cyclohexanone, using the $I^+/I^-$ redoxcycle catalysts.

The major drawback as yet of SPE cells for organic synthesis is the stability of the membrane/electrode combination but, to date, little development work has been done on this aspect.

## 3. Concluding remarks

It is to be expected that electrochemical processes will only be applied industrially in rather special cases. As has again recently been pointed out[61,62], however, it is imperative not to turn to electrochemistry only when other methods have failed. Considering electrosynthesis at an earlier stage might well have unexpected advantages.

## References


1. H. Wendt, Electrocatalysis in organic electrochemistry, *Electrochim. Acta* **29** (1984) 1513.
2. E. Steckhan, Organic syntheses with electrochemically regenerable redox systems, *Topics in Current Chemistry* **143** (1987) 1.
3. H.-J. Schafer, Oxidation of organic compounds at the nickel hydroxide electrode, *Topics in Current Chemistry* **143** (1987) 101.
4. D. Degner, Organic electrosyntheses in industry, *Topics in Current Chemistry* **148** (1988) 1.
5. A. Merz, Chemically modified electrodes, *Topics in Current Chemistry* **152** (1990) 49.
6. A. Pletcher and F.C. Walsh, *Industrial Electrochemistry* 2nd ed. (Chapman and Hall, 1990).
7. A.M. Couper, D. Pletcher and F.C. Walsh, Electrode materials for electrosynthesis, *Chem. Rev.* **90** (1990) 837.

8. F. Beck, B. Wermeckes and E. Zimmer, Kann eine Elektrode die Rolle eines Redoxsystems in der Lösung ubernehmen?, *Dechema-Monographien, Band 112* (VCM Verlagsgesellschaft, 1988) p. 257.
9. K. Weissermel and H.-J. Arpe, *Industrielle Organische Chemie* (Verlag Chemie, 1976).
10. K. Otsuka et al., *Electrochim. Acta* **34** (1989) 1485; *Chem. Lett.* (1990) 509; *Nature* **345** (1990) 697.
11. *Catalytica Highlights*, **16**, No. 3 (1990) 3.
12. F. Frusteri, A. Iannibello, A. Parmaliana, A. Cannizarro and N. Giordani, in: *New Developments in selective Oxidation* (G. Centi and F. Trifiro, Eds.) (Elsevier 1990) p. 733; A. Parmaliana et al., paper presented at the 11th NAM Catalysis Soc., Dearborn, May 1989
13. D. Hughes, *The Chemical Engineer*, September (1987) p. 17.
14. J.M. van der Eijk, Th.J. Peters, N. de Wit and H.A. Colijn, *Catal. Today* **3** (1988) 259.
15. K.-H. Grosse Brinkhaus, E. Steckhan and D. Degner, *Tetrahedron* **42** (1986) 553.
16. (a) R.P. Kreh, R.M. Spotnitz and J.T. Lundquist, *J. Org. Chem.* **54** (1989) 1526;
(b) *Electrochem. Processing — A Quarterly Commentary*, ICI Appl. Electrochem., August 1993.
17. P. Verbrugge, J. de Waal and D. Sopher, EP 199413 (1987).
18. J. Kaulen and H.J. Schäfer, *Tetrahedron* **38** (1982) 3299.
19. H. Schulz and F. Beck, *Angew. Chem.* **97** (1985) 1047 (Int. Ed. Engl. 24 ~ 1049).
20. H.H. Horowitz et al., US Patent 4203871 (1980), Eur. Pat. Appl. 813.0574.8 (1971), US Patent 4434031 (1984); H.H. Horowitz, H.S. Horowitz and J.M. Longo, in: *Electrocatalysis, Proceedings, Vol. 82-2* (The Electrochemical Society, 1982) p. 285; H.S. Horowitz, J.M. Longo and H.H. Horowtz, *J. Electrochem. Soc.* **130** (1983) 1851; H.S. Horowitz, J.M. Longo, H.H. Horowitz and J.T. Lewandowski, *ACS Symp. Ser.* **279** (1985) 143.
21. J.A.R. van Veen, J.M. van der Eijk, R. de Ruiter and S. Huizinga, *Electrochim. Acta* **33** (1988) 51.
22. T.R. Felthouse, *J. Am. Chem. Soc.* **109** (1987) 7566.
23. D. Danly, in: *Emerging Opportunities for Electroorganic Processes* (Marcel Dekker, 1984) p. 229.
24. H.-J. Schafer, Ref. 5, p. 91.
25. M.P.J. Brennan and R. Brettle, *J. Chem. Soc., Perkin Trans. I* (1973) 257.
26. D.C. Johnson, J.A. Polta, T.Z. Polta, G.C. Neuburger, J. Johnson, A.P.-C. Tang, I.-H. Yeo and J. Baur, *J. Chem. Soc., Faraday Trans. I* **82** (1986) 1081; I.-H. Yeo and D.C. Johnson, *J. Electrochem. Soc.* **134** (1987) 1973; H. Chang, Ph.D. Thesis, lowa State University (1989); I.-H. Yeo, S. Kim, R. Jacobson and D.C. Johnson, *J. Electrochem. Soc.* **136** (1989) 1395.
27. J.M. Bobbitt, *ACS Symp. Ser.* **390** (1989) 176.
28. F. Beck, *Elektroorganische Chemie* (VCH Verlagsgesellschaft, 1974).
29. D.F. Steele, D. Richardson, J.D. Campbell, D.R. Craig and J.D. Quinn, *Transl. Chem E* **68** (1990) 115.
30. D.K. Kyriacou and D.A. Jannakoudakis, *Electrocatalysis for Organic Synthesis* (John Wiley and Sons, 1986).
31. *New Horizons in Electrochemical Science and Technology* (National Academy Press, 1986).
32. M. Michman and M. Weiss, Ref. 12, p. 667.
33. S. Torii, *Electroorganic Synthesis. Part 1: Oxidations* (VCH Verlagsgesellschaft, 1985).
34. K. Yoshida, *Electrooxidation in Organic Chemistry* (John Wiley and Sons, 1984).
35. M.M. Baizer and H. Lund, *Organic Electrochemistry*, 2nd ed. (Marcel Dekker, 1985).
36. J. Heinze, Ref. 5, p. 1.
37. J.A.R. van Veen and J. F. van Baar, *Rev. Inorg. Chem.* **4** (1982) 293.
38. B. van Wingerden, J.A.R. van Veen and C.T.J. Mensch, *J. Chem. Soc., Faraday Trans. I* **84** (1988) 65; J.A.R. van Veen, H.A. Colijn and J.F. van Baar, *Electrochim. Acta* **33** (1988) 801.
39. D. Wöhrle et al., *Adv. Polym. Sci.* **50** (1983) 45; *J. Mol. Catal.* **21** (1983) 255; *Angew. Makromol. Chem.* **117** (1983) 103; *J. Electrochem. Soc.* **132** (1985) 2144; *Makromol. Chem., Macromol. Symp.* **8** (1987) 195.
40. A. Bettelheim, B.A. White and R.W. Murray, *J. Electroanal. Chem.* **217** (1987) 271.

41. J.F. van Baar, J.A.R. van Veen and N. de Wit, *Electrochim. Acta* **27** (1982) 57; J.F. van Baar, J.A.R. van Veen, J.M. van der Eijk and N. de Wit, *Electrochim. Acta* **27** (1982) 1315.
42. H. Nishihara, K. Pressprich, R.W. Murray and J.P. Collman, *Inorg. Chem.* **29** (1990) 1000.
43. M.S. Freund, J.A. Labinger, N.S. Lewis and J.E. Bercan, *J. Molec. Catal.* **87** (1994) L11.
44. A. Nanthakumar and H.M. Goff, *J. Am. Chem. Soc.* **112** (1990) 4047.
45. J.T. Groves and T.J. Murry, Am. Chem. Soc. Abstracts of Papers, 186th Meeting, April 1984.
46. R.R. Adzic, *Adv. Electrochem. Electrochem. Eng.* **113** (1984) 159.
47. M.M.P. Janssen and J. Moolhuysen, *J. Catal.* **46** (1977) 289; *Electrochim. Acta* **21** (1976) 861, 869; B.D. McNicol, *Electrocatalysis, Specialist Periodical Reviews: Catalysis, Vol. 2* (Royal Society of Chemistry, 1978) Ch. 10.
48. T. Iwasita, F.C. Nart and W. Vielstich, *Ber. Bunsenges. Phys. Chem.* **94** (1990) 1030.
49. S. Motoo, in: *The Chemistry and Physics of Electrocatalysis, Proceedings, Vol. 84-12* (The Electrochemical Society, 1984) p. 331.
50. G. Kokkinidis et al., *Electrochim. Acta* **30** (1985) 493; *Electrochim. Acta* **30** (1985) 1611; *J. Electroanal. Chem.* **257** (1988) 239; *Electrochim. Acta* **34** (1989) 803; *J. Electroanal. Chem.* **192** (1985) 375.
51. I.T. Bae, X. Xing, C.C. Liu and E. Yeager. *J. Electroanal. Chem.* **284** (1990) 335.
52. S.J. Langer and J.C. Card, *J. Mol. Catal.* **42** (1987) 331; see also: T.D. Tran, I. Londner and S.H. Langer, *Electrochim. Acta* **38** (1993) 221.
53. V.D. Neff, *J. Electrochem. Soc.* **125** (1978) 886; D. Ellis, M. Eckhoff and V.D. Neff, *J. Phys. Chem.* **85** (1981) 1225; K. Itaya et al., *J. Appl. Phys.* **53** (1982) 804; *J. Am. Chem. Soc.* **104** (1982) 4767; K. Itaya, l. Uchida and V.D. Neff, *Acc. Chem. Res.* **19** (1986) 162.
54. C.G. Murray, R.G. Nowak and D.R. Rolison, *J. Electroanal. Chem.* **164** (1984) 205; P.K. Ghosh and A.J. Bard, *J. Am. Chem. Soc.* **105** (1983) 5691; P.K. Ghosh, A.W.-H. Mau and A.J. Bard, *J. Electroanal. Chem.* **169** (1984) 315; H.Y. Liu and F.C. Anson, i*J. Electroanal. Chem.* **184** (1985) 411.
55. B. Keita and L. Nadjo, *J. Electroanal. Chem.* **230** (1987) 267; B. Keita, L. Nadjo and J.M. Savéant, *J. Electroanal. Chem.* **243** (1988) 267.
56. P.J. Kulesza and L.R. Faulkner, *J. Am. Chem. Soc.* **110** (1988) 4905; *J. Electroanal. Chem.* **259** (1989) 81.
57. P.J. Kulesza and L.R. Faulkner, *J. Electroanal. Chem.* **248** (1988) 305.
58. C.M. Castro-Acuna, F.R. Fan and A.J. Bard, *J. Electroanal. Chem.* **234** (1987) 347; W.E. Rudzeinski and D. Root, *J. Electroanal. Chem.* **243** (1988) 367.
59. K.E. Creasy and B.R. Shaw, *Electrochim. Acta* **33** (1988) 551; B.R. Shaw et al., *J. Electrochem. Soc.* **135** (1988) 869; D.R. Rolison, *Chem. Rev.* **90** (1990) 867.
60. K. Prater, *J. Power Sources* **29** (1990) 239.
61. P.M. Bersier, L. Carlsson and J. Bersier, Electrochemistry for a better environment, *Topics in Current Chemistry* **170** (1994) 113.
62. J.H.P. Utley, *Chemistry and Industry*, 21 March (1994) 215; *New Scientist*, 31 July (1993) 24.

# INDUSTRIAL HETEROGENEOUS GAS-PHASE OXIDATION PROCESSES

P. L. MILLS, M. P. HAROLD AND J. J. LEROU
*DuPont Central Research and Development*
*Wilmington, DE 19880-0262, USA*

## ABSTRACT

**An overview of heterogeneous gas-phase oxidation processes is given from a reactor technology viewpoint. Existing and emerging catalytic reactors are described and advantages and disadvantages are discussed.**

## 1. Introduction

Oxidation processes using air and oxygen are generally used for the synthesis of various inorganic or organic chemicals having utility as intermediates or final products. These processes can be conducted in either the liquid phase or vapor phase, depending upon such factors as the reactant volatility, thermal stablity of the reactants and products, specific reaction rate, and overall process economics. Examples of inorganic chemicals produced include nitric acid, sulfuric acid, hydrogen cyanide and the oxychlorination of HCl. Typical organic chemicals that are commercially manufactured using vapor-phase oxidation processes include ethylene oxide, acrolein and acrylic acid, methacrolein, methacrylic acid, maleic anhydride, and phthalic anhydride. The products of organic oxidation processes can be further distinguished by selectivity or reaction type, such as those obtained from non-selective oxidation to carbon oxides and water, partial oxidation to oxygenated products, and oxidative dehydrogenation to desired products containing no oxygen. A related category is ammoxidation, in which a mixture of air and ammonia reacts catalytically with a hydrocarbon to form a nitrile. The emphasis in this chapter is placed upon vapor-phase catalytic selective oxidation processes that have achieved commercialization or under development towards commercialization.

### *1.1 Distinguishing Features*

All selective oxidation reactions have a number of common distinguishing features. Some key points on these features are summarized below.

Because oxidation reactions often involve breaking of saturated or unsaturated carbon-carbon or carbon-hydrogen bonds, the reactions are highly exothermic. The degree of exothermicity can be quite significant when non-selective combustion reactions occur in series or parallel with the selective reactions. This is the primary reason why reactor selection and design are of critical importance for industrial oxidation processes. Unless adiabatic operation is feasible, the reactor system must be capable of controlling the temperature within certain safely-defined limits by proper management of the high heat load. This places some specific requirements on their design.

Since oxidation processes are based upon contacting hydrocarbons or intermediate oxygenates in the presence of air or oxygen in a given reactor type, the resulting mixture composition corresponds to a fixed point on the composition versus flammability map. Depending on the composition, the hydrocarbon/air or hydrocarbon/oxygen mixture can spontaneously ignite, so that safe operation of oxidation reactors requires avoiding the flammability region. A recent review on the explosion limits of hydrocarbon/air mixtures by Westerterp[1] suggests that the flammability limits determined under static conditions, which are often used in determining safe limits for reactor operation, are too conservative since they do not account for mixing effects. Newer emerging processes based upon recirculating solids reactors[2] are operated with a hydrocarbon-rich feed gas whose composition is above the upper limit of the flammability region so that safe operation is ensured.

Oxidation processes that are economically viable require development of catalyst and reactor technology where the yield of the desired product is maximized. Optimal yields of the desired product can only be obtained if the product does not undergo thermal decomposition in the given reaction environment and in any downstream unit operations associated with product recovery. However, many products of selective oxidation reactions are highly unstable since they are based upon anhydrides or epoxides and may contain reactive organic functional groups, such as aldehydes or acids. Some newer reactor concepts, such as membrane reactors, involve simultaneous reaction and removal of the reaction product and have potential for significant improvement in product yield due to isolation of the product in a more favorable reaction environment.

Notwithstanding the inherent hazards of oxidation processes, only 2% of chemical plant accidents occurred in oxidation process, as shown in Table 1.

Table 1. Causes of Chemical Plant Accidents.[3]

| Process | % of Incidents |
|---|---|
| Polymerization | 48% |
| Nitration | 11 |
| Sulfonation | 10 |
| Hydrolysis | 7 |
| Salt formation | 6 |
| Halogenation | 6 |
| Friedel-Crafts | 4 |
| Amination | 3 |
| Diazonation | 3 |
| Oxidation | 2 |
| Esterification | 1 |

### *1.2 Oxidation Catalysts*

Catalysts used for vapor-phase selective oxidation processes are quite numerous and have been brought to commercialization realization after thousands of man-years of development efforts in industrial and academic research laboratories. It can be generally concluded that most catalysts can be placed into one of the following two classifications: (1) transition mixed-metal oxides where oxygen has some mobility in the lattice, such as multicomponent bismuth molybdates for the selective oxidation of propylene to acrolein; (2) metals with added promoters or dopants onto which molecular oxygen is chemisorbed as a first reaction step, such as supported silver catalysts used in the selective oxidation of ethylene to ethylene oxide.

In some cases, the same reaction can be commercially realized using catalysts from both classes. For example, two unique industrial processes exist for the oxidation of methanol to formaldehyde. One process is based upon a silver gauze catalyst, while another uses an iron molybdate oxide catalyst. Another special class of reactions occurs when the catalyst pores contain a liquid melt that participates in the catalysis corresponding to a supported liquid-phase system. This is the case for the potassium-promoted $V_2O_5$ used in the oxidation of $SO_2$ to $SO_3$, and potassium-acetate promoted silver catalysts used in the selective oxidation of ethylene and acetic acid to vinyl acetate.

### *1.3 Process Overview*

A non-exhaustive list of the major industrial vapor-phase oxidation processes is given in Table 2 along with typical values for reactant conversions and product selectivities. The reactants used in these processes include the important commodity $C_1$ to $C_6$ range hydrocarbons as well as several key aromatic feedstocks.

Table 2. Heterogeneous Vapor-Phase Catalytic Oxidation Processes.

| Monomer | Reactants | % Conv. | % Sel. |
|---|---|---|---|
| Formaldehyde | Methanol/air | 99 | 94 |
| Ethylene oxide | Ethylene/oxygen | 15 | 80 |
| 1,2-Dichloroethane | Ethylene/air/HCl | 95+ | 95+ |
| 1,2-Dichloroethane | Ethylene/oxygen/HCl | 95+ | 97+ |
| Acrolein | Propylene/air | 90+ | 80-85 |
| Acrylic acid | Acrolein/air | 95+ | 90-95 |
| Acrylonitrile | Propylene/air/$NH_3$ | 99+ | 73-77 |
| Methacrolein | Isobutene/air | 97+ | 85-90 |
| Methacrylic acid | Methacrolein/air | 70-75 | 80-90 |
| Maleic anhydride | Benzene/air | 98 | 75 |
| Maleic anhydride | n-Butane/air | 75-90 | 67-72 |
| Phthalic anhydride | Naphthalene/air | 99+ | 84 |
| Phthalic anhydride | o-Xylene/air | 99+ | 79 |

Table 3 lists some of the key monomers that are used as chemical intermediates and form the basis for a large variety of engineering plastics and polymer. A comparison between the reactants used in most existing processes and newer reactants from recent patent and open literature publications is also provided. This comparison clearly points to the direction where future oxidation process development is being emphasized. The trend is obviously to replace more expensive olefin feedstocks by cheaper alkanes. This effort was actually initiated at least ten years ago and has continued. Recent examples are the synthesis of acrylonitrile from propane versus propylene, and the synthesis of methacrylic acid from isobutane versus isobutylene.

Table 3. Emerging Oxidation Processes.

| Monomer | Current Reactants | New Reactants |
|---|---|---|
| Formaldehyde | Methanol/air | Methane/air or $O_2$ |
| Ethylene | Ethane | Methane/air or $O_2$ |
| Vinyl chloride | Ethylene/air/HCl | Methane/HCl/air or $O_2$ |
| 1,2-Dichloroethane | Ethylene/air/HCl | Ethane/HCl/air or $O_2$ |
| Acetic acid | Methanol/CO | Ethane/air or $O_2$ |
| Acrylic acid | Acrolein/air | Propane/air or $O_2$ |
| Acrylonitrile | Propylene/air/$NH_3$ | Propane/air/$NH_3$ |
| Methacrylic acid | Methacrolein/air | Isobutane/air |
| Phthalic anhydride | Benzene, o-Xylene | n-Pentane/air |

### *1.4 Chapter Objectives*

The primary objective of this chapter is to review industrial vapor-phase heterogeneous oxidation processes from the perspective of reactor and process technology. Particular reactor types that will be discussed include: (1) Single and multi-stage adiabatic fixed-bed reactors, including reverse-flow reactors (section two), (2) multi-tubular reactors (section three), (3) fluidized-bed reactors (section four), (4) recirculating solids reactors (section five), and (5) moving bed and chromatographic reactors (section six). Although the development of novel catalysts and associated processes for both existing and emerging processes is an active area of research, only those processes that have achieved commercial status will be discussed. Details on reactor modeling are not included for brevity, but will be presented elsewhere[4].

## 2. Fixed-Bed Reactors

### *2.1. Introduction*

Compared to other types of vapor-phase catalytic reactors, fixed-bed reactors have received the most attention from the perspective of application and reaction engineering analysis. For this reason, the level of understanding for this configuration is perhaps the greatest when compared to other reactor types. This is primarily due to the development of a hierarchy of fixed-bed reactor models. A combination of experimental data on intrinsic and apparent reaction kinetics, engineering correlations for transport coefficients, and detailed reactor modeling allows rational design and scale-up procedures to be followed. Even though the knowledge base in fixed-bed reactor engineering is very good, potential pitfalls in reactor design or analysis of performance are still present.

Figure 1 lists some of the key issues must be considered when modeling fixed-bed reactors for vapor-phase catalytic systems.[5] A wide range of length scales must be traversed in a realistic model of the fixed-bed reactor. The individual catalyst sites represent the microscale where the catalytic reactions occur. Reaction kinetic measurements using state-of-the-art laboratory reactors and information derived from catalyst characterization instrumentation provide the basis for development of kinetic models and identification of kinetic parameters based upon a sequence of elementary steps.

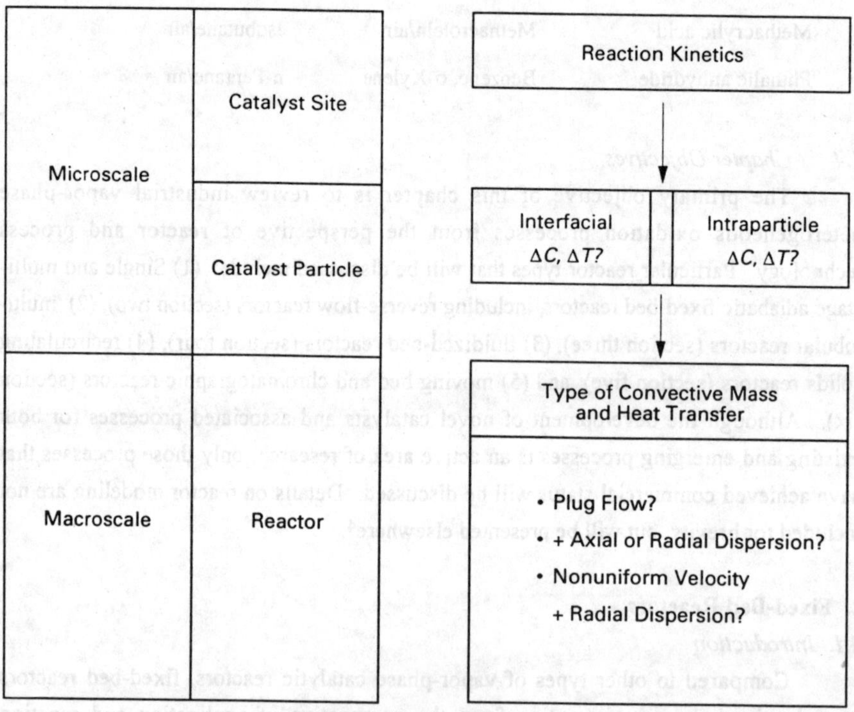


Figure 1 Modeling issues for fixed-bed reactors.

The collection of catalytic sites represent the next greater dimension on the microscale since the characteristic dimension now corresponds to the shape factor for the catalyst particle. Here, both intraparticle and interparticle transport processes and their interactions with local intrinsic kinetics must be considered. Moreover, a realistic description of the catalyst particle morphology, such as the distribution of pore sizes, is needed.

The next larger length scale is the bed of catalyst particles and corresponds to the macroscale level. Here, transport processes that occur on characteristic dimensions of the reactor, such as the reactor diameter and overall length of the fixed-bed are the focus. This includes a description of deviations of the gas flow pattern from ideal plug-flow, and deviations of the temperature within the catalyst particles and in the gas bulk from an ideal isothermal condition.

As illustrated in Figure 2, mathematical models for fixed-bed reactors are typically classified as being either the pseudo-homogeneous or heterogeneous type. Pseudo-homogeneous models do not explicitly distinguish between the fluid and solid phases during formulation of the mass, energy balances and momentum balances. On the contrary, heterogeneous models explicitly account for transport-kinetic interactions for both the gas and solid phases. The advantage of the heterogeneous models is their ability to describe intraparticle and interparticle transport-kinetic interactions on a single particle level. Heterogeneous models should be used for reactions with moderate to fast rates and moderate to high heats-of-reaction.

| | Pseudo Homogeneous Models<br>$T = T_s$, $C = C_s$ | Heterogeneous Models<br>$T \neq T_s$, $C \neq C_s$ |
|---|---|---|
| 1-D | basic, ideal | + interfacial gradients |
| | + axial mixing | + intraparticle gradients |
| 2-D | + radial mixing | + radial mixing |

Figure 2. Classification of fixed-bed reactor models.

Both pseudo-homogeneous and heterogeneous models can be developed so that they describe transport-kinetic interactions in either one or two-dimensions. One-dimensional models explicitly account for axial gradients of concentration, temperature, and total pressure. Radial transport phenomena are often treated in an approximate fashion by using overall transport coefficients. Two-dimensional pseudo-homogeneous models explicitly account for radial gradients in the differential material and energy balances.

*2.2 Adiabatic Fixed-Bed Reactors*

The least complex class of reactors used for gas-phase catalytic oxidation reactions are adiabatic fixed-bed reactors. The single-stage or single-bed adiabatic fixed-bed reactor has the simplest design. The number of processes that employ adiabatic reactors for oxidation processes is limited. Reactions carried out in the single-stage adiabatic reactor include: (1) the oxidation of ammonia to nitric oxide over a Pt/Rh gauze, which is the key step in the manufacture of nitric acid, and (2) the oxidation of a mixture of methane and ammonia to HCN, which is an important intermediate needed in the synthesis of several polymers, over a Pt/Rh gauze. Also included in the adiabatic reactor class is the multi-stage design, even though heat is exchanged between adjacent stages. Reactions carried out in this multi-reactor system include: (1) the oxidation of $SO_2$ to $SO_3$ on vanadium pentoxide catalyst, which is the key step in the manufacture of sulfuric acid, and (2) the silver-catalyzed oxidation of methanol to formaldehyde. Finally, the most recently developed reactor type in the adiabatic class is the reverse-flow reactor. This reactor has been used in the former Soviet Union to oxidize $SO_2$ to $SO_3$.

Unlike catalytic reactors with heat exchange within the reactor, the distinguishing feature of the adiabatic reactor is that the heat generated by an exothermic reaction goes unchecked. The steady-state temperature profile that results is a monotonically increasing function of distance from the inlet. The temperature increase asymptotically approaches the adiabatic temperature rise. Temperatures can exceed the adiabatic temperature rise in the unsteady state adiabatic reverse-flow reactor, as described in section 2.3.

Because of the potential severity of the heat effects, oxidation reactions carried out in adiabatic reactors form a special sub-class. Not only must the catalyst tolerate a temperature rise of several hundreds of degrees, but the desired product selectivity must also not be negatively impacted by the high temperature. If these two requirements are met, then one can take advantage of the high rate of reaction at the high temperature without significant penalty. Moreover, the relative simplicity of the adiabatic reactors, compared to their cooled counterparts, can be exploited in design, scale-up, and operation.

Some typical adiabatic reactors are shown in Figures 3 and 4. These include the simple catalyst bed [6], the disk or shallow-bed reactor [6,7], and the radial-flow reactor [7]. The shallow-bed reactor is used in fast reactions where complete conversion can be achieved in a very short contact time. The shallow-bed affords minimal pressure drop, although flow distribution can be an imposing problem[7]. The radial-flow reactor is used to reduce pressure drop as well.

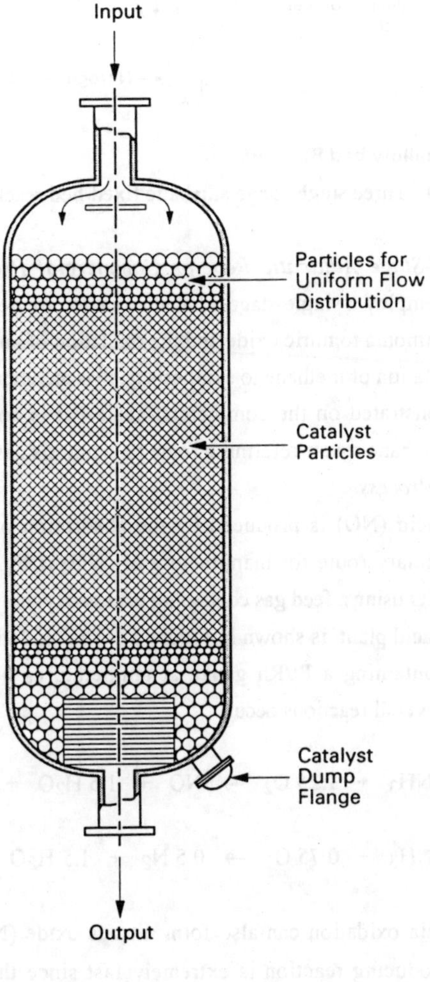


Figure 3. A typical adiabatic catalytic fixed-bed reactor

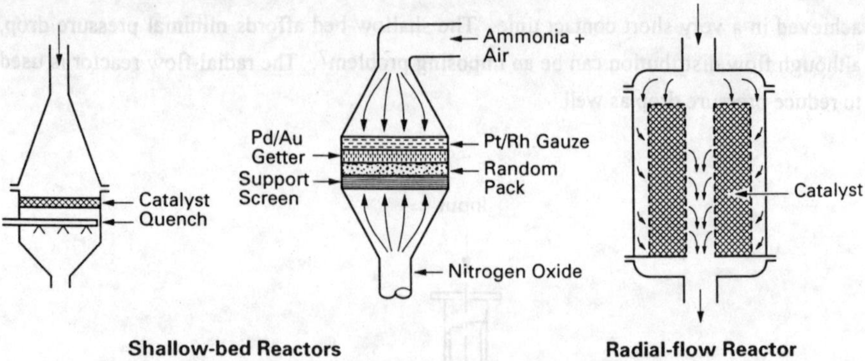


Figure 4. Three single-stage adiabatic fixed-bed reactor types.

*2.2-1. Single-Stage Adiabatic Reactor Commercial Processes.* Two commercial processes that employ a single-stage adiabatic reactor are described below. These are the oxidation of ammonia to nitric oxide and the oxidation of ammonia and methane to HCN. The partial oxidation of methane to synthesis gas[8] is an emerging reaction system that has not been demonstrated on the commercial scale. It is an attractive alternative to the conventional methane steam-reforming route to synthesis gas.

*(i) Nitric Acid Process*

Nitric acid (NO) is produced by the absorption of nitrogen dioxide ($NO_2$) in water. The primary route for manufacture of $NO_2$ is by the sequential oxidations of ammonia and NO using a feed gas containing excess oxygen. A schematic of a typical high pressure nitric acid plant is shown in Figure 5 . The oxidation of ammonia is carried out in a reactor containing a Pt/Rh gauze at temperatures between 850 to 950 °C. The following two overall reactions occur:

$$NH_3 + 1.25\, O_2 \rightarrow NO + 1.5\, H_2O + 226\text{ kJ} \quad (1)$$

$$NH_3 + 0.75\, O_2 \rightarrow 0.5\, N_2 + 1.5\, H_2O + 317\text{ kJ} \quad (2)$$

Ammonia oxidation can also form nitrous oxide ($N_2O$) in small amounts. The desired NO producing reaction is extremely fast since the contact time is about one millisecond and is limited by gas-solid mass transfer. For this reason, the gas linear velocity and flow uniformity are critical issues. The yield to NO is an increasing function of temperature and a slight decreasing function of total pressure. At atmospheric pressure and 850 °C, the NO yield is 98%, while yields of ca. 96% are obtained at 8 atm and 900 °C. The loss of precious metal catalyst becomes more important at higher temperatures.

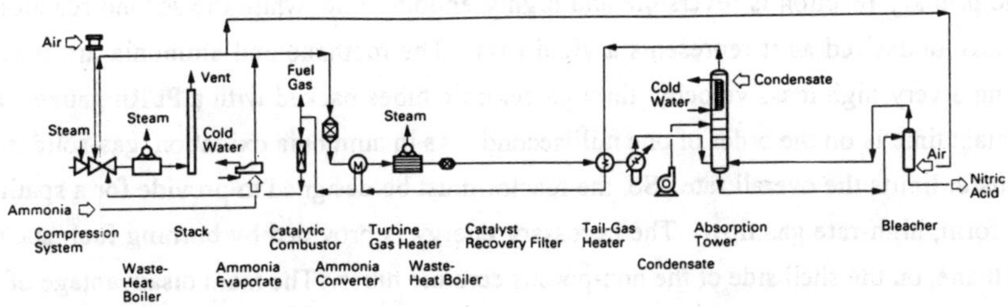


Figure 5. High pressure ammonia oxidation process. [9]

The high pressure process reaction operates at 110 psig and offers reduced equipment size at the expense of slightly lower NO selectivity. In the split-pressure process, the ammonia conversion is carried out at an intermediate pressure of about 30 psig, while the absorption is carried out at higher pressure. The NO that is formed in the ammonia converter oxidizes further to $NO_2$:

$$NO + 0.5\,O_2 \rightarrow NO_2 + 57\text{ kJ} \quad (3)$$

This reversible exothermic reaction is thermodynamically favored at low temperature. Finally, the mixture of NO and $NO_2$ pass through a condenser and then a cooled absorption column to produce the nitric acid. NO that is formed during the reaction between $NO_2$ and water is reoxidized to $NO_2$.

*(ii) HCN Process*

The type of catalytic reactor used in the oxidation of NO is also encountered in the production of HCN. It involves a reaction between methane and ammonia at high temperature.

The main reactions that occur in the Degussa process in the absence of oxygen in the feed gas are

$$CH_4 + NH_3 \leftrightarrow HCN + 3\,H_2 \text{ -251 kJ/mole} \tag{4}$$

$$NH_3 \leftrightarrow 0.5\,N_2 + 1.5\,H_2 \tag{5}$$

The primary reaction is reversible and highly endothermic, while the second reaction is clearly undesired as it represents a yield loss. The methane and ammonia are reacted using a very high mass velocity through ceramic tubes packed with a Pt/Rh gauze. The contact time is on the order of one millisecond. As in ammonia oxidation, gas-solid mass transfer limits the overall rate. So, the reactor must be designed to provide for a spatially uniform, high-rate gas flow. The necessary energy is provided by burning fuel, such as methane, on the shell side of the non-porous ceramic tubes. The main disadvantage of the Degussa process is the degradation of the tubes at the high reaction temperatures (1000 - 1100 °C).

In the Andrussow process, oxygen is present in the feed gas. The main and side reactions are

$$CH_4 + NH_3 + 1.5\,O_2 \rightarrow HCN + 3\,H_2O \tag{6}$$

$$NH_3 + 0.75\,O_2 \rightarrow 0.5\,N_2 + 1.5\,H_2O \tag{7}$$

$$CH_4 + 0.5\,O_2 \rightarrow CO + 2\,H_2 \tag{8}$$

$$CH_4 + 2\,O_2 \rightarrow CO_2 + 2\,H_2O \tag{9}$$

$$HCN + H_2O \leftrightarrow NH_3 + CO \tag{10}$$

The main role of oxygen is to oxidize methane to provide the necessary energy for the HCN formation.

There are three primary points of distinction in comparing the Andrussow process to the Degussa process. First, the Andrussow converter is an adiabatic reactor, which considerably simplifies the reactor design. Second, since oxygen is fed with the ammonia and methane, and additional side reactions occur that reduce the overall yield. As shown above, these include the oxidation of ammonia to nitrogen and the hydrolysis of HCN to ammonia and CO. To minimize the latter reaction, the reactor effluent must be rapidly cooled. Third, an additional constraint is the need to operate outside the zone of flammability.

*2.2-2. Reactor Performance and Modeling.* As described earlier, modeling and analysis of a single adiabatic fixed-bed reactor requires knowledge of the catalytic kinetics, intraparticle and interparticle transport processes, and the macroscopic flow and transport phenomena[5]. The modeling of adiabatic reactors is more straightforward than nonadiabatic reactors because the complexities introduced by intrareactor heat exchange are avoided. Modeling of multi-stage adiabatic reactors is much more difficult because of the thermal coupling that occurs between the stages.

Scale-up of the single-stage adiabatic reactor from laboratory-scale is simple in principle. For example, suppose the production rate in a laboratory-scale adiabatic reactor with a cross-sectional area for flow of A ($m^2$) and bed-depth of d (m) is equal to P (moles/hr). Moreover, suppose that complete conversion of the limiting reactant is achieved in the laboratory reactor. Suppose further that the desired commercial-scale production rate is $10^4 \cdot P$. Then the commercial-scale reactor should have the same bed depth as the laboratory reactor but a cross-sectional area equal to $100 \cdot A$. This area ensures that the same linear velocity, and hence contact time, is maintained in both reactors. Moreover, for fast reactions external transport processes are typically rate-limiting. By maintaining a constant velocity in the lab and pilot-scale reactors, the external transport rates are held constant as well. Factors that could complicate this simple scale-up procedure include non-adiabatic operation in the laboratory reactor, or flow maldistribution in the commercial-scale reactor.

The conversion and product distribution data from a laboratory reactor study of Pt-catalyzed methanol oxidation are shown in Figure 6. These data display the important requirement that a reaction system must maintain a highly desired product selectivity at high temperature, if it is to be carried out adiabatically. As the bed temperature is increased for a feed that is rich in methanol, the formaldehyde selectivity actually increases from less than 50% at 250 °C to over 90% at 600 °C. At reduced methanol to oxygen feed ratios, the selectivity does not exhibit such favorable features. It is for this reason that more than one adiabatic reactor can be used to avoid the need to feed all the required oxygen to attain a desired methanol conversion per pass to a single reactor. This operational strategy is discussed in the next section.

*2.2-3. Multi-Stage Adiabatic Reactor.* There are two classes of reaction systems for which the single-stage adiabatic reactor is incapable of satisfying the conversion and/or selectivity demands. The first class is reversible exothermic reactions. Multiple stages with interstage cooling are required for these reactions in order to achieve an acceptable conversion level with a reasonable reactor volume. The classical oxidation example is $SO_2$

to $SO_3$, which is described in more detail below. The second class is hydrocarbon partial oxidation reaction systems in which the desired product selectivity is sufficiently sensitive to temperature and oxygen concentration. More specifically, yield losses to carbon oxides, which result from sequential or parallel side reactions, undermine the goal of achieving high, intermediate partial oxidation product selectivity at a reasonable hydrocarbon conversion per pass. Most hydrocarbon partial oxidations have this feature, two examples of which include the silver-catalyzed oxidations of methanol to formaldehyde and the catalytic oxidation of monomethylformamide to methyl isocyanate.

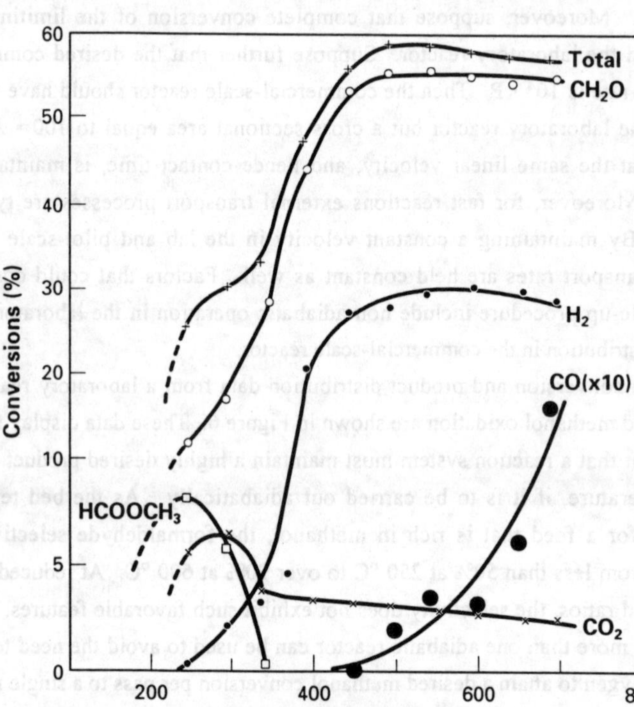


Figure 6. The influence of temperature on the conversion of methanol over a silver catalyst[11]. The feed composition is 2% $O_2$, 8.8% methanol, and the superficial gas velocity is 8.2 cm/s.

*2.2-4. Commercial Processes*

*(i) Sulfur Dioxide to Sulfur Trioxide Process*

The manufacture of sulfuric acid involves the oxidation of elemental sulfur to $SO_2$, followed by the catalytic oxidation of $SO_2$ to $SO_3$ over vanadium pentoxide, which is followed by the absorption of $SO_3$ with water. These two reactions are given by:

$$SO_2 + 0.5\,O_2 \leftrightarrow SO_3 \tag{11}$$

$$SO_3 + H_2O \rightarrow H_2SO_4 \tag{12}$$

The key attribute of the first reaction is that it is both exothermic and reversible. As shown in the conversion versus temperature plot in Figure 7, low temperature favors

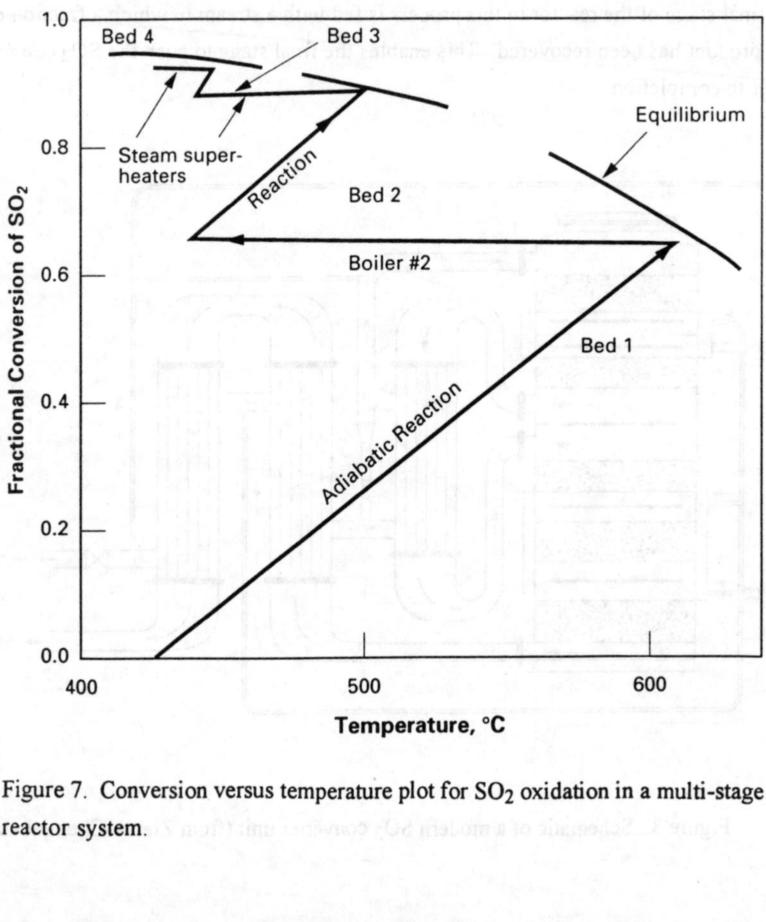


Figure 7. Conversion versus temperature plot for $SO_2$ oxidation in a multi-stage reactor system.

equilibrium. This is problematic, since the rate of reaction is an increasing function of temperature. For this reason, the reaction is carried out in a multi-stage adiabatic reactor. In the first stage, a large fraction of the $SO_2$ conversion is carried out. A high temperature is desirable in the first reactor stage because the feed is free of reaction product $SO_3$ so that the beneficial kinetic effect of a higher temperature can exploited. The reactor temperature increases linearly with conversion and becomes sufficiently high in the first bed that the reaction mixture must be cooled in order to confront the equilibrium constraint. In practice, several stages are employed with interstage cooling using heat exchangers or "cold shots" of air.

Figure 8 shows a $SO_2$ multi-stage reactor system (of Zieren-Chemiebau). The particular one shown has feed-effluent heat exchangers and cold-shot interstage heating. More elaborate multi-stage reactors have been developed due to environmental pressures to reduce sulfur emissions in sulfuric acid plants. One such process is shown in Figure 9. The final stage of the reactor in this process is fed with a stream in which a fraction of the $SO_3$ product has been recovered. This enables the final stage to push the $SO_2$ conversion closer to completion.

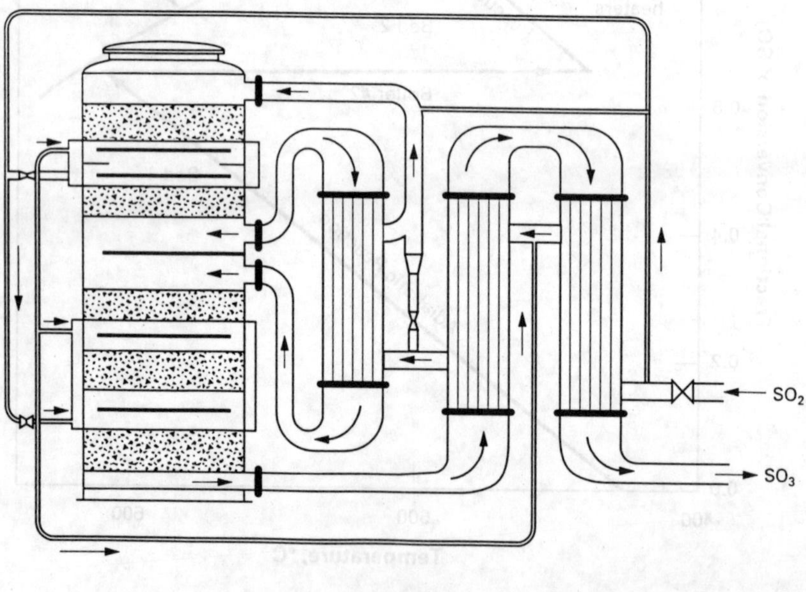


Figure 8. Schematic of a modern $SO_2$ converter unit (from Zieren-Chemiebau).[6]

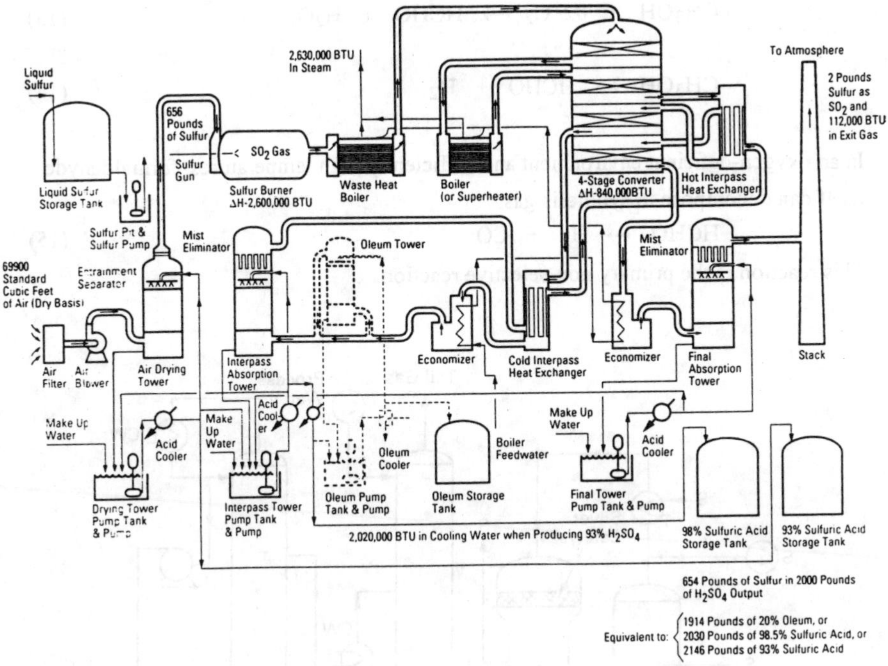


Figure 9. Process flowsheet for a sulfuric acid process.[12]

*(ii) Methanol to Formaldehyde over a Silver Catalyst*

Formaldehyde is another important chemical intermediate that is produced by oxidizing methanol in a hydrocarbon-rich feed mixture at 550-650 °C over a silver catalyst in a single-stage or two-stage adiabatic reactor. Alternatively, an iron molybdate catalyst can be used. In this case, a hydrocarbon-lean mixture is fed to a cooled multi-tubular reactor. The former case is relevant in this section.

Figure 10 shows a schematic flowsheet for the silver-catalyzed methanol to formaldehyde process. Fresh and recycled methanol, after being combined with air and vaporization, pass through an adiabatic reactor containing a thin layer (1 to 5 cm) of metallic silver crystallites or gauze. The feed to the bed is rich in methanol. The two desirable overall reactions are the oxidative dehydrogenation and straight dehydrogenation of methanol to formaldehyde:

$$CH_3OH + 0.5\,O_2 \rightarrow HCHO + H_2O \quad (13)$$

$$CH_3OH \leftrightarrow HCHO + H_2 \quad (14)$$

In an oxygen-deficient environment and sufficiently high temperatures, formaldehyde itself can decompose to synthesis gas:

$$HCHO \leftrightarrow H_2 + CO \quad (15)$$

This reaction is the primary non-selective reaction.

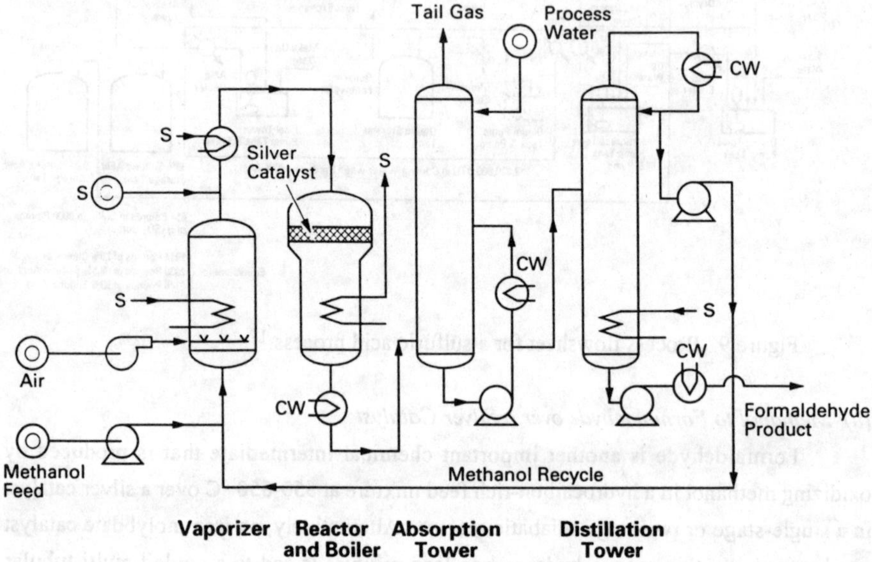


Figure 10. Process flowsheet for a methanol-to-formaldehyde process.

Although not shown in Figure 10, the oxygen addition is distributed between two adiabatic reactors in some cases. In the first stage, the feed oxygen is completely converted and most of the methanol converts to formaldehyde. Additional air and cooling is carried out between the first and second reactors. The feed temperature and feed compositions to each stage must be selected to maximize the formaldehyde yield and methanol conversion. Moreover, recall that high temperatures in this system are beneficial (see Figure 6). This means that the oxygen feed concentration and feed temperature to each reactor must be sufficiently high to ensure light-off of the catalyst. On the other hand, too high an oxygen concentration or feed temperature may mean a reduction in formaldehyde selectivity or sintering of the catalyst.

The inherent advantage of the silver-catalyzed, adiabatic reactor formaldehyde process over the metal oxide catalyzed, cooled reactor counterpart is the smaller process vessel sizes. The smaller volumes are possible because temperature control is not critical and a hydrocarbon-rich feed mixture is used. Thus, a high diluent flow rate is not needed. The disadvantage is that methanol must be recovered and recycled. The alternative process is discussed in section three.

*2.2-5. Reactor Performance and Modeling.* In multi-stage adiabatic reactor systems, heat exchange is accomplished with a standard heat exchanger or with a "cold shot" of one of the reactants. The reactor design can be an imposing optimization problem that involves a large number of manipulated variables. The simplest objective function consists of two terms: (1) the revenue generated by the production of desired product, and (2) the annualized cost of the reactor system and catalyst. The difference between these is a measure of the net profit. Obviously, this profit value is a liberal estimate given that there are other costs in the overall process. Solution for the optimal set of variables requires a detailed description of the multi-reactor and heat exchanger system.[13,14]

For the partial oxidation reaction system in particular, the multi-stage reactor optimization problem is especially challenging given that multiple steady-states and parametric sensitivity are the rule in exothermic reaction systems. In fact, to our knowledge, this problem has not yet been solved in the open literature. Multi-stage adiabatic reactors are impractical for these reaction systems if the number of stages is too large. Moreover, radial temperature gradients may become too large from the standpoint of catalyst durability or desired product selectivity. In these cases, the logical design choice is a cooled multi-tubular reactor. This class of reactors is discussed in section three.

Consider the oxidation of $SO_2$ as a representative example of a single exothermic reversible reaction. In this situation, the interstage cooling is accomplished with standard heat exchangers. To achieve complete conversion in a single stage, the reaction would have to be performed at a low temperature. However, the required reactor volume would be prohibitively large. Instead, several reactors in series are used in practice that have progressively decreasing feed and reactor temperatures. The feed temperatures to the first and second stages are sufficiently high to ensure that a reasonable conversion is achieved. The final stage has a sufficiently low temperature so that remaining reactant is converted. Additional measures may be needed to push the conversion to 100%; one measure is shown below.

*2.3. Reverse-Flow Reactor*

The reverse-flow reactor is an intriguing reactor that has been conceived by Boreskov and Matros[15,16] and Matros and co-workers.[17] As shown in Figure 11, this reactor is a standard adiabatic fixed-bed where the feed gas enters opposing ends of the reactor in a deliberate periodic fashion. The main idea is to utilize the exothermic heat of reaction as efficiently as possible within the catalyst bed itself.

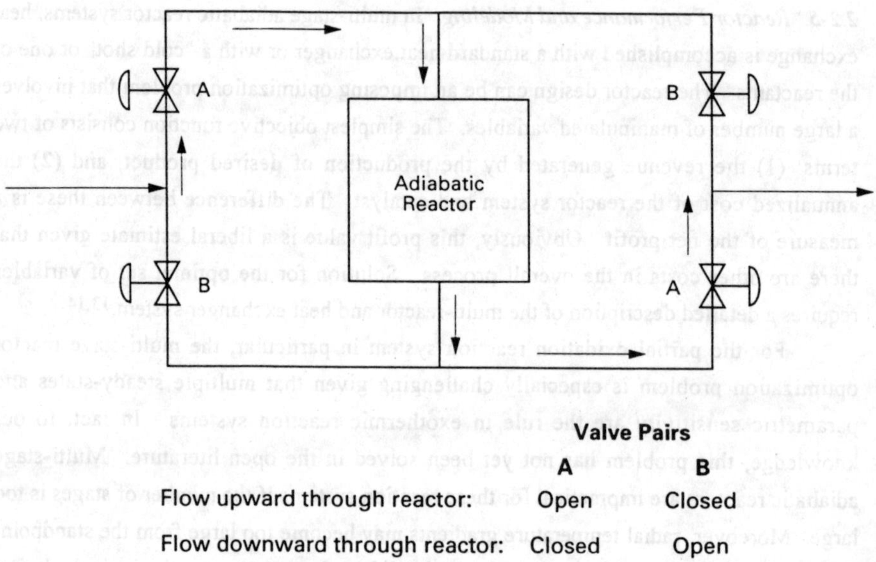


| | Valve Pairs A | Valve Pairs B |
|---|---|---|
| Flow upward through reactor: | Open | Closed |
| Flow downward through reactor: | Closed | Open |

Figure 11. Schematic of the reverse-flow reactor.

The reactor is operated as follows, as described lucidly by Eigenberger and Nieken:[18] "... After a sufficient portion of the packing has been heated to temperatures higher than the ignition temperature, the burner can be turned off. Thereafter, cold polluted air enters into the packing, where the air is heated by the hot bed so that catalytic oxidation takes place. The introduction of cold air leads to a progressive cooling of the inlet portion of the bed; and, as a result, to a continuous displacement of the temperature front, a so-termed migrating combustion zone. Without further action, the temperature front would move out of the reactor after a certain time, thereafter, the reaction would be extinguished. To prevent extinction, the direction of flow through the fixed bed is periodically reversed with the help of valves. As a result, the portion of packing which has cooled down is heated up again by the combustion zone moving in the opposite direction. Hence, the two end regions of the packing act as regenerative heat exchangers.

After a considerable number of flow reversals, a periodically steady state is established in the fixed bed in which, in one-half period, the zone of reaction moves a distance upward, whereas, in the next half period, the zone moves the same distance downward. ..."

The ability to capture the hot spot within the bed by flow reversal relies on the large difference in the characteristic time for convective mass flow and conductive energy transport. Flow switching can be easily accomplished on a time period that is much shorter than the characteristic time of transit of a creeping hot spot to traverse the entire length of the reactor.

The reverse-flow reactor can be used to carry out the complete combustion of pollutants contained in air streams and to carry out reversible exothermic reactions. For the former application, the efficient exchange of energy enables the complete combustion of pollutants in small concentrations in air streams with low feed temperature. For the latter application, the pseudo steady-state temperature profile that is established within the bed is close to the optimal profile for achieving high conversion in an equilibrium limited situation. It is this application that has relevance for oxidation to useful chemicals.

Figure 12 shows representative temperature profiles within a bed of vanadium oxide particles during flow reversal for $SO_2$ oxidation to $SO_3$.[15] Profile one was measured during flow from left to right through the bed, while profiles two, three, and four during flow from right to left. The profiles reveal the very large temperature rise that can be sustained (ca. 400 °C) compared to the adiabatic temperature rise (ca. 70-80 °C) for the 1.7% $SO_2$ mixture in air.

## 3. Multi-tubular Reactors

### *3.1. Introduction*

A larger fraction of selective oxidation reactions require finer temperature control than the staged adiabatic reactor types can provide. In these reaction systems, the selectivity is a decreasing function of temperature. This implies that the activation energy of the desired reaction is exceeded by that of one or more of the non-selective reactions.

In such cases, ineffective control can lead to hot spots and a subsequent loss in desired product selectivity. The multi-tubular catalytic reactor consists of a bank of tubes arranged in parallel that are immersed in a vessel through which a heat transfer fluid, such as Dowtherm®, steam, or molten salt, is circulated. The schematic in Figure 13 does not display the intricacy of the state-of-the-art multi-tubular reactor, which may contain up to 25,000 individual tubes. The multi-tubular arrangement provides sufficient heat transfer area and reduces the effective radial heat transfer distance. A large number of parallel reactors are sometimes used instead of a single unit to prevent hot spot formation and desired product losses.

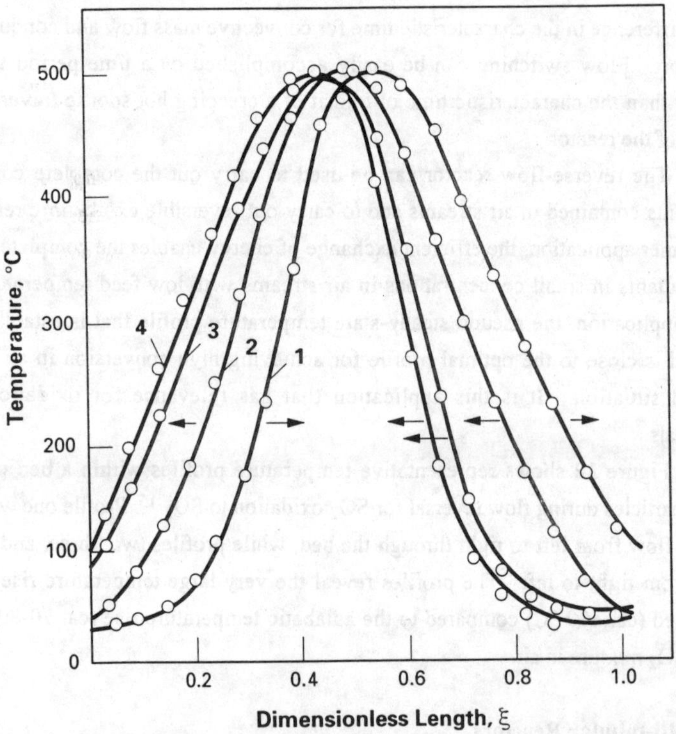


Figure 12. Measured temperature profiles during $SO_2$ oxidation in a reverse-flow reactor.

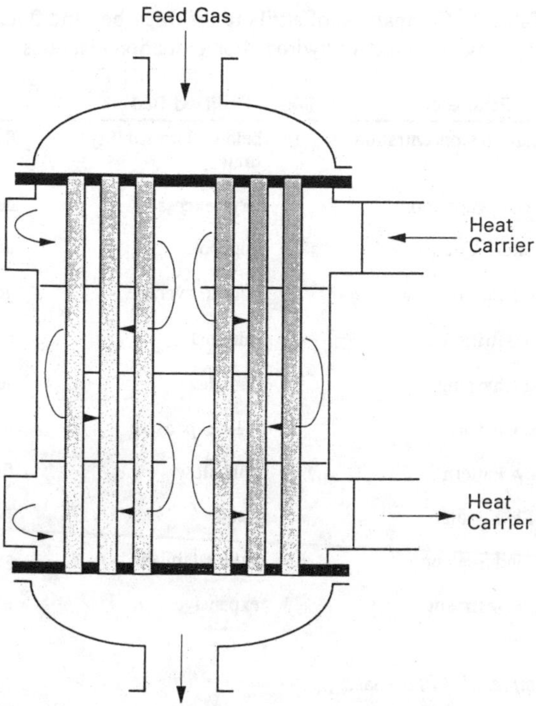


Figure 13. Schematic of a multitubular reactor.[9]

Some of the key attributes of the multi-tubular fixed-bed are provided in Table 4. Shown also for comparison are the corresponding attributes of the fluidized bed, which is an alternative reactor type for carrying out oxidation reactions that will be described in section four.

The choice between the multi-tubular and alternative reactor types must be considered on a case-by-case basis, taking all of these issues into account. Additional considerations on reactor selection are discussed in section four.

Table 4. Comparison of attributes of fixed-bed and fluidized-bed reactors for carrying out selective hydrocarbon oxidation reactions.

| Parameter | Fixed-Bed | Fluidized-Bed |
|---|---|---|
| Hydrocarbon Concentration | below flammability limit | flammable region possible |
| Oxygen Concentration | large excess | near stoichiometric |
| Temperature Control | hot spot | nearly isothermal |
| Catalyst Effectiveness | poor to average | good |
| Catalyst Attrition | minimal | possible problem |
| Catalyst Charging | complex | straightforward |
| Catalyst Cost | least expensive | more expensive |
| Gas Flow Pattern | plug-flow | flow-regime dependent |
| Solids Flow Pattern | fixed | flow-regime dependent |
| Design and Scaleup | well-established | system dependent |
| Capital Investment | expensive | less expensive |

### *3.2. Commercial Processes*

*3.2.1. Methanol to Formaldehyde on Metal Oxide Catalyst.* As discussed in section 2.2-4, formaldehyde can be produced by oxidizing methanol over a metal oxide catalyst, such as iron molybdate, under hydrocarbon lean conditions. This lower temperature approach requires the use of a multi-tubular reactor to avoid the non-selective production of carbon oxides. A schematic flowsheet of the metal oxide process is shown in Figure 14. The process is considerably different than the alternative high temperature, methanol-rich process that employs a silver catalyst. The most notable advantage of the metal oxide process is the elimination of a methanol recycle loop. The notable disadvantage is the need for larger equipment to accommodate the high flow rates of the heat-carrying diluent.

*3.2.2. Ethylene to Ethylene Oxide.* Ethylene oxide capacity is on the order of 10 million tons worldwide because of it's use in the production of many important chemicals, including ethylene glycol, ethanolamines, and polymers containing the hydroxyethyl group. Ethylene oxidation to ethylene oxide is a prime example of the need for a rational catalyst and reactor design to maximize yield of a partial oxidation product.

The desired reaction is

$$C_2H_4 + 0.5\,O_2 \rightarrow C_2H_4O \qquad (16)$$

Undesired reactions are primarily the parallel and consecutive complete combustion reactions:

$$C_2H_4 + 3\,O_2 \rightarrow 2\,CO_2 + 2\,H_2O \qquad (17)$$

$$C_2H_4O + 2.5\,O_2 \rightarrow 2\,CO_2 + 2\,H_2O \qquad (18)$$

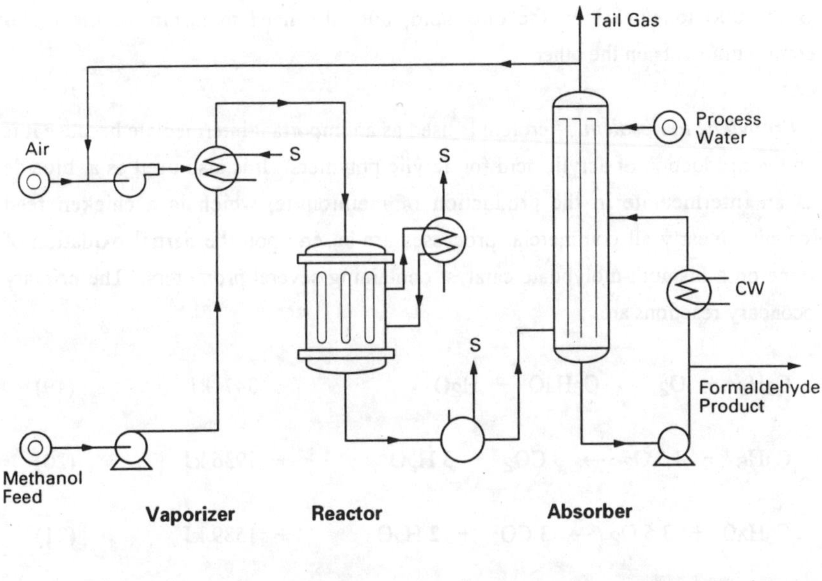


Figure 14. Process flowsheet for methanol to formaldehyde or an iron molybdate catalyst.[19]

All the reactions are exothermic. The heat of reaction of the complete oxidation is 1419 kJ per mole of ethylene. Moreover, the activation energies of the undesired reactions are higher than that of the desired reaction. This temperature sensitivity of the ethylene oxide selectivity requires very good temperature control. In order to suppress the consecutive oxidation of ethylene oxide, ethylene conversions per pass are low, ca. 10 to 20%. Selectivity to ethylene oxide is in the range of 60 to 80%. Reaction temperatures and total pressures are typically 200 to 300 °C and 10 to 30 atm, respectively. Either oxygen-enriched air or pure oxygen are used as the oxidant stream.

Supported catalysts with promoters are used in all commercial processes. The catalyst is typically a low surface area $\alpha$-$Al_2O_3$ bead impregnated with silver. The large pore support is used in order to reduce pore diffusion limitations that serve to promote the non-selective combustion of ethylene oxide. Finally, catalyst promoters are essential to achieve high intrinsic selectivity and catalyst life.

Because of the demand for fine temperature control, the multi-tubular reactor is preferred. A heat-transfer fluid is circulated on the shell-side of tubes containing the supported silver catalyst. Many tubes are needed because of the large capacity plants (up to 150,000 tons/year) on the one hand, but the need to minimize the radial temperature gradients on the other.

*3.2.3. Propylene to Acrolein.* Acrolein is used as an important intermediate because it is used in the production of acrylic acid for acrylic polymers. It is also used as a biocide and as an intermediate in the production of methionine, which is a chicken feed supplement. Nearly all commercial processes are based upon the partial oxidation of propylene on a bismuth-molybdate catalyst containing several promoters. The primary and secondary reactions are

$$C_3H_6 + O_2 \rightarrow C_3H_4O + H_2O \quad + 347\ kJ \qquad (19)$$

$$C_3H_6 + 4.5\,O_2 \rightarrow 3\,CO_2 + 3\,H_2O \quad + 1936\ kJ \qquad (20)$$

$$C_3H_4O + 3.5\,O_2 \rightarrow 3\,CO_2 + 2\,H_2O \quad + 1589\ kJ \qquad (21)$$

$$C_3H_6 + 2\,O_2 \rightarrow C_2H_4O + CO_2 + H_2O \quad + 814\ kJ \qquad (22)$$

Carbon dioxide and acetaldehyde are the primary carbon-containing by-products. The commercial bismuth-molybdate catalyst has several promoters, that result in an acrolein selectivity as high as 85% at a propylene conversion of about 90%.[20] Given the exothermic nature of the reaction system, a multi-tubular reactor is used to minimize temperature nonuniformities and the resultant loss in acrolein selectivity.

*3.2.4. Butane to Maleic Anhydride*

Maleic anhydride (MAN) is an unsaturated diacid that is used in the manufacture of several key products. Its' greatest use is in the formation of unsaturated polyester resins. The polymeric resin is formed from the reaction between MAN, ethylene glycol, and a vinyl monomer. The production of MAN is carried out in most existing processes by the selective oxidation of n-butane over a vanadium-phosphorous oxide (VPO) catalyst.

The main selective and non-selective reactions that occur are

$$C_4H_{10} + 3.5\,O_2 \rightarrow C_4H_2O_3 + 4\,H_2O \quad + 1236\text{ kJ} \qquad (23)$$

$$C_4H_{10} + 6.5\,O_2 \rightarrow 4\,CO_2 + 5\,H_2O \quad + 2656\text{ kJ} \qquad (24)$$

$$C_4H_{10} + 4.5\,O_2 \rightarrow 4\,CO + 5\,H_2O \quad + 1521\text{ kJ} \qquad (25)$$

The ability to achieve a 60 to 70% yield of MAN at 80 to 90% butane conversion is remarkable given the complexity of the selective reaction since it involves the abstraction of fourteen hydrogen atoms, the incorporation of three oxygen atoms, and ring closure. Moreover, the selective and non-selective reactions are quite exothermic.

A schematic of the Huntsman fixed-bed MAN process is shown in Figure 15. The process consists of a fixed-bed multitubular reactor, energy recovery units, a MAN absorber/stripper section, and a refinery section. As in other processes discussed in this section, thousands of tubes are used in the fixed-bed reactor to decrease undesirable radial temperature gradients.

Recent advances in fixed-bed based MAN processes include the development of reactors that can operate in the flammable regime, and of total butane recycle processes that can give overall process yields between 65 and 75%.

*3.2.5. o-Xylene to Phthalic Anhydride*

Another anhydride of paramount significance in the chemical industry is phthalic anhydride (PAN). The reaction chemistry used to synthesize PAN is based on the oxidation of o-xylene over a vanadium oxide on titanium oxide catalyst. The complexity of the chemistry rivals that of maleic anhydride. The primary and secondary reactions are

$$C_6H_4CH_3CH_3 + O_2 \rightarrow C_6H_4CHOCH_3 + H_2O \qquad (26)$$

$$C_6H_4CHOCH_3 + O_2 \rightarrow C_6H_4COOHCH_3 \qquad (27)$$

$$C_6H_4COOHCH_3 + O_2 \rightarrow C_6H_4COOCH_2 + H_2O \qquad (28)$$

$$C_6H_4COOCH_2 + O_2 \rightarrow C_6H_4COOCO + H_2O \qquad (29)$$

$$C_6H_4CH_3CH_3 + 10.5\ O_2 \rightarrow 8\ CO_2 + 5\ H_2O \qquad (30)$$

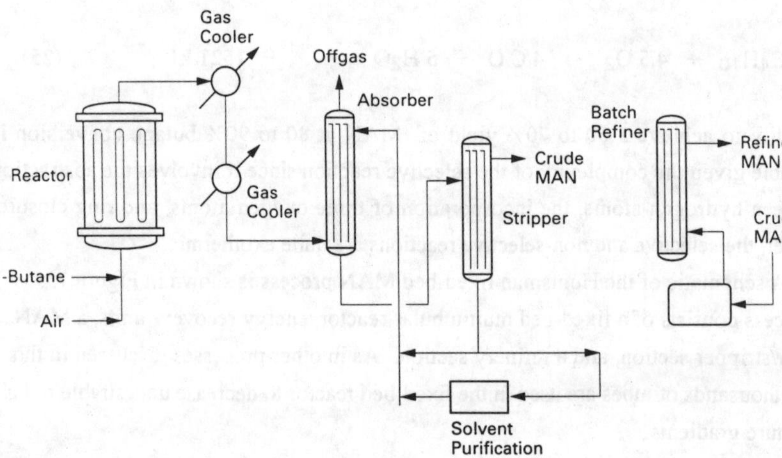


Figure 15. Process flowsheet for n-butane oxidation to maleic anhydride using multitubular reactors.[21]

As indicated above, PAN ($C_6H_4COOCO$) is the final product in a sequence of partial oxidations. PAN itself is difficult to oxidize further, but the parallel oxidation of o-xylene to carbon oxides is an important reaction pathway leading to selectivity losses. The selective catalytic chemistry relies on the interaction between the surface vanadium oxide and the underlying titanium oxide support.

A schematic of a typical o-xylene to PAN process is shown in Figure 16. The process consists of a feed and preheat section, the multi-tubular catalytic reactor, effluent stream heat exchangers, and a separation system. In practice, the multi-tubular reactor consists of a very large number of tubes because of the potential for both yield losses and reactor runaway. Various measures are taken to reduce the magnitude of the hot spot, some of which are described in the next section.

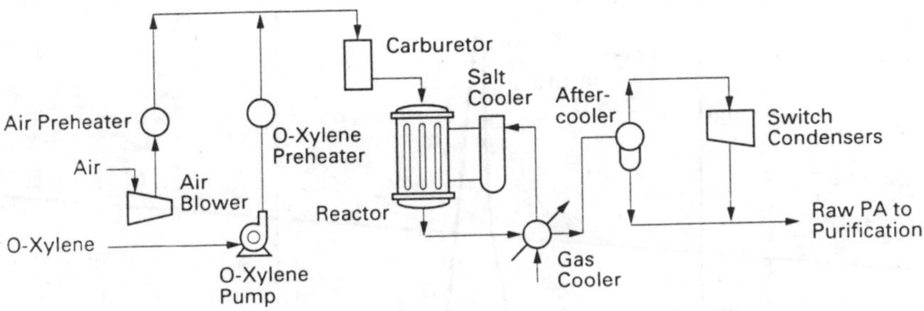


Figure 16. Process flowsheet for o-xylene oxidation to phthalic anhydride using multitubular reactors.[22]

### 3.3. *Reactor Performance and Modeling*

Control of the exothermic heat of reaction is the most critical issue in the design and operation of the nonadiabatic fixed-bed reactor.[22,23] Figure 17 shows the simulated temperature profiles in a cooled fixed-bed within which a single exothermic reaction occurs.[23] Sensitivity to the feed temperature is apparent. A 1 °C increase in the feed temperature from 343 to 344 °C results in an increase in the maximum temperature from approximately 480 to 900 °C. These hot spots must be avoided in temperature-sensitive selective oxidation reactions because of the detrimental effects on the desired product yield.

There are about five measures that can be taken to reduce the severity of the hot spot or to eliminate hot spots completely. These include: (1) reduce coolant temperature, (2) increase carrier gas flow rate, (3) increase bed thermal conductivity, (4) reduce tube diameter, and (5) dilute the catalyst bed, which corresponds to activity profiling. The first three measures are the simplest to employ, but adversely affect the process economics. For example, an increased carrier gas flow rate means that the volume of the reactor and downstream units increases. Moreover, the separation demands are increased. An increase in the bed thermal conductivity can be accomplished by using a different support material or by diluting the catalyst with catalytically inert particles that have a higher conductivity. The final two measures are considered in more detail.

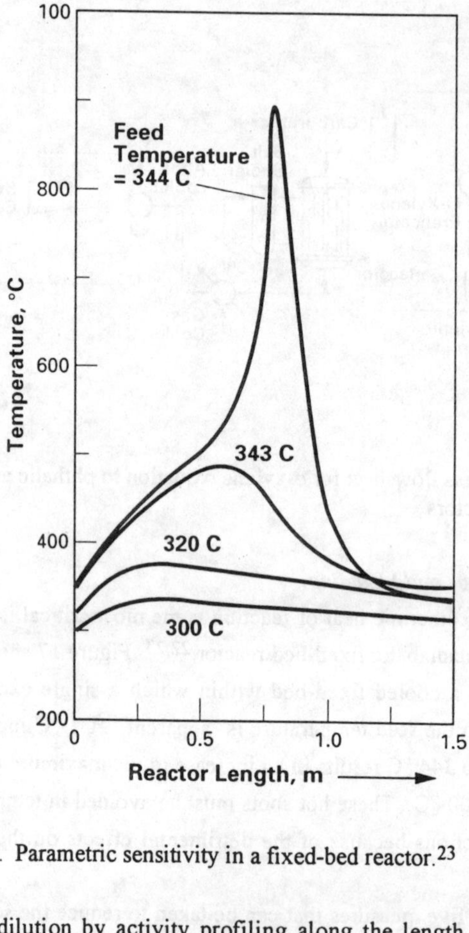


Figure 17. Parametric sensitivity in a fixed-bed reactor.[23]

Catalyst dilution by activity profiling along the length of the fixed-bed is an effective means of suppressing hot-spot formation. Selected results of Eigenberger[23] are shown in Figure 18 for a model exothermic reaction. The simplest dilution technique is to use a activity step function along the reactor length, as in Figure 18a. The effect of reducing the activity by a factor of two over the first 20% of the bed length is not impressive. More elaborate profiling, as shown in Figure 18b, is more effective in suppressing the hot spot. The best profile consists of saw-tooth-type depression in the activity profile that effectively flattens the temperature rise in the region of temperature sensitivity. Obviously, catalyst dilution has the drawback of requiring a larger reactor to achieve the desired conversion. However, the investment in a larger reactor may be worthwhile if the desired product yield is improved in a multiple reaction system. The final decision should be guided by process economics considerations.

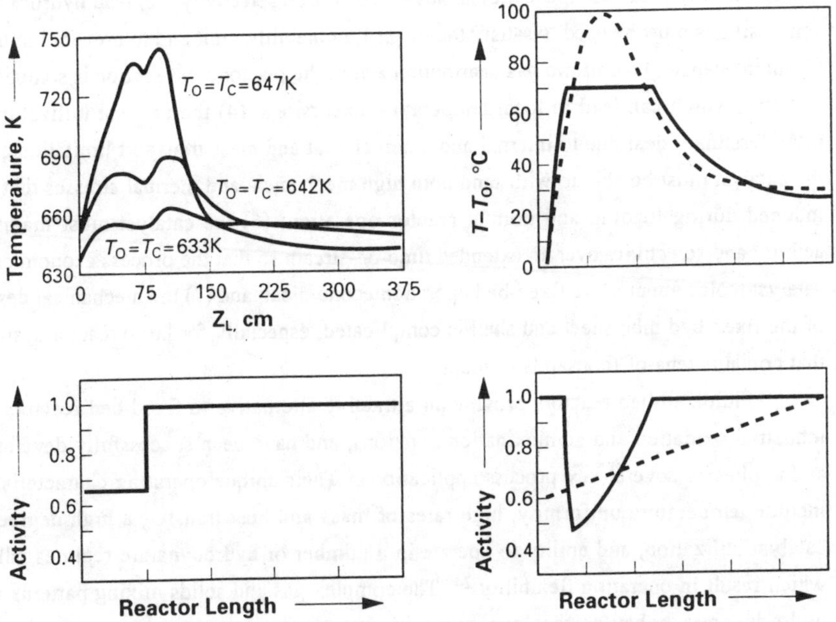


Figure 18. Catalyst activity profiling to reduce the magnitude of hot spots in a fixed-bed reactor.[23] Figure 18a (left). Straightforward step-function activity profiling. Figure 18b (right). More elaborate saw-tooth type of activity profiling.

While multi-tubular reactors are by now considered a classical catalytic reactor, there are challenges that, if overcome, have the potential to improve multi-tubular reactor performance and expand their domain of application. These challenges include: (1) pushing the envelope by using hydrocarbon/oxygen feed ratios in a fuel-rich regime, improving heat transfer capabilities, increasing operating temperatures, and improving reactor control strategies; (2) improve understanding of catalytic kinetic, and (3) improved reactor models by using computational fluid dynamic modeling of tube and shell sides, and realistic descriptions of the bed structure

## 4. Fluidized-Bed Reactors

### 4.1. Introduction

While fixed-bed reactors have been widely utilized for industrial oxidation and ammoxidation processes with a significant degree of success, they have several non-ideal characteristics that have prompted numerous investigations on development of alternate reactor types and operating modes. The most notable non-ideal characteristics of fixed-beds include: (1) non-isothermal temperature profiles in both the axial and radial direction that can lead to local hot spots with an adverse effect on selectivity; (2) lean hydrocarbon compositions must be used to satisfy the lower flammability limit and to prevent ignition; (3) maintenance of a uniform gas distribution across the reactor cross section is sometimes difficult, which can lead to local temperature excursions; (4) the catalyst utilization is often less than ideal due to internal and external heat and mass transport limitations; (5) the catalyst must be able to withstand both high mechanical and thermal stresses that are induced during loading and normal reactor operation; (6) the catalyst must maintain activity and selectivity over an extended time-on-stream so that the process economics of catalyst replacement make fixed-bed operation economical; and (7) the mechanical design of the fixed-bed tube sheet and shell is complicated, especially for large reactor systems that contains tens-of-thousands of tubes.

Fluidized-bed reactors provide an attractive alternative to fixed-bed reactors for industrial oxidation and ammoxidation reactions, and have been successfully developed and applied to several key process applications. Their unique operating characteristics include temperature uniformity, high rates of mass and heat transfer, a high degree of catalyst utilization, and ability to operate in a number of hydrodynamic regimes, all of which result in operating flexibility.[24] The complex gas and solids mixing patterns and hydrodynamic behavior that can occur in this reactor type provides a number of challenges from a process and reaction engineering perspective. These difficulties, along with some important issues related to the development of an attrition-resistant catalyst that can withstand the rigors of gas-particle fluidization, provide a partial explanation why fixed-bed reactors continue to be the dominant reactor type.

The primary objective of this section is to provide an overview of various industrial applications of fluidized-bed reactors in industrial oxidation and ammoxidation processes with an emphasis upon chemical transformations involving light hydrocarbons and aromatics. The particular ones given here are illustrations of catalytic reactions that have been commercialized in fluidized-bed reactors and are actually in operation today. If this objective were extended to include reaction systems that have commercial potential and are the subject of active research in major laboratories of academic and industrial institutions, then the number of potential processes would be significantly greater. Details on reactor modeling are not included, since these are covered in a number of recent reviews and monographs on the subject, as well as in a future review.[4]

*4.2. Origin of Fluidized-Bed Reactors*

The origin of fluidized-bed reactors can be traced back to the discovery that a Friedel-Crafts aluminum chloride catalyst could be used to catalytically crack heavy petroleum oil.[25] This invention, which was made by A. M. McAfee of Gulf Refining Company (now part of Chevron) in 1915, was originally practiced in fixed-bed reactor systems. The fixed-bed processes that were commercially introduced in 1937 were a significant improvement over the classical thermal cracking methods, since they produced gasoline with a higher octane rating and less low-value heavy fuel-oil by-product. A chronology of these developments is illustrated in Figure 19.

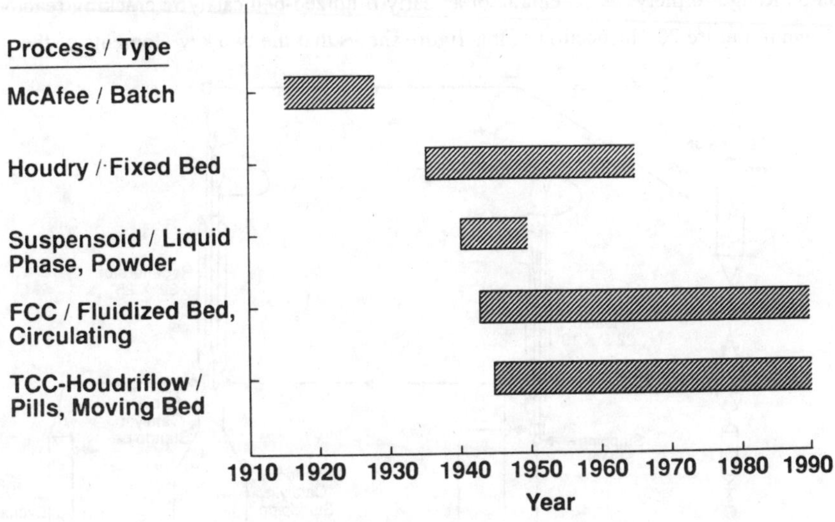


Figure 19. Key developments in fluidized-bed reactor processes.

The demands placed upon the petroleum refining industry as a result of World War II provided the impetus for development of more efficient processes with increased production rates than that available with existing fixed-bed processes. Innovations to the fixed-bed processes included the so-called *alternating-bed* reactors. In these systems, the regenerator and reaction beds were connected in series, and the catalyst was continuously moved from one reactor to the other. Superficial gas velocities less than 1 m/s were used to minimize erosion of the reactor hardware and catalyst attrition. This led to the observation that dense beds of catalyst powder could be maintained in place with relatively small losses due to elutriation, even at superficial gas velocities, that were orders of magnitude greater than the calculated settling velocity of the individual particles that comprised the bed. It was observed that at these gas velocities the particles were considerably agitated by gas bubbles that flowed upward through the mixed particles. The measured pressure drop through this gas-solid suspension was equivalent to that calculated using the weight of catalyst in the bed as the basis. Hence, the weight of the catalyst bed was buoyed by the force induced by the upward flowing gas bubbles. Industrial application of fluidized-bed technology was born from these observations.

The above efforts resulted in the development and operation of the first truly fluidized-bed cracking process in 1942 at the Standard Oil Company of New Jersey's Baton Rouge refinery. A schematic of an early fluidized-bed catalytic cracking reactor is shown in Figure 20. Inspection of this figure shows that the two key elements of the

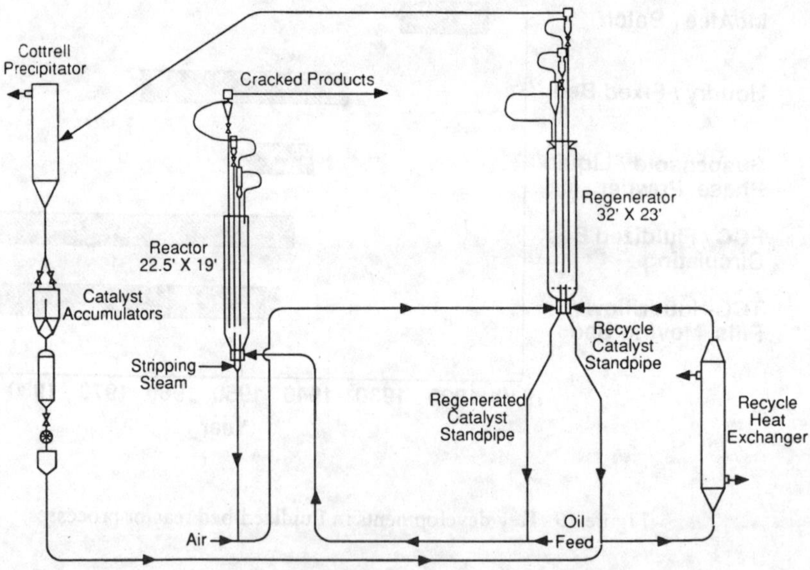


Figure 20. Early fluidized-bed catalytic cracking reactor.[25]

system are the reactor and regenerator which are connected so that fresh and regenerated catalyst can be sent both to and from these two units. The vaporized oil was cracked in a dense, fluidized bed of catalyst with solids residence times on the order of two to three minutes. Additional details on the design and operation of these early units are given in the above reference and those cited therein. Today, there are more than 250 commercial fluidized catalytic cracking units in operation, with more than half of these located in the United States.[25] Most of these have been designed by either Kellog or UOP.

*4.3. Fluidized-Bed Reactors in Chemical Processing*

*4.3-1. Reactor Description.* While the greatest application of fluidized-bed technology is in petroleum refining, a number of key products based upon catalytic oxidation or ammoxidation are also manufactured in fluidized-bed reactors. A generalized schematic of a fluidized bed reactor that might be used in a particular chemical processing application is illustrated in Figure 21.

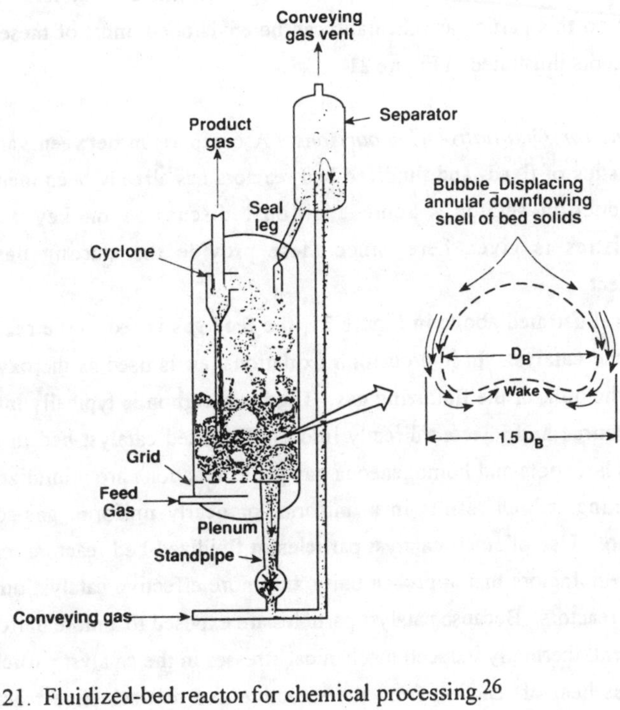


Figure 21. Fluidized-bed reactor for chemical processing.[26]

In nearly all processes involving fluidization, the solid is transported from one stage or processing step through pipelines that may contain various types of valves. Upward transport of solids is usually achieved by contacting the solids with a flowing gas stream with a sufficiently high velocity so that the gas-to-particle drag force exceeds the downward drag force of the particles. Transport of solids in the downward direction is performed by allowing the solids to settle through a pipeline. Examining Figure 21, the process gas enters the plenum below a grid that supports the bed where it forms bubbles that flow upward and fluidized the solids. When the bubbles reach the upper region of the gas-solid fluidized zone, they burst into the freeboard or disengagement region and entrain some particles. The exiting gas passes through a cyclone that separates the gas-solid suspension into a product gas stream and a solids stream where the latter is returned to the bed. If the catalyst particles undergo deactivation or require some type of post-reaction treatment, they can be withdrawn through a standpipe and returned to the fluidized bed via a pneumatic lift and separator. Once in the separator, the solids will flow downward through the standpipe seal leg back into the reactor. Although several variations on this particular schematic can be envisioned, most of these incorporate the basic concepts illustrated in Figure 21.

*4.3-2. Reactor Qualitative Comparisons.* A comparison between various qualitative characteristics of fixed- and fluidized-bed reactors has already been mentioned above in the introduction section. Some additional discussion on key aspects of these characteristics is given here, since these provide the starting basis for process development.

As illustrated above in Figure 21, the feed gas is fed to the reactor bottom and fluidizes the catalyst. In hydrocarbon oxidations, air is used as the oxygen source and provides the bulk of the fluidizing gas. The hydrocarbon is typically introduced from a separate source and injected directly into the fluidized catalyst bed in the grid region. Localized hot spots and homogeneous gas-phase reactions are minimized by rapid gas-solids mixing, which results in a uniform, or nearly uniform, gas-solid suspension temperature. Use of small catalyst particles in fluidized-bed reactors results in catalyst effectiveness factors that approach unity and more effective catalyst utilization than in fixed-bed reactors. Because catalyst particles are exposed to a more uniform temperature environment, thermally induced mechanical stresses in the catalyst particles are reduced. The excess heat of reaction from the selective and non-selective reactions is readily removed in fluidized-bed reactors by cooling coils located in the gas-solid suspension. Both the internal and external heat transfer resistances in these coils are smaller than those obtained in either internal or external heat exchanger used in fixed-bed reactors, which results in a more effective use of the heat transfer surface.

The ability to use hydrocarbon concentrations that are in the flammable region is another attractive feature of fluidized-bed reactors over fixed-bed reactors.[27] This is accomplished by keeping the fuel and oxygen sources separate until they are contacted in the fluidized bed of catalyst. The high heat capacity of the catalyst and the rapid heat transfer between the gas and solid particles minimizes or prevents the formation and movement of a stable flame front that is needed for ignition. In butane oxidation, for example, the hydrocarbon concentration is ca. 4 mole % in air for certain fluidized-bed processes, while fixed-bed processes contain less than 2 mole % of hydrocarbon in the feed.[27] The use of a higher butane concentration results in a lower air compressor duty, lower utility costs, and produces a more concentrated product stream. This, in turn, reduces the size of the product gas cooler, reduces the amount of nitrogen that is recycled, and reduces the size of the absorbers used for product recovery. This last point is noteworthy, because it suggests that any comparisons of process technology involving different reactor types must be performed using a model of the entire process, and not just direct comparisons of reactor performance variables.

Charging and removal of the tableted catalyst used in fixed-bed reactors requires that the catalyst have a high mechanical strength and that the characteristic pressure drop versus gas-flow rate response function for each tube fall within a certain tolerance to minimize preferential gas flow. By contrast, loading and unloading of the catalyst in a fluidized-bed reactor is much easier and can be accomplished through well-known pneumatic transport techniques. However, the catalyst must possess sufficient attrition resistance so that the process economics associated with attrition and elutriation of fines is not adversely affected.

Catalyst costs associated with fluidized-bed processes can be more expensive than their fixed-bed counterparts because of the additional steps involved in catalyst manufacturing that are associated with imparting the required attrition resistance. For applications where catalyst deactivation occurs, fluidized beds may be the only economical option, since catalyst replacement is simpler than fixed-beds.

One significant complication of fluidized beds is generally associated with the complexities associated with their gas-solid hydrodynamics and mixing characteristics. The ideal mixing states of plug-flow of gas and perfectly backmixed solids are often not readily achieved in fluidized beds. More importantly, these mixing states may not even be desirable for certain types of reaction networks where the selectivity to the desired product may be adversely affected. In addition, achievement of a high degree of reactant conversion may be difficult in a fluidized bed and may require the use of staging or internals to reduce solids backmixing or minimize imperfect gas-solid contacting or gas bypassing. By contrast, reaction engineering of fixed-bed reactors is better understood and amenable for scale-up using intrinsic kinetic data and existing knowledge on transport effects.

*4.4. Commercial Processes*

Fluidized bed reactors have been used in a number of commercial processes involving the oxidation and ammoxidation of light hydrocarbons. A summary of the monomers produced from these processes, the hydrocarbon reactant used, and the corresponding company that developed the process technology is given in Figure 22.

| Monomer | Reactants | Company |
|---|---|---|
| Acrylonitrile | Propylene/air/$NH_3$ | British Petroleum<br>Montedison - UOP |
| Maleic Anhydride | Butylenes/air<br>n-Butane/air<br>n-Butane/air | Mitsubishi Chemicals<br>Alusuisse Italia<br>British Petroleum |
| Phthalic Anhydride | Naphthalene/air | Sherwin Willaims/Badger |

Figure 22. Commercial fluidized-bed processes for selective oxidation and ammoxidation.

Most of the world's production of acrylonitrile (AN) is produced using Sohio's (now British Petroleum) fluidized-bed process from the vapor-phase catalytic air oxidation of propylene and ammonia.[28] Although the first report of producing AN from propylene occurred in a patent by Allied Chemical and Dye Corporation in 1947, a commercial catalyst system was not developed until the late 1950's by researchers at Sohio using a composition based upon a bismuth-phosphomolybdate oxide. Acrylonitrile is an important chemical intermediate that is used in a variety of applications in the fields of fibers, synthetic resins, elastomers, and as intermediates in organic synthesis. More than forty-five commercial plants based upon this process have either been licensed or are operated by British Petroleum using this technology.

Commercial fluidized-bed processes for production of maleic anhydride have been developed in the laboratories of several major industrial companies.[29] The first commercial fluidized-bed process was developed by Mitsubishi Chemical in 1970, and was based upon the oxidation of the mixed butenes produced from naphtha crackers. Both Alusuisse Italia and British Petroleum have developed dense-phase fluidized-bed processes, but these are based on n-butane as the hydrocarbon feedstock. They also use proprietary methods for imparting attrition resistance to the vanadium-phosphorus oxide catalyst. Both of these processes are available for licensing and have been recently commercialized in Europe and the Far East. The other differences that exists in these processes are primarily in the downstream processing associated with recovery and purification of the crude maleic anhydride product stream. Details associated with preparation of the catalyst precursor and the methods used for imparting attrition resistance are not available, and are kept as proprietary information.

Another application of fluidized-bed reactors is the selective oxidation of naphthalene to phthalic anhydride.[30] The original process design was developed by Sherwin-Williams, while the catalyst technology was developed by Davison Chemical. The first plant was commercialized in 1945, which led the way for development of fourteen additional plants based upon this technology. The original catalyst was vanadium oxide on a silica carrier.

Some more specific details on each of the above processes are given below.

*4.4-1. Sohio-BP Acrylonitrile Process.* A process flow diagram of the Sohio-BP process for the manufacture of AN from the air oxidation of propylene and ammonia is shown in Figure 23. The overall reaction is

$$2\ CH_2{=}CH{-}CH_3 + 2\ NH_3 + 3\ O_2 \rightarrow 2\ CH_2{=}CH{-}CN + 6\ H_2O \qquad (31)$$

The main reaction by-products are hydrogen cyanide, acetonitrile, and carbon oxides. The reaction mechanism has been studied in detail, and involves a complex sequence of oxidation-reduction reactions involving various metal oxides.[20]

The reaction is typically conducted using stoichiometric ratios of the reactants with mean-residence times of the gas being on the order of a few seconds. The operating conditions may vary, but can range between 20 to 200 kPa gauge (2.9 to 29 psig) total pressure and 400 to 500 °C. The process is highly selective, and does not require recycling to produce high AN yields of approximately 0.8 to 0.9 kilogram per kilogram of propylene. The excess heat of reaction is recovered in the form of steam. Commercially recoverable quantities of both hydrogen cyanide and acetonitrile (0.10 to 0.2 kg per kg of propylene fed) are also produced.

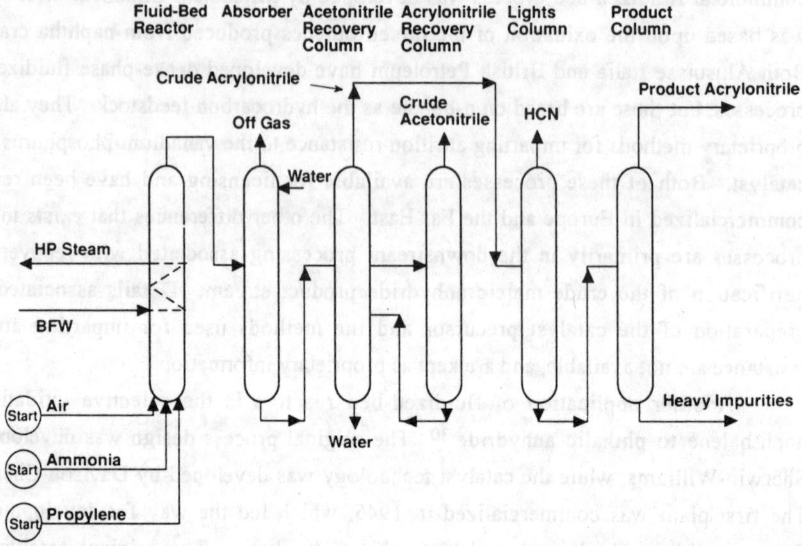


Figure 23. Process flowsheet for propylene ammoxidation to acrylonitrile using a fluidized bed reactor.[28]

The bismuth-phosphomolybdate catalyst used in the early Sohio process has been steadily improved and replaced over the years by more highly active and selective catalysts. Sohio introduced Catalyst 21 in 1967, which was an antimony-uranium based oxide. Catalyst 41, which was based upon ferrobismuth-phosphomolybdate, was introduced in 1972, while Catalyst 49 was introduced in 1978. All of these give improved process efficiency and a reduction in by-products.

In the process-flow diagram shown above in Figure 23, the fluid-bed reactor crude effluent is cooled and scrubbed with water in a countercurrent absorber. The off-gas, which consists primarily of nitrogen, is vented. The reaction products remain in the absorber aqueous phase. By-product acetonitrile is removed from the absorber bottoms by extractive distillation in the acetonitrile recovery column. In the next column, the crude acrylonitrile and hydrogen cyanide are removed in the distillate, while water and residual acetonitrile are removed in the column bottoms product. In the remaining two columns, the hydrogen cyanide is separated from the wet acrylonitrile, the water content of the product is then reduced, and any remaining nonvolatile impurities are removed in the bottoms product.

The major by-products of this process, hydrogen cyanide and acetonitrile, normally are incinerated as their supply often exceeds demand. Unused ammonia can be recovered as ammonium sulfate and then disposed, but it is often vented to the atmosphere. Aqueous wastes containing cyanides, sulfates, and various organic by-products must be incinerated, by deepwell injection, or be pre-treated for subsequent biological waste treatment. Recent state and government regulations involving plant emissions have resulted in major reductions and handling of these waste products.

*4.4-2. Maleic Anhydride Processes.* Maleic anhydride is an important raw material in the manufacture of alkyd and polyester resins, surface coatings, lubricant additives, plasticizers, copolymers, and agricultural chemicals. For this reason, a number of industrial processes have been developed that are based upon fixed-bed reactors, fluidized-bed reactors, and recirculating solids reactors. These are summarized in Figure 24. Those developed by Mitsubishi Chemicals, Alma/Alusuisse, and Sohio/UCB are based upon classical fluidized-bed reactors and have been implemented on the commercial

| Company | Reactor |
|---|---|
| ALMA (Alusuisse) | Fluidized Bed |
| Alusuisse Italia | Fixed Bed |
| Amoco | Fixed Bed |
| Mitsubishi | Fluidized Bed |
| Monsanto | Fixed Bed |
| Scientific Design | Fixed Bed |
| Sohio-UCB | Fluidized Bed |
| Dupont | Recirculating Solids |

Figure 24. Commercial-scale maleic anhydride processes.

scale. Another process developed by Badger and Denka has also been developed and implemented on the pilot-scale, but has not yet been realized on a commercial scale. Some noteworthy features of these processes are discussed individually below.[27]

*(i) Mitsubishi process.* A process-flow diagram of the fluidized-bed maleic-anhydride process developed by Mitsubishi Chemicals is shown in Figure 25. The hydrocarbon source is the crude $C_4$ fraction from naphtha crackers, so it represents a departure from the more classical benzene-based processes and more recent processes based upon n-butane. Although recent information is not available, the capacity of the plant is at least 18,000 metric tons per year of purified maleic anhydride.

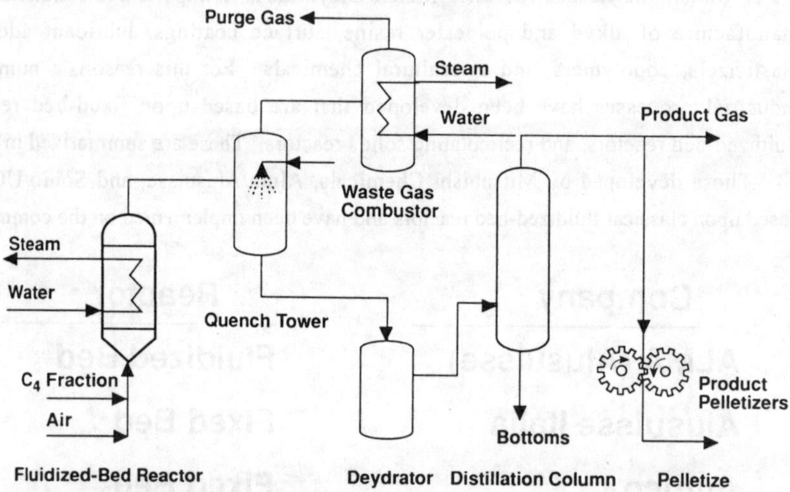


Source: S. Ushio, Chemical Engineering, September 20 (1971) 107.

Figure 25. Mitsubishi fluidized-bed process for maleic anhydride.

The fluidized-bed reactor is fed with the crude $C_4$ fraction and air where the hydrocarbon concentration is high enough that it exists in the explosive range (i.e., between 1.8 to 8.9 mole %). Because of the rapid gas-solids mixing and associated hydrodynamics in the grid region, an ignition cannot occur, so that safe operation is possible. The excess heat of reaction is removed by internal cooling coils that results in the generation of high pressure (430 to 710 psig) steam. The hot reactor off-gas, which contains the crude maleic anhydride, is quenched in a spray tower so that an aqueous solution of crude maleic acid is produced. The off-gas from the spray tower overhead contains a small fraction of reaction by-products and is incinerated in the waster gas combustion chamber to generate additional high pressure steam.

Dehydration and purification of the crude maleic acid stream obtained from the quench tower bottoms is performed by evaporation and distillation. The purified maleic anhydride vapors are taken overhead and condensed for subsequent pelletization and packaging.

A clear advantage of this process is the ability to utilize a relatively cheap $C_4$ source versus a more expensive n-butane feedstock, and the generation of high pressure steam that can be integrated into other near-by processes located on the same manufacturing site. The attrition-resistant vanadium-phosphorus catalyst is manufactured using classical spray-drying techniques with sufficient silica to impart hardness.

(*ii*) *Alusuisse/Alma process.* A process-flow diagram of the Alusuisse/Alma fluidized-bed process is shown in Figure 26. Unlike the Mitsubishi Chemicals process, n-butane is used as the hydrocarbon source. The n-butane and air are fed separately to the grid zone so that the concentration of hydrocarbon of the mixed stream would be about 4 mole %.

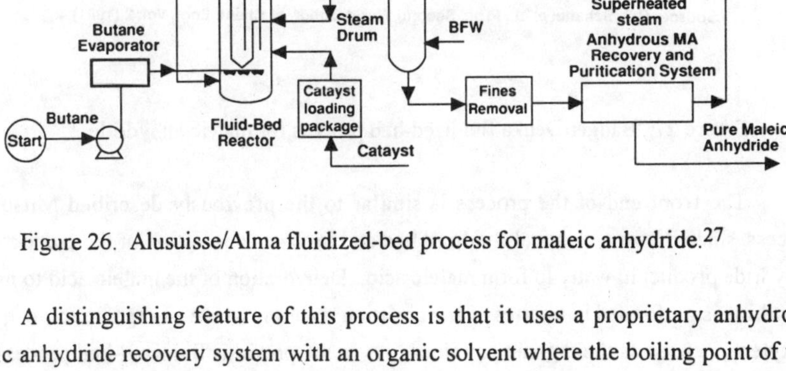

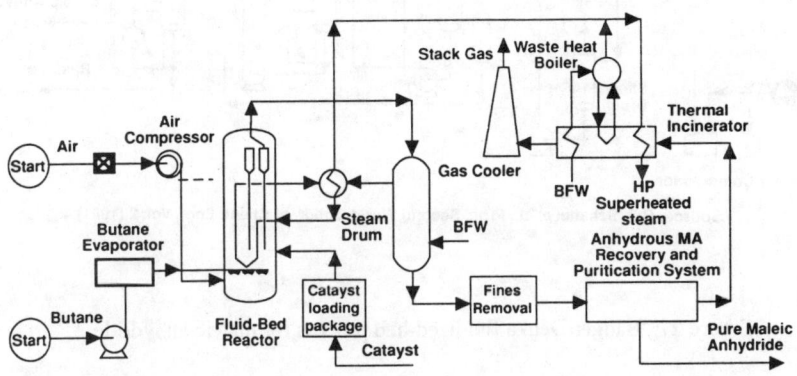


Figure 26. Alusuisse/Alma fluidized-bed process for maleic anhydride.[27]

A distinguishing feature of this process is that it uses a proprietary anhydrous maleic anhydride recovery system with an organic solvent where the boiling point of the latter exceeds that of maleic anhydride (ca. 202 °C at 1 atm). This is claimed to minimize the formation of unwanted by-products that are otherwise formed in aqueous recovery processes involving the evaporation of aqueous maleic acid.

The vanadium-phosphorus catalyst is transformed into an attrition-resistant form using a proprietary commercial-scale spray-drying process. It is claimed that the catalyst experiences negligible losses due to activity and attrition so that economical operation can be maintained. Both the catalyst and process are available for licensing, and a commercial facility was recently started up in Europe.[29]

*(iii) Badger/Denka process.* A process-flow diagram of the Badger/Denka maleic anhydride fluidized-bed process is shown in Figure 27. To the authors' knowledge, this process has not been commercialized, but it was demonstrated on the pilot-scale in the early 1980's.

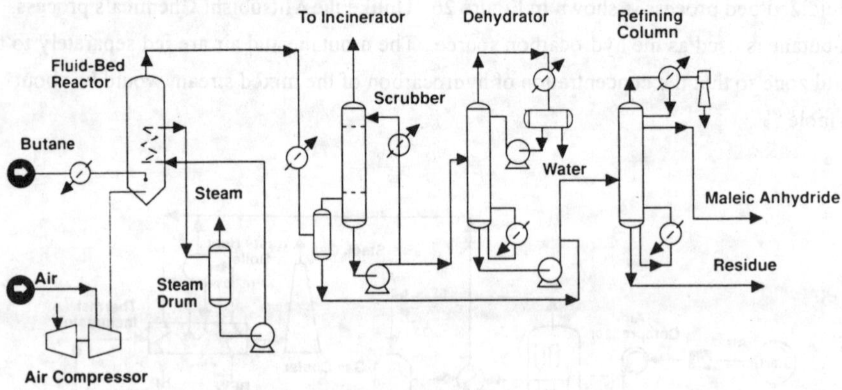

Source: G.S. Schaffel et al., Proc. Second World Congr. of Chem. Eng., Vol. 2 (1981) 4.

Figure 27. Badger/Denka fluidized-bed process for maleic anhydride.

The front end of the process is similar to the previously described Mitsubishi process, since it uses a classical fluidized bed and involves absorption of the crude maleic anhydride product in water to form maleic acid. Dehydration of the maleic acid to maleic anhydride is performed in a series of fractionation columns where an organic solvent, such as xylene, is used as the organic entrainer. As in the previous fluidized-bed processes, a catalyst having excellent attrition resistant and stable activity over an economical life is claimed.

*(iv) Sohio/UCB process.* A process-flow diagram of the Sohio/UCB maleic anhydride fluidized-bed process is shown in Figure 28. A maximum of 50 mole % of the crude maleic anhydride vapor in the reactor off-gas is continuously condensed out by cooling it below the dew point of maleic anhydride while maintaining the dew point above that of water. The partial condenser used for this step must be periodically cleaned, due to accumulation of solid. The remaining product is absorbed in water and recovered as crude maleic acid.

Unlike the Badger/Denka process, no organic solvent entrainer is used in the maleic acid dehydration step. The aqueous maleic acid steam is first evaporated under vacuum, which is followed by dehydration to the anhydride using a specially developed thermal dehydration reactor system that minimizes isomerization.

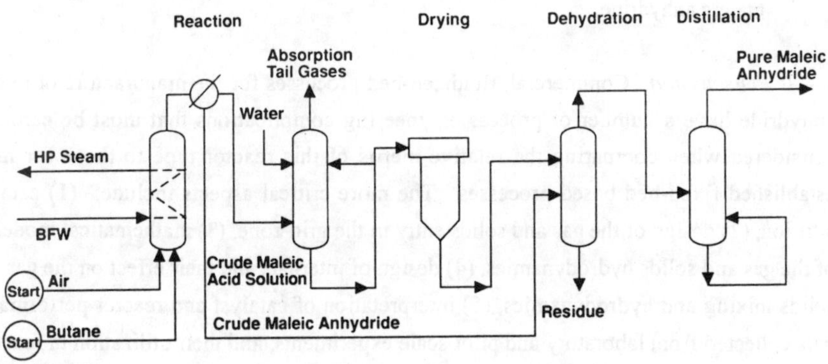


Figure 28. Sohio/UCD fluidized-bed process for maleic anhydride.[27]

The catalyst used in this process is based upon a proprietary attrition-resistant manufacturing technique developed at Sohio/UCP. Given their extensive experience in the development of acrylonitrile fluidized bed and other related processes, this is not surprising. It is claimed that this proprietary catalyst gives a 50% yielu of maleic anhydride on a once-through basis. As in the case of the previous processes, the catalyst and process technology is available for licensing.

*4.4-3. Process comparisons.* A qualitative comparison of the above maleic anhydride fluidized-bed processes is provided in Table 5. The key differences are in the hydrocarbon source used, the methods used to transform the crude maleic anhydride to the purified polymer-grader product, and particulars related to the manufacture of the attrition-resistant vanadium-phosphorus catalyst.

| | Mitsubishi | ALMA | Badger | Sohio/UCB |
|---|---|---|---|---|
| HC Source | crude $C_4$ | n-Butane | n-Butane | n-Butane |
| $O_2$ Source | air | air | air | air |
| % HC in Feed | unknown | 4% | unknown | unknown |
| Capacity, mt/hr | 18,000 | 40,000 | unknown | unknown |
| Recovery system | aqueous | organic | aqueous | aqueous |
| Entrainer | none | proprietary | o-xylene | none |

Table 5. Qualitative comparison of fluidized-bed processes for production of maleic anhydride.

*4.4-4. Discussion.* Commercial, fluidized-bed processes for the manufacture of maleic anhydride have a number of process engineering complications that must be carefully considered when comparing the relative merits of this reactor type to those for more established fixed-bed based processes. The more critical aspects include: (1) catalyst attrition, (2) design of the gas and solids entry in the grid zone, (3) mathematical modeling of the gas and solids hydrodynamics, (4) design of internals and their effect on the gas and solids mixing and hydrodynamics, (5) interpretation of catalyst and reactor performance data collected from laboratory and pilot-scale experiments, and their utilization in reaction engineering models for scale-up to commercial units, and (6) assessment of the process economics of fluidized-bed processes, especially in relation to other competing processes based upon other reactor types. A detailed treatment of these issues is complicated enough to be the subject of a separate study, and lies outside the objectives of this section. Several of these are considered, at least in part, as subjects of several review papers, chapters, and monographs on fluidized-bed reaction.[24-30] The remaining issues, such as catalyst attrition, scale-up of data, and process economics analysis, are part of industrial process development and often kept as proprietary in-house knowledge within industry and are not necessarily relevant to the objectives of academic research. In fact, all of the issues given above are not specific to maleic anhydride processes, but generally apply to any application where a fluidized-bed reactor is being considered.

*4.5. Phthalic Anhydride Process*

A schematic diagram of the Sherwin-Williams/Badger fluidized-bed process for production of phthalic anhydride by the air oxidation of naphthalene is given in Figure 29. The primary reaction is

$$C_{10}H_8 + 9/2\ O_2 \rightarrow C_8H_4O_3 + 2\ H_2O + 2\ CO_2 \tag{32}$$

The non-selective reactions produce small quantities of naphthoquinone and maleic anhydride. In addition, CO and $CO_2$ are produced from the combustion of naphthalene and phthalic anhydride. The reaction network (not shown) suggests that naphthoquinone may be an intermediate oxidation product of naphthalene and that maleic anhydride may be an over-oxidation reaction product.

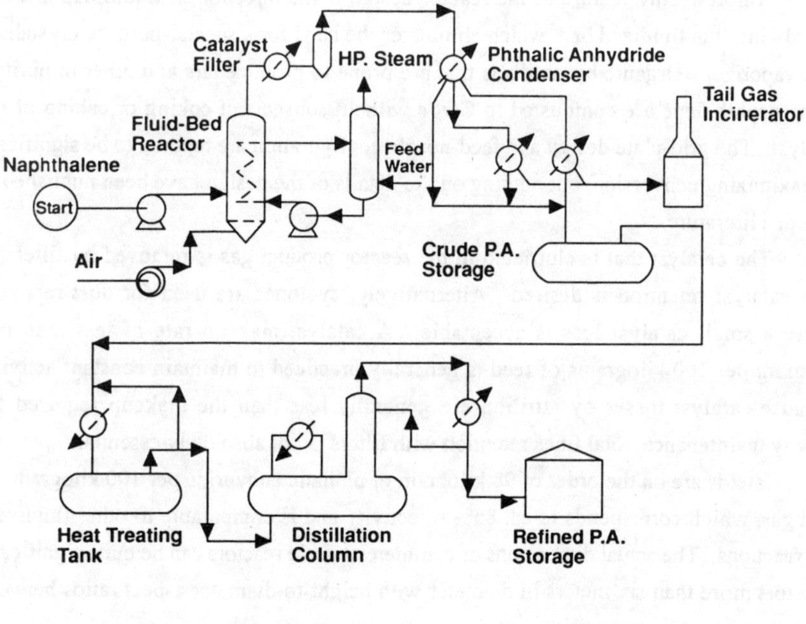


Figure 29. Sherwin-Williams fluidized-bed process for phthalic anhydride.

The hydrodynamics of the gas-solid mixing in large-scale reactors typically results in uniform temperatures with a maximum temperature gradient of +/- 1 °C, with resulting maleic anhydride by-product yields of less than 0.5%. Fixed-bed reactors typically produce maleic anhydride yields between 5 to 10%, with temperature gradients between 10 to 40 °C. Incomplete gas-solid contacting in the fluidized-bed reactor can result in yields of under-oxidation products, such as naphthoquinone, on the order of 2%. Nearly complete conversion of naphthalene is achieved, since it is hardly detectable in the crude product gas.

The catalyst used in commercial processes is principally vanadium oxide on a silica-gel base, and has a particle-size distribution between 5 to 300 microns.

The reactor operates at temperatures between 340 to 385 °C, and the superficial velocity of the gas is 0.3 to 0.6 m/s, so it operates in the bubbling flow regime. The reaction is highly exothermic, and the excess heat of reaction is removed by internal cooling coils with steam generation. The enormous internal rate of catalyst circulation provides a uniform gas-solid suspension temperature, even though the coolant is several hundred degrees below the average suspension temperature.

A noteworthy feature of the reactor design is the injection of liquid naphthalene directly into the fluidized bed, which eliminates the need for a special-purpose crystalline feed vaporizer. Organic by-products that are prone to produce tars and other impurities in the naphthalene are combusted to $CO_X$'s without subsequent coking or caking of the catalyst. The grid-plate design and feed-nozzle configuration are reputed to be significant in maximizing conversion, but nothing on the details of the design have been published in the open literature.

The catalyst that is elutriated in the reactor product gas is removed by filters, if total catalyst retention is desired. Alternatively, cyclones are used for dust recovery where a small catalyst loss is acceptable. A catalyst makeup rate of less than one kilogram per 100 kilograms of feed is generally practiced to maintain constant activity. Because catalyst losses by attrition are generally less than the makeup required for activity maintenance, total fines retention with filters is not absolutely essential.

Yields are on the order of 98 kilogram of phthalic anhydride per 100 kilograms of feed gas, which corresponds to ca. 85% selectivity and is comparable to other fluidized-bed reactions. The actual dimensions of commercial-scale reactors can be quite significant. Reactors more than six meters in diameter with height-to-diameter aspect ratios between one to two are still in operation.

*4.6. Summary*

In this section, an overview of commercial applications of fluidized-bed reactors for industrial vapor-phase oxidation and ammoxidation processes was given. The particular ones described here included acrylonitrile from propylene and ammonia, maleic anhydride from butylenes or n-butane, and phthalic anhydride from naphthalene. Newer, emerging reaction systems that have been practiced in fixed-bed reactors or which take advantage of cheaper alkane feedstocks were not reviewed, although these represent the basis for the next generation of possible commercial processes later in this decade. Despite the apparent advantages of this reactor type for this class of reactions, major challenges exist in the commercialization of new catalytic chemistries. Some of these challenges are due to insufficient economic growth in the markets where the chemical intermediates from these processes might be used as feedstocks. The other challenges are technical in nature and represent concerns with issues such as catalyst attrition, activity and selectivity, catalyst manufacture, and reactor design and scale-up. A small sampling of the pros and cons of fluidized-beds for oxidation and ammoxidation reactions that include these and other related issues is given in Figure 30. These represent some areas where further research is needed, and provide the impetus for investigation of other reactor types and operating modes.

| Pros | Cons |
|---|---|
| Temperature Control | Lack of Temperature Gradients |
| Heat Management | Higher Pressure Drop |
| Continuity of Operation | Catalyst Elutriation |
| Less Active Catalysts | Catalyst Attrition |
| Deactivating Catalysts | Incomplete Gas-Solid Contacting |
| Higher Hydrocarbon Concentrations | Selectivity - Backmixing |
| Simpler Mechanical Design | Homogeneous Gas-Phase Reactions |

Figure 30. Advantages and disadvantages of fluidized-bed reactors for oxidation and ammoxidation.

## 5. Recirculating Solids Reactors

*5.1. Introduction*

Gas-solid fluidization is generally characterized by various hydrodynamic regimes that span the fixed-bed or delayed-bubbling regime to the fast-fluidization regime, as illustrated in Figure 31. For a fixed geometry and particle characteristics, as defined according to the classification of Geldart, the transition from the fixed-bed regime to successive regimes occurs as the superficial gas velocity is increased beyond the minimization fluidization velocity $U_{mf}$. Typical fluidized catalytic-cracking riser reactors with solids that belong to the Geldart powder group A can operate in the fast, fluidized-bed regime with superficial gas velocities between 8 to 18 m/s.[32] This particular regime occurs when the gas velocity is increased within the turbulent fluidization regime so that the overall bed voidage increases and the top surface between the dense bed and the freeboard region becomes less and less distinct. The gas velocity is large enough that solids elutriate from the top of the reactor so that it becomes necessary to either add fresh solids at the reactor inlet, or to capture the elutriated solids and recycle them to the reactor, if the solids inventory is to be kept constant.

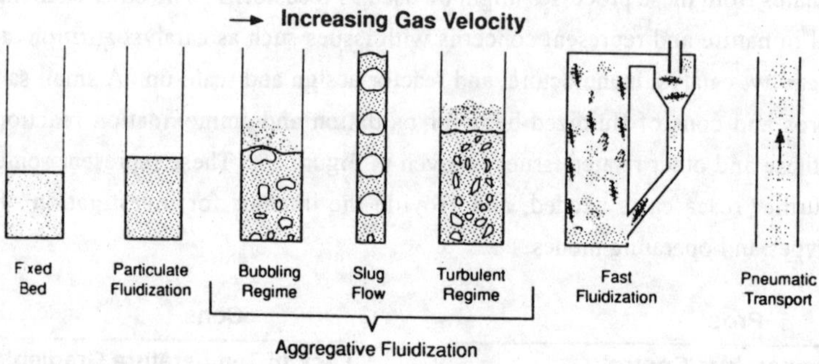


Figure 31. Hydrodynamic regimes in fluidized systems.[31]

A more quantitative method for distinguishing between various hydrodynamic flow regimes in gas-solid fluidization is to use a flow-regime map. An illustration of such a map is shown in Figure 32. In this map, the flow regimes for a given Geldart powder type are shown in terms of the Archimedes number raised to the one-third power on the abscissa and a dimensionless superficial gas velocity U* on the ordinate. Practical ranges of these two parameters where the various hydrodynamic regimes exist are also illustrated along with specific types of reactors, e.g., circulating beds and transport reactors. Most of the fluidized-bed processes discussed in the previous section, such as that used for manufacture of acrylonitrile, typically operate in the turbulent regime.

Figure 33 provides a qualitative comparison between the mean-gas velocity and the mean-solids velocity as a function of voidage for selected hydrodynamic regimes encountered in gas-solid fluidized bed reactor systems. These regimes correspond to the classical bubbling bed to the transport bed reactor. It shows that the difference between the mean-gas velocity and mean-solids velocity, or the so-called *slip velocity*, is greatest for Type C circulating fluidized-bed reactors. The remaining reactor types, such as the classical bubbling fluidized-bed, and the transport-bed reactor, result in gas-solid hydrodynamics with lower slip velocities. Based upon these comparisons, the circulating fluid bed would be the preferred mode of operation for systems where high rates of interparticle mass transfer and rapid gas-solids mixing might lead to an overall improvement on performance.

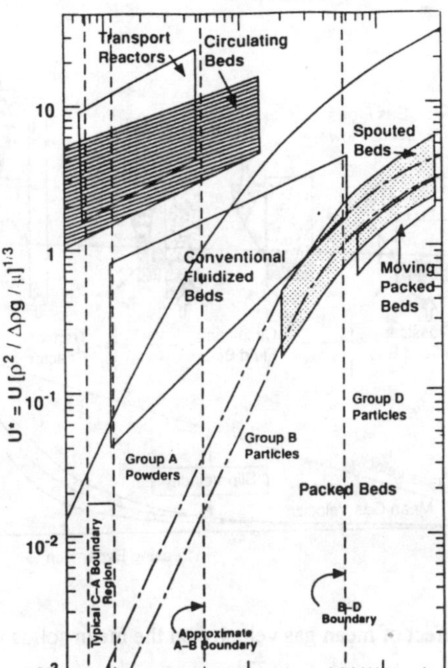


Figure 32. Flow regime map for fluidized systems.[31,32]

The above discussion suggests that the advantages of operating a fluidized bed in the fast-fluidization regime for gas-solid catalyzed reactions, relative to the other lower-level hydrodynamic regimes, include: (1) higher gas throughput per unit cross-sectional area of the reactor, (2) adjustable retention time of the catalyst, (3) the gas flow pattern approaches plug-flow with negligible axial dispersion, (4) high rates of heat and mass transfer between the gas and solids on a local level, (5) uniform or nearly uniform gas-solid suspension temperature, (6) reduced tendency of the solids to cake together or agglomerate, (7) possibility of staged addition of gas along the reaction zone, (8) addition and removal of catalyst in the recycle leg is simplified, and (9) the catalyst can be exposed to two different reaction environments. From the perspective of industrial oxidation and ammoxidation reactions, these advantages can translate into superior process operation and economics if the proper combination of kinetics, transport effects, and catalyst performance can be identified or developed.

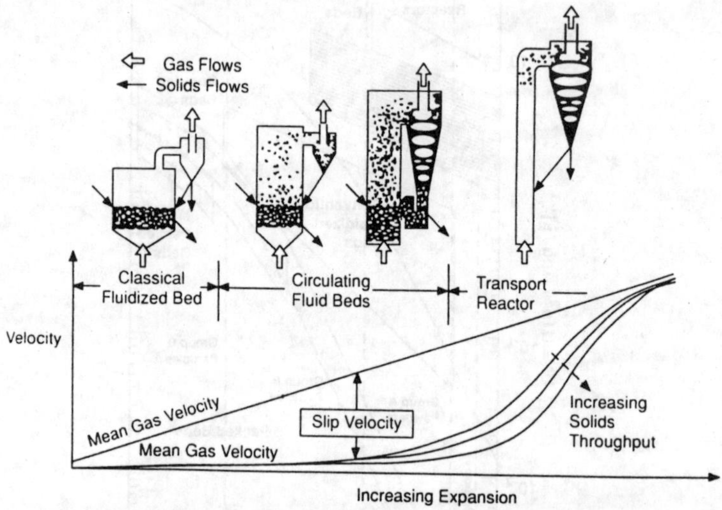


Figure 33. Effect of mean gas velocity on the mean solids velocity for various fluidized-bed operating regimes.[33]

The primary objective of this section is to present various industrial applications where fluidized-bed reactors in industrial oxidation and ammoxidation processes are operated with continuous recirculation of the catalyst corresponding to the fast-fluidized or transport-bed hydrodynamic regime. As in the previous section, the emphasis here is placed upon illustrations of commercial processes where this mode of operate is utilized to gain operating advantage. Because this particular mode of operation has been commercially realized in this particular application area in only a few instances, some additional emphasis will be given to reactor performance data where available.

*5.2 Origin of Recirculating Solids Reactors*

Early commercial applications of fluidized-bed reactors for catalytic cracking of petroleum feedstocks during the early 1940's were based upon superficial gas velocities of at least 1.5 m/s, and represented the first sustained attempt at operating a fluidized bed of catalyst beyond the classical bubbling regime.[34] A number of technical difficulties, such as increased rates of catalyst attrition, resulted in a reduction of these gas velocities to less than 1 m/s. Fluidized-bed processes for acrylonitrile that were developed by researchers at Sohio in the late 1950's, along with other gas-solid noncatalytic fluidized bed processes, re-kindled an interest in high gas velocity fluidized beds.[32] The coupling of high gas velocity with rapid solids recycle in gas-solid noncatalytic applications of fluidized beds, such as that encountered in high temperature metallurgical applications, provided the basis for further development of the circulating fluidized-bed reactor concept.

Recent work at DuPont led to the development and application of a circulating fluidized bed for the selective oxidation of n-butane to maleic anhydride using a novel attrition-resistant vanadium-phosphorus metal oxide catalyst.[35] This concept was used in the first reaction step of a two-step process to manufacture tetrahydrofuran, and is currently being commercialized in Asturias, Spain.[36] This represents the first commercial application of this particular reactor type to an industrial vapor-phase selective oxidation process. Some particular details about this process and a related process for the selective ammoxidation of aromatics are presented in the sections that follow.

*5.3 Reactor Description*

A schematic diagram of the recirculating solids reactor system developed by DuPont for the selective oxidation of n-butane to maleic anhydride is shown in Figure 34. The system consists of a fluidized-bed riser reactor, a gas-solid separator, a catalyst stripper, a fluid-bed regenerator, and a standpipe. Although the description given below applies to this particular reacting system, the working concept is general and could be applied, in principle, to any related oxidation or ammoxidation system.

The riser or transport bed is basically a vertical pipe where the catalyst particles are injected at the bottom from the standpipe and fluidized upwards at a high velocity with the vaporized n-butane plus dilution gas, if desired. The entrained catalyst is discharged into a gas-solids separator where the solid catalyst is directed into the stripper. The reaction between the butane-rich reactant gas and the catalyst is very fast and occurs in the riser within a matter of seconds. The attrition-resistant vanadium-phosphorus metal oxide catalyst undergoes a reduction in the net oxidation state as a result of the reaction with butane, since the oxygen incorporated into the maleic anhydride is obtained from the catalyst lattice. The ratio of the gas to solids flow rates, the riser inside pipe diameter and overall length, the particle characteristics according to the Geldart classification, the thermodynamic and transport properties of the reaction gas, and the suspension recirculation rate must be carefully considered to ensure that the optimum yield of maleic anhydride is obtained. A comparison between the gas velocities and solids concentration for transport-bed and classical bubbling fluidized-bed modes of operation is given in Table 6.

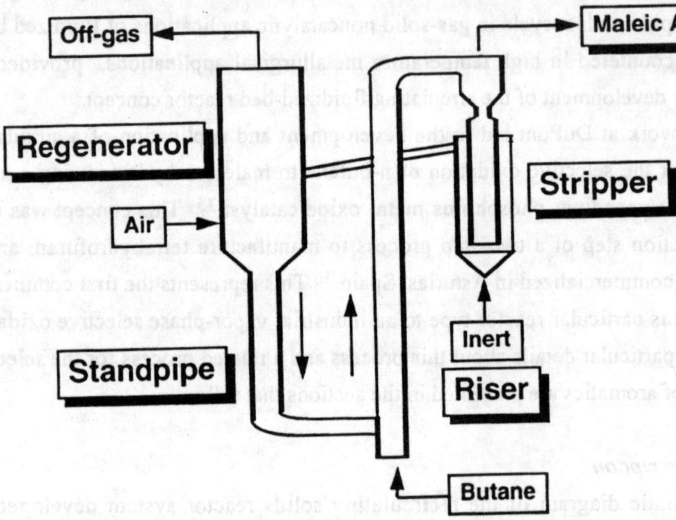


Figure 34. Recirculating solids reactor for n-butane oxidation to maleic anhydride.[35]

The reduced catalyst, once it enters the stripper, is contacted with inert gas to reduce the concentration of any adsorbed maleic anhydride and other species on the catalyst surface before being introduced into a conventional fluidized-bed regenerator that is operated in the bubbling hydrodynamic regime. Here, the reduced catalyst is contacted with air for several minutes so that any lattice oxygen removed during the reduction step can be replenished. The heat of combustion and reaction is removed by internal cooling coils with steam generation. The off-gases from the regenerator, which contain $CO_x$'s and nitrogen, are directed to a CO converter for generation of additional steam. Because the catalyst is exposed to butane-rich and oxygen-rich reaction gases in the riser and regenerator, respectively, the average oxidation state of the metal oxide is cycling at the characteristic frequencies associated with the gas-solid contact time distribution in these two reaction vessels.

Table 6. Comparison between characteristic operating ranges for transport-bed and fluidized-bed reactors.

| | **Transport bed** | **Fluidized bed** |
|---|---|---|
| Gas velocity, ft/s | 5 to 40 | < 3 |
| Solids concentration, lb/cu ft | 1 to 10 | 20 to 45 |

The advantages of using a transport-bed reactor system for vapor-phase oxidations and ammoxidations are summarized in Figure 35. A particularly important advantage is two separate reactors allows the reaction temperature, inlet gas composition, gas-solid contact time and flow patterns to be optimized for the catalyst reduction zone and the re-oxidation zone. It is generally known that metal oxides undergo re-oxidation at significantly slower rates when compared to those for reduction. Hence, a longer contact time between the reduced catalyst and the air can be maintained in the fluid-bed regenerator, whereas a short contact time can be established between the re-oxidized catalyst and the n-butane in the riser. In fixed-bed and fluidized-bed reactors operated without catalyst recirculation, both the catalyst re-oxidation and reduction occur in the same reactor so that the reaction cannot be optimized and the maleic anhydride yield will be adversely affected.

Another advantage of the transport-bed reactor is that a high concentration of the product in the reactor off-gas can be established because the butane concentration in the reactor feed gas can, in principle, be set at any desired value. The butane is not mixed with any oxygen, so the potential for ignition and explosions is minimized or non-existent. Furthermore, the maleic anhydride is not exposed to any gas-phase oxygen so that homogeneous gas-phase combustion reactions are not possible. Because the butane concentration is high in the riser feed gas, the need to handle large volumetric flow rates of inert gas is reduced, which results in a significant reduction in the capital investment for the downstream processing equipment, such as absorbers, compressors, blowers, heat exchangers, and other associated equipment.

The use of high gas velocities in the riser section, along with high catalyst recirculation rates, can result in the development of a core-annular type of flow structure.[37] The gas has a net upward velocity, while the solids exist as a dense layer at the wall with a net downward velocity component and a lean suspension in the core with a net upward velocity component. Vigorous exchange of gas and solids occurs at any given local axial position along the riser height. The gas-phase flow pattern approaches plug-flow, while the solids approaches plug-flow with some degree of internal recycle before it escapes the riser section and enters the stripper and catalyst regenerator. Such a description of the gas and solids flow patterns have been observed in gas and solids tracer experiments, as well as in other non-invasive techniques, such as computer-aided tomography and computer-aided radioactive particle tracking.[37] Taken collectively, this type of hydrodynamic behavior reduces local temperature gradients and is more readily scaled up to larger vessels, since the production rate per unit volume of reactor is greater than either fixed beds or fluidized beds without any recycle.

- **Riser Reactor Zone**
  - **—High selectivity . Plug flow. No hot spots.**
  - **—No free board burning**
  - **—High turndown ratio**
  - **—Ease of scale-up**
- **Fluidized Bed Regenerator Zone**
  - **—High heat transfer coefficient**
  - **—Good temperature control**
- **Separate Catalyst Redox Zones**
  - **—Independent control of two zones**
  - **—High Selectivity**
- **Concentrated product stream**
  - **—High hydrocarbon concentration in feed**
  - **—Product gas and regen gas are separate**
- **High throughput**
- **Low catalyst inventory**
- **Reduced explosion risk**

Figure 35. Advantages of transport-bed reactors for selective oxidations and ammoxidations.[27]

The use of a conventional bubbling bed for the catalyst regeneration step permits well-known techniques for design and scale-up of this section of the transport-bed system to be applied. The hydrodynamics of the gas-solid mixing are such that high heat transfer coefficients between the internal coils and gas-solid suspension can be obtained, which translates into good temperature control.

Possible limitations with this reactor system include the usual complexities associated with the design and operation of coupled reactor vessels and separators where transport of gas-solid suspensions are involved. These difficulties can be overcome with fundamental process models for the reactor system that account for kinetics, transport effects, and hydrodynamics, as well as a well-designed and proven process control systems. In addition, an attrition-resistant catalyst that can maintain long life, high activity and high selectivity within the process economic constraints is essential. Key to the development of the DuPont riser process was the development of novel technology for a catalyst that met these constraints that was proven in laboratory-scale and pilot-plant performance evaluations. Given the special nature of this particular challenge, this is discussed in detail below.

*5.4 Attrition-Resistant Catalyst Technology*

A key part of the recirculating reactor process for n-butane oxidation to maleic anhydride is the development of an attrition-resistant VPO catalyst which can maintain both high activity and selectivity. Since the mean gas velocities in transport-bed reactors are typically between 5 to 40 ft/s, the particle momentum is significantly greater than that encountered in fluidized bed reactors where the gas velocities are often 3 ft/s or less. A cross-sectional view of a typical fluidized-bed catalyst is shown in Figure 36a. Here, the attrition resistance is obtained by spray drying the active catalyst microspheres with silica so that the matrix contains between 30 to 50 wt % silica. The effective surface area of the active catalyst is decreased relative to the total surface area; and more importantly, the product selectivity can be adversely affected. DuPont's new catalyst technology, which is shown in Figure 36b, encapsulates the active VPO catalyst in a porous silica shell whose pore openings permit reaction species to readily diffuse into and out of the inner region of the catalyst particle without significantly affecting the maleic anhydride selectivity.

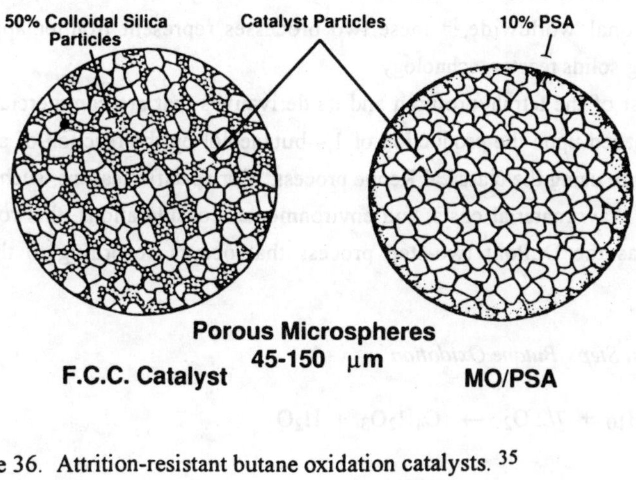


Figure 36. Attrition-resistant butane oxidation catalysts. [35]
a (left): conventional spray-dried catalyst; b(right): porous hardened shell catalyst.

The shell of the DuPont attrition-resistant catalyst typically contains 5 to 10 wt % silica, which is significantly less than that present in the conventional catalyst. One advantage of using this method is that the active VPO catalyst phases are relatively unchanged, owing to their minimal exposure to silica when compared to the traditional hardening process on a per particle or other equivalent basis. Additional details are provided by Bergna.[38,39]

Despite recent advances in development of attrition-resistant shells for the vanadium-phosphorous metal oxide catalyst used in butane oxidation, many technical opportunities and challenges will continue to exist. These include: (1) the development of novel catalyst compositions, (2) new methods for imparting and controlling attrition resistance, (3) techniques for characterization of the spray-dried shell catalysts, (4) new and improved processes for catalyst manufacturing that are based on state-of-the-art process concepts and process control schemes, and (5) more basic understanding of the factors that affect catalyst recipe scale-up from the lab to commercial-scale production.

*5.5 Commercial Processes*

Recirculating solids reactors have been used in two industrial oxidation and ammoxidation processes that have already been commercialized, or are in the process of being commercialized. These include: (1) the DuPont process for manufacture of tetrahydrofuran[36], and (2) the Lummus process for production of certain aromatic nitriles.[40] In comparison to fluidized catalytic-cracking processes where over 250 units are operational worldwide,[34] these two processes represent frontier applications of recirculating solids reactor technology.

Most of the tetrahydrofuran and its derivatives that are commercially produced today are based upon the production of 1,4-butanediol by the reaction of acetylene and formaldehyde using the classical Reppe process. For specific reasons, such as acetylene availability, raw material costs, and environmental considerations, this route is not as attractive as the DuPont two-step process that occurs according to the following reactions:

*First Step: Butane Oxidation*

$$C_4H_{10} + 7/2\, O_2 \rightarrow C_4H_2O_3 + H_2O \qquad (33)$$

*Second Step: Maleic Acid Hydrogenation*

$$CO_2H(CH{=}CH)CO_2H + 6\,H_2 \rightarrow C_4H_4O + 3\,H_2O \quad (34)$$

Maleic anhydride (MAN) is an important intermediate used in the production of various polymers, resins, and other specialty chemical products. Examples of products that are based upon MAN include γ-butyrolactone, tetrahydrofuran, polybutyleneterethalates (PBT), polyurethanes, copolyester elastomers (COPE), pyrrolidones, polytetramethylene ether glycol (PTMEG), and tetrahydrothiophene. These products are used in a wide variety of applications that include various engineering plastics, automotive products, solvents for the manufacture of pharmaceuticals and video tape, and specialty fibers.[36] Within DuPont, the principal end use of MAN is in the production of PTMEG for which tetrahydrofuran (THF) is a precursor. The main outlet for PTMEG is in the manufacture of Spandex fibers and copolyester elastomers (COPE) for products such as Lycra®, Hytrel®, and other consumer products. A 100 million lb/yr plant based on maleic anhydride from n-butane is scheduled to start up in 1996 in Asturias, Spain.[36]

Although most commercial maleic anhydride processes in operation today use n-butane as the hydrocarbon source, other feedstocks, such as benzene, butene, and butadiene, can be used. Prior to the 1960's, benzene was exclusively used as the raw material for commercial MAN processes. Early in the 1970's, increases in the cost of benzene lead to a raw material advantage for $C_4$ hydrocarbons, especially n-butane, as the preferred feedstock. Increasing environmental hazards associated with benzene, the lower carbon efficiency of benzene when compared to n-butane since one-third of the benzene carbon is lost due to combustion, and potentially higher MAN yields from n-butane, lead to the commercial dominance of n-butane as the preferred feedstock. Most commercial processes in operation today are based upon multitubular fixed-bed reactors with n-butane as the feedstock. The largest single reactor is capable of producing about 40,000 tons/yr. The DuPont THF process represents the first commercial application of transport-bed reactor technology to the area of selective oxidation.

Aromatic nitriles, such as benzonitrile, phthalonitrile, isophthalonitrile, terephthalonitrile, and nicotinonitrile, are important chemical intermediates that were most recently produced by conventional ammoxidation in fluidized-bed or fixed-bed reactors using air, ammonia, and the hydrocarbon reactant in the presence of a suitable metal oxide catalyst. Nicotinonitrile may be hydrolyzed to nicotinamide or nicotinic acid or niacin, either of which may be used as a component of vitamin B complex.

Isophthalonitrile is used industrially for the manufacture of herbicides and fungicides, such as tetrachloro-1,3-dicyanobenzene. Phthalonitrile is an intermediate used in the manufacture of phthalocyanine pigments. Generally speaking, aromatic nitriles are used in organic synthesis for a wide variety of applications.

Lummus developed commercial nitrile processes for the above aromatic nitriles, using oxidative ammonolysis in the absence of molecular oxygen, by using lattice oxygen from a metal-oxide catalyst that is transported between reaction vessels so the catalyst has a low and high average oxidation state.[40] The overall reactions are given in Figure 37, where the oxygen source is from the catalyst lattice. The feedstocks for these reactions include ortho-xylene, meta-xylene, para-xylene, and 2-methyl-5-ethylpyridine. Detailed reaction schemes that show the major and minor reactions, including those that occur in the riser section and the regenerator, are provided elsewhere.[40] The advantages of using a recirculating solids reactor for these reactions are similar to those given explained above for the butane oxidation to maleic anhydride.

Some additional details on each of these processes are given below.

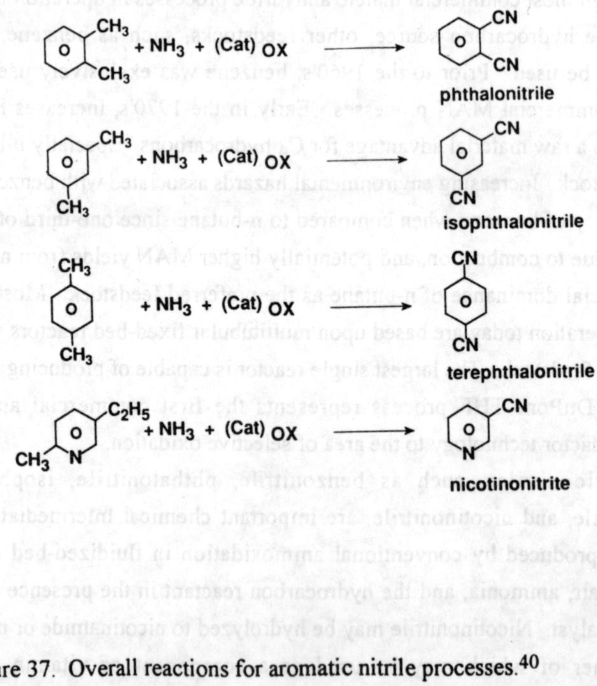


(35)

(36)

(37)

(38)

Figure 37. Overall reactions for aromatic nitrile processes.[40]

*5.5-1 DuPont Tetrahydrofuran Process.* A schematic diagram of the proposed DuPont process is given in Figure 38. In the first step of the process, n-butane is oxidized to maleic anhydride using the attrition-resistant VPO catalyst described above in section 5.4 in a transport-bed reactor. The spent catalyst is carried overhead where it is separated and regenerated with air in a conventional bubbling fluidized-bed reactor. After regeneration, it is recycled to the transport-bed reactor where the re-oxidized catalyst is exposed again to the n-butane rich process feed gas. The maleic anhydride vapors that exit the cyclone separator are adsorbed in water to form maleic acid (MAC) with unreacted n-butane recycled to the transport bed. In the second step of the process, the crude MAC is hydrogenated to THF using a palladium-rhenium catalyst.[41] The crude THF is purified by extractive distillation of the THF-water azeotrope with unreacted hydrogen being recycled. By adjusting process conditions, an alternate or co-reaction product γ-butyrolactone can be produced.

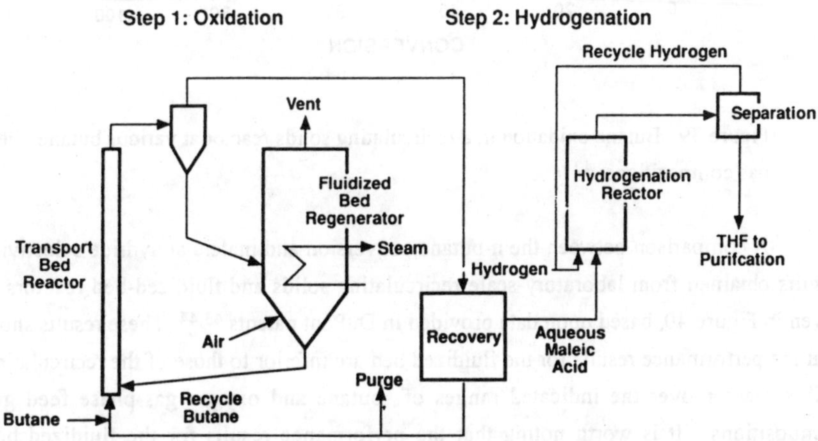


Figure 38. DuPont process for production of tetrahydrofuran.[36]

Figure 39 shows the maleic anhydride selectivity versus n-butane conversion data extracted from the DuPont riser reactor system shown in Figure 34 with the patented attrition-resistant VPO catalyst. These data show that a wide range of n-butane feed compositions can be used in the absence of gas-phase oxygen without affecting the maleic anhydride selectivity to any significant degree. The selectivity losses are due to the formation of $CO_X$ that is associated with the non-selective oxygen species that is present on the catalyst surface after re-oxidation in the fluidized-bed catalyst regenerator shown earlier in Figure 34.

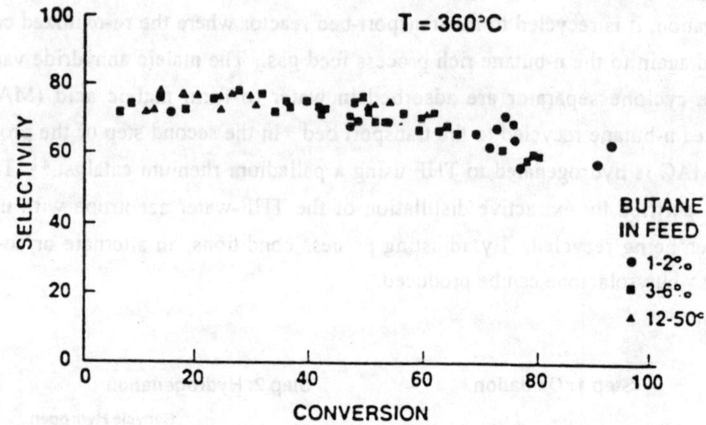


Figure 39. Butane oxidation in a recirculating solids reactor at various butane feed gas compositions.[42]

A comparison between the n-butane conversion and maleic anhydride selectivity results obtained from laboratory-scale recirculating solids and fluidized-bed reactors is given in Figure 40, based upon data provided in DuPont patents.[43,44] These results show that the performance results for the fluidized bed are inferior to those of the recirculating solids reactor over the indicated ranges of butane and oxygen gas-phase feed gas compositions. It is worth noting that the performance results for the fluidized bed generally appear to decrease as the composition of n-butane in the feed is increased when the feed composition of oxygen is held almost constant.

The formation of maleic anhydride from n-butane involves the consumption of lattice oxygen in the catalyst. Regeneration of the catalyst with air results in an increase in the adsorbed surface oxygen, which is thought to participate in the formation of non-selective reaction products. Several approaches to reduce the concentration of surface oxygen have been suggested and include: (1) stripping of the regenerated catalyst with an inert gas, and (2) reduction in the mean residence time of the catalyst in the fluidized-bed regenerator. Experimental evidence that removal of excess surface oxygen gives an increase in the maleic anhydride selectivity versus n-butane conversion performance in a riser reactor is shown in Figure 41. Here, the re-oxidized catalyst is stripped before being directed into the riser and compared with the performance obtained without stripping of excess surface oxygen. An increase of about 7 to 10% in the selectivity to maleic anhydride is observed.

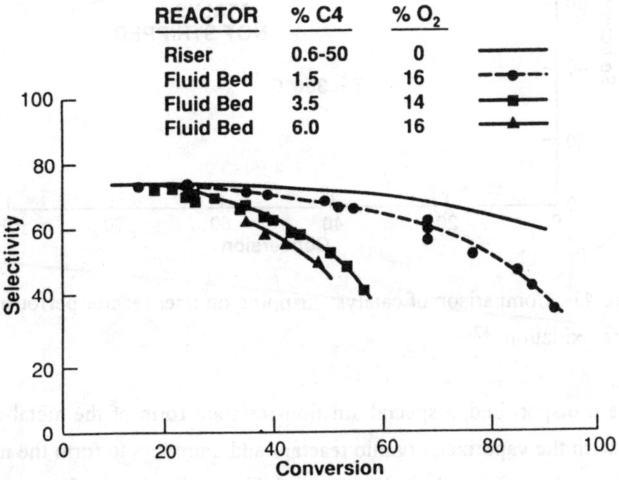


Figure 40. Butane oxidation in a recirculating solids and fluidized-bed reactors.[43,44]

*5.5-2. Lummus Aromatic Nitrile Process.* A schematic of the process flow diagram for the manufacture of aromatic nitriles by oxidative ammonlysis is shown in Figure 42. The process can be divided into a reaction section, a product-recovery section, and an ammonia-recovery section.

The reactor section is very similar to that used in the DuPont THF process, and contains separate reactors for reduction of the metal oxide catalyst, stripping of the catalyst to remove adsorbed organics, and regeneration of the metal oxide with air.

The reactor feed contains the organic reactant, recycle gas and organic intermediates, and ammonia supplied from the recovery section and storage. The partially vaporized feed is introduced to the reactor through the gas distributor along with the liquid heel. The use of partial vaporization avoids the problem of fouling the feed evaporator with residues having a low volatility. If any heavies are formed that deposit on the catalyst, they can be removed in the fluidized-bed regenerator by total oxidation with air.

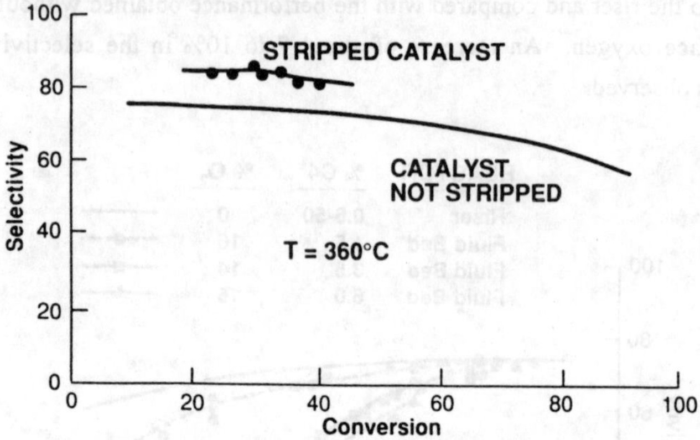


Figure 41. Comparison of catalyst stripping on riser reactor performance for butane oxidation.[42]

In the transport bed, a special attrition-resistant form of the metal-oxide catalyst is contacted with the vaporized organic reactant and ammonia to form the nitrile product. Generally, the reaction is conducted with a stoichiometric excess of ammonia so that the ammonia also functions as a carrier gas. The spent catalyst then flows to the stripper where it is contacted with steam or other inert gas to remove any adsorbed material. It is then conveyed to the regenerator using air. The air used as a lift gas is usually only a small part of the total air required for regeneration. In addition to re-oxidizing the catalyst and combusting some of the adsorbed organics, the regenerator also serves as a catalytic oxidizer for disposal of any plant gaseous wastes.

The crude reactor effluent contains the desired nitrile product, plus any intermediates and unreacted feed gas for recycle, as well as ammonia, water vapor, nitrogen, carbon oxides, and a small amount of hydrogen cyanide. The hot reactor effluent is first cooled in a quench tower as a single-stage separation of the condensibles, e.g., organics and water vapor, from the non-condensibles such as nitrogen, carbon oxides, and ammonia. Isolation of the desired aromatic nitrile may be performed by a variety of combined unit operations and is dependent upon the particular nitrile being manufactured owing to their different physical properties. Typically, the unit operations involved will include crystallization, fractionation, washing, and centrifugation. Additional details are available in the paper of Sze and Gelbein.[40]

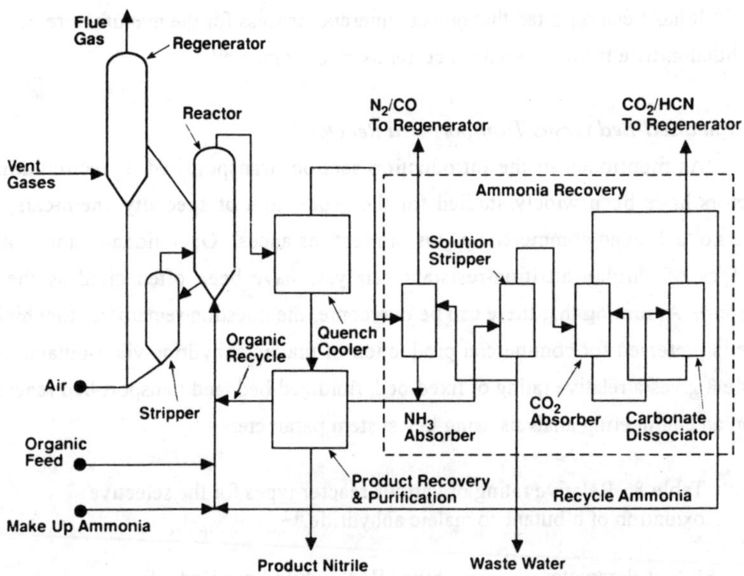


Figure 42. Lummus aromatic nitrile process. [40]

The use of excess ammonia in the riser dictates that it must be recovered and recycled for economical plant operation. This is performed using the classical approach through ammonia absorption, solution stripping, $CO_2$ absorption, and ammonia carbonate dissociation.

Table 7 gives a summary of the aromatic nitrile yields reported by Lummus[40] that are based upon operation of a pilot-plant. Overall yields of at least 75%, with most being between 85 to 90%, are claimed.

Table 7. Overall yields of aromatic nitriles from the Lummus aromatic oxidative ammonlysis process.

| Feedstock | Product | Product Yield, % |
|---|---|---|
| Toluene | Benzonitrile | ca. 90 |
| *p*-xylene | Terepthalonitrile | ca. 90 |
| *m*-xylene | Isophthalonitrile | 80 - 85 |
| *o*-xylene | Phthalonitrile | 75 - 80 |
| β-picoline | Nicotinonitrile | 80 - 90 |

It has been reported that one commercial process for the manufacture of isophthalonitrile from *m*-xylene is currently in operation.[40]

## *5.6 Fluidized-Bed versus Transport-Bed Reactors*

As mentioned in the Introduction section, transport-bed or recirculating-solids reactors have been widely studied for the production of specialty chemicals, but they have not achieved commercial status in most instances. Operational complications and the lack of highly attrition-resistant catalysts have been often cited as the primary reason.[42] Assuming that these can be overcome, the question remains as to which reactor type is preferred for commercial production of maleic anhydride via n-butane oxidation. Table 8 gives a relative rating of fixed-bed, fluidized-bed, and transport-bed reactors based upon an engineering analysis using key system parameters.

Table 8. Relative rating of various reactor types for the selective oxidation of n-butane to maleic anhydride.[42]

| System Parameter | Fixed Bed | Fluidized Bed | Riser |
|---|---|---|---|
| Heat Removal | – | + | + |
| Temperature Control | – | + | + |
| Maximum Feed Conc., % | 1.8–2.1 | 4 | >10 |
| Capital Investment | – | + | ++ |
| Selectivity | + | – | ++ |
| Catalyst Attrition | + | – | – |

Although the specifics associated with the above ratings are omitted, inspection of the results indicates that the ranking for this particular application is riser reactor > fluidized bed > fixed bed. This assumes that a suitable attrition-resistant catalyst can be identified which has been successfully performed in the current application.[39,40]

## *5.7. Summary*

A review of commercial processes that use recirculating solids reactors in industrial oxidations and ammoxidations has been presented. The motivation for using this particular hydrodynamic regime is primarily associated with various advantages of reactor operation where the catalyst oxidation-reduction cycle can be optimized through control of reaction and other process conditions. In addition, the use of high superficial gas velocities and catalyst recirculation permits flexibility of operation and enhances local rates of heat and mass transport.

Particular processes that are being commercialized, or are in the process of being commercialized, include the DuPont tetrahydrofuran process and the Lummus aromatic-nitrile process. The DuPont process combines a novel application of the recirculating solids reactor to the selective oxidation of n-butane to maleic anhydride with a novel invention for imparting catalyst attrition resistance. In the Lummus process, ammoxidation of aromatics to nitriles is conducted in a fashion that is analogous to the DuPont process, although precise details on the attrition resistant are lacking. The benefit of separating the selective reaction that uses lattice oxygen from the catalyst from the catalyst re-oxidation is also apparent in this application.

Reaction engineering of recirculating solids reactors for these particular applications is perhaps more complicated than that for classical fixed-bed or fluidized-bed reactors without catalyst recycle. Part of this is due to the lack of published data on the oxidation-reduction kinetics and availability of fundamental models for the gas-solids hydrodynamics and transport effects. These, along with other issues, provide the basis for future fundamental and practical challenges associated with this technology.

## 6. Emerging Reactor Types

Two promising unconventional catalytic reactors for carrying out selective oxidation reactions are in the early stages of research and development. These are inorganic membrane reactors (section 6.1) and moving bed/chromatographic reactors (section 6.2). Given the increased environmental pressures to develop chemical processes with reduced waste, emerging reactor types that increase the selective oxidation product yield deserve close scrutiny. However, modifications of conventional reactor types should also be considered as alternative approaches. In addition to assessing economic and environmental performance of new reactors, safety and reliability should also be factored into the reactor comparisons.

### *6.1. Membrane Reactors*

#### *6.1.-1. Introduction*

The catalytic inorganic-membrane reactor consists of a thin film of mesoporous or microporous inorganic material supported by a macroporous material. The thin film may simultaneously serve as a catalyst and as a permselective membrane or nonpermselective diffusion barrier. The successful design and implementation of inorganic membrane reactors requires skills in materials science, catalytic chemistry and kinetics, chemical reactor engineering, and mathematical modeling. The reader is referred to several reviews of inorganic membranes; in particular, Zaspalis and Burggraaf,[45] Armour,[46] Gellings and Bouwmeister,[47] Hsieh,[48] Tsotsis et al.,[49] and Harold et al.[50] Several applications of catalytic inorganic membrane reactors have been demonstrated in the literature:

- Control of rapid gas phase catalytic reactions requiring strict stoichiometric feeds with nonpermselective porous membrane reactors[51-53]
- Overall rate enhancement of volatile reactant limited multiphase reactions with nonpermselective porous membrane reactors[54]
- Improvement of conversion in equilibrium-limited reactions with permselective dense and porous membrane reactors[48-50,55-59]
- Improvement of desired product yield in consecutive-parallel gas phase catalytic reaction systems in permselective membrane reactors, and nonpermselective membrane reactors (described below)

Figure 43 shows four common catalytic membrane reactor types. The catalytic membrane reactor (CMR) consists of a support layer and a permselective layer (membrane) with a catalytic function. The catalytic nonpermselective membrane reactor (CNMR) consists of a support layer and a nonpermselective catalytic layer. In the packed bed catalytic membrane reactor (PBCMR), the catalyst is located external to the supported permselective membrane. Finally, the packed- bed membrane reactor (PBMR) consists of a inactive permselective layer with catalyst located in the flow stream.

Of pertinence to the current review is the application of membrane reactors in selective hydrocarbon oxidation reactions. Consider the multiple reaction network

$$A + B \rightarrow R \tag{39}$$

$$A + R \rightarrow P + Q \tag{40}$$

$$A + B \rightarrow P + Q \tag{41}$$

where, without any loss in generality, A is oxygen, B is the hydrocarbon, R is the desired selective product, and P and Q are undesired total oxidation products.

It is instructive from a catalytic membrane design standpoint to examine the reaction structure for two limiting cases.[60] Under conditions in which component A is the limiting reactant and the rate of the third reaction is negligible, the network has the parallel structure given by

$$A \rightarrow R \tag{42}$$

$$A \rightarrow P + Q \tag{43}$$

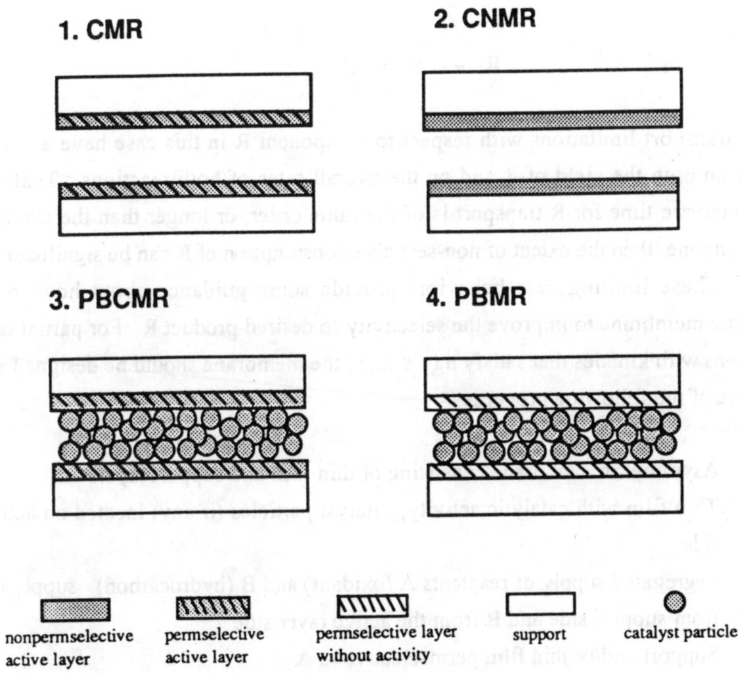


Figure 43. Four common catalytic membrane reactor types.

Under these conditions, the point selectivity to R, the ratio of the rates of the first and second reactions at a local point (•) within the reactor, depends on the reaction orders with respect to A in each reaction, given by $n_{A1}$ and $n_{A2}$ for the first (selective) and second (non-selective) reactions, respectively. If both $n_{A1}$ and $n_{A2}$ are positive, then the following rules apply to the impact of mass transport limitations with respect to A:

- If $n_{A1} > n_{A2}$, then mass transport limitations are detrimental to the yield of R
- If $n_{A1} = n_{A2}$, then mass transport limitations have no effect on the yield of R
- If $n_{A1} < n_{A2}$, then mass transport limitations are beneficial to the yield of R
- For positive $n_{A1}$ and $n_{A2}$ values, the overall rate of A consumption decreases as the limitation due to mass transport increases

On the other hand, if component B is the limiting reactant in the reaction network above, the reaction system has the consecutive structure given by

$$B \rightarrow R \tag{44}$$

$$R \rightarrow P + Q \tag{45}$$

Mass transport limitations with respect to component R in this case have a detrimental effect on both the yield of R and on the overall rates of both reactions. That is, if the characteristic time for R transport is of the same order, or longer than the characteristic reaction time, then the extent of non-selective consumption of R can be significant.

These limiting case behaviors provide some guidance about how to tailor a catalytic membrane to improve the selectivity to desired product R. For partial oxidation reactions with kinetics that satisfy $n_{A1} < n_{A2}$, the membrane should be designed with one or more of the following properties:

(i) Asymmetric membrane consisting of thin film and support layers
(ii) Thin film with catalytic activity; catalyst particles (if any) located on active layer side
(iii) Segregated supply of reactants A (oxidant) and B (hydrocarbon): supply of A from support side and B from the active layer side
(iv) Support and/or thin film permselective to A
(v) Support and/or thin film permselective to R

The combination of properties (i) - (iv) implies a controlled supply of the oxidant (A) through the support layer to the active layer. By feeding B from the active layer side, a low effective A/B ratio is achieved in the reaction zone. A segregated feed condition is best accomplished if the thin film is permselective to A. Another way to achieve a segregated feed is to use a support layer with a sufficiently large thickness or sufficiently small pores to limit the flux of A to the active layer. Both of these approaches are described below.

The combination of properties (i) - (iii) and (v) means that the intermediate R that is formed is selectively removed from the catalytic zone. The resulting reduction in the concentration of R, and hence the rate of the non-selective reaction, increases the selectivity of R. This concept has been advanced theoretically by Lund and coworkers.[61,62] This approach is described below.

*6.1-2 Permselective Membranes for Partial Oxidation.* A membrane reactor concept developed by Agarwalla and Lund[61] focuses on the following consecutive reaction network:

$$B \rightarrow R \rightarrow P. \tag{46}$$

Their concept is to employ a PBMR in order to maximize the yield of the desired intermediate species R. The hypothetical PBMR consists of catalyst on the tube side wherein reaction occurs. The key feature of the reactor is the membrane that is assumed to be permselective to the intermediate R. As reaction occurs on the tube side, the desired intermediate is selectively removed to the shell side where it is swept away by an inert component. The model shows that for an intermediate range of conversion of main reactant B, the yield of R is higher for the permselective compared to nonpermselective situation.

These simulations provide a materials challenge: Develop an inorganic membrane through which, for example, a partially oxidized hydrocarbon permeates at a higher rate than the unoxidized hydrocarbon and oxygen. Membranes that come closest to meeting this challenge are of the Knudsen variety. Selectivity ratios for R/B, for example, do not exceed 3 for such membranes. Thus, while intriguing, the concept remains untested.

A second type of permselective membrane specifically applied to partial oxidation reactions involve the selective permeation of oxygen through a solid electrolyte. Solid electrolytes used for oxygen transport fall into two main categories. The first includes mixtures of divalent and trivalent cations with tetravalent metals, an example being yttria-stabilized zirconia ($Y_2O_3$-$ZrO_2$) or solid mixtures containing $Bi_2O_3$ (e.g., $(Bi_2O_3)_{0.85}(La_2O_3)_{0.15}$). The transport of oxygen proceeds through a lattice that has anion vacancies created by the doping material. The second category is the mixed-conducting (ionic and electronic) materials. One example is the family of perovskite materials, denoted by the general formula $ABO_3$, such as $(La_{1-x}Sr_x)MnO_{3-d}$.[63-65] The oxygen transport in this case proceeds by a vacancy diffusion mechanism, with the driving force being an imposed chemical potential gradient. Another example is YSZ doped with an oxide of a metal that has both ion and electron conductivity, such as $TiO_2$ or $CeO_2$.[66]

The concept of the dense oxide membrane reactor for partial oxidation is as follows: Air is flowed on one side of a membrane device that consists of a macroporous support and a dense oxide layer that can selectively permeate oxygen. Hydrocarbon is flowed on the other side. The permeation of oxygen through the oxide lattice provides a controlled supply of lattice oxygen to the other side where the catalyst and flowing hydrocarbon are located. The catalyst is either the permselective oxide layer itself or is another material that is deposited in particulate form on top of the membrane layer. The key to the idea is that the lattice oxygen may be the more selective form of oxygen for hydrocarbon partial oxidation reactions.[67,68] On the other hand, gas phase oxygen and adsorbed oxygen react with the hydrocarbon to form undesirable carbon oxides in addition to the partially oxidized hydrocarbon. Thus, if the supply of lattice oxygen can be properly balanced with its rate of consumption to the desired species, this approach enables a sustained oxidation of the hydrocarbon without mixing of hydrocarbon and air.

*6.1.-3. Nonpermselective Membranes for Partial Oxidation.* Selected experimental and modeling results which demonstrate the concept of the nonpermselective membrane reactor with reactant segregation are presented in this section. Compared to the dense oxygen-permeable membrane design, this approach offers higher oxygen flux but limited permselectivity.

Keizer et al.[68] carried out the partial oxidation of ethylene to acetaldehyde on an alumina-supported vanadium oxide ($Al_2O_3$-$V_xO_y$) membrane. The membranes consisted of mesoporous (average pore diameter of approximately 4 nm), thin $\gamma$-$Al_2O_3$ films (thickness of approximately 5-10 mm) supported on top of macroporous $\gamma$-$Al_2O_3$ substrates (thickness of 2 μm, average pore diameter of approximately 160 nm). The substrates were in the form of flat circular plates with a diameter of 39 mm. Microprobe analysis indicated that most of the vanadium was located within the $\gamma$-$Al_2O_3$ top layer. Membrane layers were prepared in both supported and non-supported forms for studies using either the membrane reactor configuration or in a packed-tube configuration. A stagnation flow configuration was used to feed gas mixtures to both sides of the asymmetric membrane.

The experimentally observed product distribution from integral operation of a packed tube of $Al_2O_3$-$V_xO_y$ catalyst indicated the following reaction network:

$$C_2H_4 + 0.5\,O_2 \rightarrow C_2H_4O \tag{47}$$

$$C_2H_4O + 2.5\,O_2 \rightarrow 2\,CO_2 + 2\,H_2O \tag{48}$$

$$C_2H_4 + 2\,O_2 \rightarrow 2\,CO + 2\,H_2O \tag{49}$$

$$CO + 0.5\,O_2 \rightarrow CO_2 \tag{50}$$

Both acetaldehyde and CO (and water) co-produced at low ethylene conversion. As the ethylene conversion (i.e., temperature) was increased, acetaldehyde and CO were fully oxidized to $CO_2$ and water.

Membrane reactor tests consisted of two flow types. In the *mixed feed* (called configuration A) experiments, a mixture of ethylene, oxygen, and helium with a prescribed composition were fed to the active layer side of the membrane reactor, while pure helium was fed to the support side. In the *segregated feed* (configuration B) experiments, a mixture of ethylene and helium were fed to the active layer side, while a mixture of oxygen and helium were fed to the support side. Comparisons of the two flow types were made by fixing the overall feed gas composition while varying the furnace temperature over a range sufficient to span a representative range of ethylene (or oxygen) conversions.

An analysis of the results from the mixed-flow experiments revealed a product distribution similar to that obtained for the packed-tube. At low temperature, the major products were CO and acetaldehyde. At high temperatures, the major product was $CO_2$. The product distribution from the segregated feed experiments was similar. Both CO and acetaldehyde were produced at lower temperatures, giving way to $CO_2$ at higher temperatures. Interestingly, the maximum selectivity to acetaldehyde was approximately 65% in the segregated feed experiments. This was considerably higher than the corresponding value in the mixed flow runs (ca. 20%). But, this maximum occurred at a higher temperature, (ca. 310 °C), than that obtained for the mixed feed (ca. 220 °C). The ability to partially oxidize ethylene to acetaldehyde depends on both the intrinsic catalytic chemistry and kinetics and the interaction between the chemical and physical rate processes. It is the latter interaction that is impacted by the use of reactant segregation with this nonpermselective membrane reactor. The observed higher acetaldehyde selectivity, at a fixed conversion level of ethylene and overall contact time, demonstrates the principle of reactant segregation. By controlling the flux of oxygen to the active layer, a higher ethylene-to-oxygen ratio is maintained. In effect, a diffusion limitation (for oxygen) is created with the reactant segregation. The oxygen diffusion limitation is beneficial to acetaldehyde selectivity, but is detrimental to overall rate. Thus, while a higher catalyst temperature is needed to achieve the desired ethylene conversion, a higher acetaldehyde selectivity is achieved. The same result might be obtained with the conventional fixed-bed reactor by maintaining a high ethylene-to-oxygen ratio along the entire reactor length.

*6.2. Moving Bed/Chromatographic Reactors*

Reaction and separation can also be accomplished in moving bed/chromatographic reactors. Whereas reaction is coupled with selective permeation through a thin film in membrane reactors, reaction is coupled with selective adsorption onto a porous material in a chromatographic reactor. Continuous operation of chromatographic units requires the contacting of a moving bed of solids with the flowing fluid. Figure 44 is a schematic representation of a moving bed reactor. Reactant is fed to an intermediate point in a downward flowing stream of solids. The solid phase consists of both the adsorbent and catalyst in a reaction/separation unit.

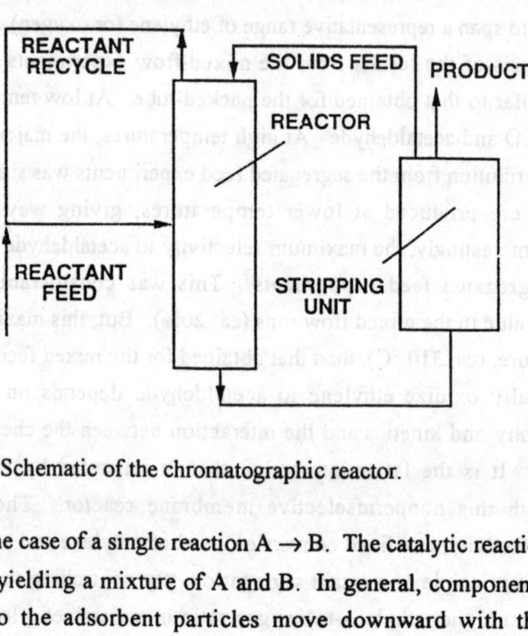


Figure 44. Schematic of the chromatographic reactor.

Consider the case of a single reaction A → B. The catalytic reaction occurs on the catalyst particles, yielding a mixture of A and B. In general, components that are more strongly bound to the adsorbent particles move downward with the solid phase. Components that bind weakly to the solids flow upward in the gas stream. In an idealized situation, weakly-bound reactant A flows out the top of the bed, while more strongly-bound reaction product B exits the bottom of the column with the solid. Thus, reaction and separation are effected in a single unit. This concept has been predicted in simulations by Viswanathan and Aris[69] and demonstrated in experiments by Takeuchi and Uraguchi.[70] Other theoretical studies of the chromatographic reactor have appeared.[71-73]

More intriguing is the ability of the chromatographic reactor to increase the conversion in an equilibrium-limited reaction or the selectivity of an intermediate in a consecutive reaction system.[74] The former concept has been experimentally demonstrated by Fish and Carr[75] using the hydrogenation of 1,3,5-trimethylbenzene to 1,3,5-trimethylcyclohexane on supported Pt as the test reaction system. In reference to the latter, Takeuchi et al.[76] have shown that the selectivity of an intermediate product in a first-order consecutive reaction system in a chromatographic reactor can be increased over selectivity levels obtained in a conventional fixed-bed reactor. This concept has direct relevance to selective oxidation reactions.

In their recent study, Tonkovich et al.[76] demonstrated the viability of the simulated countercurrent moving-bed chromatographic reactor (SCMCR) in the methane coupling reaction to ethane and ethylene. The SCMCR consists of a tandem of several beds of stationary solids with a moving feed. This configuration "simulates" a moving bed. In the methane coupling experiment the catalyst ($Sm_2O_3$) and adsorbent (activated charcoal) were contained in separated columns because of the vast difference in operating temperatures (900 to 1100 K for the former and 373 K for the latter). Considerable improvement in $C_2$ yields and methane conversion were achieved over levels obtained in fixed-bed or fluidized-bed reactors. More specifically, methane conversion exceeded 60% and $C_2$ yield of >50%. These figures compare to typical $C_2$ yields of 20% in conventional reactors. Three key operating advantages were achieved in the study. First, suppression of $C_2$ intermediate product oxidation to carbon oxides was achieved by a chromatographic separation of the $C_2$ species from the oxygen. By binding more strongly to the charcoal, the $C_2$ species were retained in the charcoal for a longer time period than the methane or oxygen. Second, better contacting of the methane with the catalyst enabled higher methane conversions. Third, the $C_2$ product stream was free of both oxygen and methane.

The results of Tonkovich et al.[76] bring to question whether other partial oxidation reactions can be carried out in a chromatographic reactor with improved performance. It would also be useful to compare the performance of the chromatographic reactor and membrane reactor in carrying out a selective oxidation reaction as part of an effort to identify the preferred reactor type.

**References**


1. N. B. Siccama and K. R. Westerterp, *Ind. Eng. Chem. Res.* **32** (1993) 1304.
2. R. M. Contractor, H. E. Bergna, H. S. Horowitz, C. M. Blackstone, B. Malone, C. C. Toradi, B. Griffiths, U. Chowdhry and A. W. Sleight, *Catalysis Today*, **1** (1987) 49.
3. G. P. Marrs, F. P. Lees, J. Barton, and N. Scilly, *Chem. Eng. Res. Des.* **67** (1989) 381.
4. P. L. Mills and M. P. Harold, *Catal. Rev.-Sci. Eng.* (to be submitted), 1995.
5. G. F. Froment and K. B. Bischoff, *Chemical Reactor Analysis and Design*, Second Edn., (Wiley, New York, 1990) pp. 475-515.
6. S. M. Walas, in *Encyclopedia of Chemical Processing and Design*, ed. J. J. McKetta (Marcel-Dekker, New York, 1994).
7. G. Eigenberger, in *Ullmann's Encyclopedia of Industrial Chemistry, Principles of Chemical Reaction Engineering and Plant Design*, Vol. B4, eds. B. Elvers, S. Hawkins, and G. Schulz (VCH Publishers, Weinheim, Germany, 1992) pp. 200-237.
8. M. P. Huff, P. M. Tornianen, and L. D. Schmidt, *Catal. Today* **21**, (1994) 113.
9. D. J. Newman, in *Kirk-Othmer Encyclopedia of Chemical Technology*, Third Edn., Vol. 15, eds. H. F. Mark, D. F. Othmer, C. G. Overberger, and G. T. Seaborg (Wiley, New York, 1981) pp. 853-871.
10. N. Waletzko and L. D. Schmidt, *AIChE J.* **34** (1987) 1146.
11. L. Lefferts, J. G. van Ommen, and J. R. H. Ross, *Appl. Catal.* **23** (1986).
12. R. W. Strickland, in *Riegel's Handbook of Industrial Chemistry*, Ninth Edn. ed. J. A. Kent (Van Nostrand Reinhold, New York, 1992) pp. 458-479.
13. K. Y.- Lee and R. Aris, *Ind. Eng. Chem. Process Des. Develop.* **2** (1963) 300.
14. J. D. Paynter., J. S. Dranoff, and S. G. Bankoff, *Ind. Eng. Chem. Process Des. Develop.* **10** (1971) 244.
15. G. K. Boreskov and Y. S. Matros, *Catal. Rev.-Sci. Eng.* **25** (1983) 551.
16. G. K. Boreskov and Y. S. Matros, in *Recent Advances in the Engineering Analysis of Chemically Reacting Systems*, ed. L. K. Doraiswamy (Halsted Press, Wiley, New York, 1984).
17. Y. S. Matros, A. Noskov, V. Chumachenko, and O. Goldman, *Chem. Eng. Sci.* **43** (1988) 2061.
18. G. Eigenberger and U. Nieken, *Int. Chem. Eng.* **34** (1994) 4.

19. W. G. Etzkorn, J. J. Kurland, and W. D. Neilsen, in *Kirk-Othmer Encyclopedia of Chemical Technology*, Fourth Edn., Vol. 15, ed. M. Howe-Grant (Wiley, New York, 1991) pp. 232-251.
20. R. T. Grasselli and J. D. Burrington, in *Adv. Catal.* **30** (1981) 133.
21. J. C. Burnett, T. R. Felthouse, S. F. Mitchell, and M. J. Mummey, to appear in *Encyclopedia of Chemical Technology*, Fifth Edn. (1995).
22 V. Nikolov, D. Klissurski, and A. Anastasov, *Catal. Rev.-Sci. Eng.* **33** (1991) 319.
23. G. Eigenberger, in *Ullmann's Encyclopedia of Industrial Chemistry, Principles of Chemical Reaction Engineering and Plant Design*, Vol. B4, eds. B. Elvers, S. Hawkins, and G. Schulz (VCH Publishers, Weinheim, Germany, 1992) pp. 200-237.
24. J. R. Grace, in *Chemical Reactor Design and Technology*, NATO Advanced Study Institute Series E: Applied Sciences, Vol. 110, ed. H. I. de Lasa, (Martinus Nijhoff, Dordrecht, 1986), pp. 245-304.
25. D. King, in *Chemical Reactor Technology for Environmentally Safe Reactors and Products*, NATO Advanced Study Institute Series E: Applied Sciences, Vol. 225, eds. H. I. de Lasa, G. Dogu and A. Ravella (Kluwer Academic, Dordrecht, 1989), pp. 17-50.
26. F. A. Zenz, in *Kirk-Othmer Encyclopedia of Chemical Technology*, Third Edn., Vol. 10, eds. H. F. Mark, D. F. Othmer, C. G. Overberger and G. T. Seaborg, (Wiley, New York, 1980), pp. 548-581.
27. R. M. Contractor, *Catalysis Today* **1** (1987) 587.
28. L. T. Groeft, in *Kirk-Othmer Encyclopedia of Chemical Technology*, Third Edn., Vol. 1, eds. H. F. Mark, D. F. Othmer, C. G. Overberger and G. T. Seaborg, (John Wiley, New York, 1980), pp. 414-426.
29. S. D. Cooley and J. D. Powers, in *Encyclopedia of Chemical Processing and Design*, Vol. 29, eds. J. J. McKetta and W. A. Cunningham (Marcel Dekker, New York, 1988), pp. 35-55.
30. J. J. Graham, in *Handbook of Multiphase Systems*, ed. G. Hetstroni (Hemisphere, Washington, 1982), pp. 8-161 to 8-162.
31. J. R. Grace, in *Chemical Reactor Design and Technology*, NATO Advanced Study Institute Series E: Applied Sciences, Vol. 110, ed. H. I. de Lasa, (Martinus Nijhoff, Dordrecht, 1986), pp. 245-304.
32. J. R. Grace, *Chem. Eng. Sci.* **45** (1990) 1953.
33. L. Reh, in *Circulating Fluidized Bed Technology III*, eds. P. Basu, M. Horio and M. Hasatari (Pergamon Press, New York, 1991), pp. 105-118.

34. D. King, in *Chemical Reactor Technology for Environmentally Safe Reactors and Products*, NATO Advanced Study Institute Series E: Applied Sciences, Vol. 225, eds. H. I. de Lasa, G. Dogu and A. Ravella (Kluwer Academic, Dordrecht, 1989), pp. 17-50.
35. R. M. Contractor, H. E. Bergna, H. S. Horowitz, C. M. Blackstone, B. Malone, C. C. Toradi, B. Griffiths, U. Chowdhry and A. W. Sleight, *Catalysis Today*, **1** (1987) 49.
36. W. E. Stadig, *Chem. Proc.* (August 1992), 27.
37. G. S. Patience, *Circulating Fluidized Beds: Hydrodynamics and Reactor Modelling*, Ph.D. Dissertation, Ecole Polytechnique de Montreal (Montreal, Canada, 1990).
38. H. E. Bergna, U. S. Patent 4,677,984 assigned to DuPont, November 11, 1985.
39. H. E. Bergna, U. S. Patent 4,769,477 assigned to DuPont, September 6, 1988.
40. M. C. Sze and A. P. Gelbein, *Hydrocarbon Processing* (February 1976) 103.
41. M. A. Mabry, W. W. Prichard and S. B. Ziemecki, U.S. Patent 4,550,185 assigned to DuPont, October 29, 1985.
42. R. M. Contractor, H. E. Bergna, H. S. Horowitz, C. M. Blackstone, U. Chowdhry and A. W. Sleight, in *Studies in Surface and Catalysis*, Vol. 38, ed. J. W. Ward (Elsevier, Amsterdam, 1987), pp. 645 - 654.
43. R. M. Contractor, U. S. Patent 4,668,802 assigned to DuPont, May 26, 1987.
44. R. M. Contractor, U. S. Patent 5,021,588 assigned to DuPont, June 4, 1991.
45. V. T. Zaspalis and A. J. Burggraaf, in *Inorganic Membranes Synthesis, Characteristics and Applications,* ed. R.R. Bhave (Van Nostrand Reinhold, New York, 1991) Chapter 7, p. 177.
46. J. Armor, *Appl. Catal.* **49** (1989) 1.
47. P. J. Gellings and H. J. M. Bouwmeister, *Catal. Today* **12** (1992) 1.
48. H. P. Hsieh, *Catal. Rev.-Sci. Eng.* **33** (1991) 1.
49. T. T. Tsotsis, R. G. Minet, A.M. Champagnie, and P. K. T. Liu, in *Computer-Aided Design of Catalysts*, eds. E. R. Becker and C. J. Pereira (Marcel Dekker, New York 1993).
50. M. P. Harold, C. Lee, A. J. Burggraaf, K. Keizer, V. T. Zaspalis, and R. S . A. de Lange, *MRS Bulletin,* Vol. XIX (April 1994) 34.
51. H. J. Sloot, Ph.D. Thesis, University of Twente, Enschede, the Netherlands (1991).
52. H. J. Sloot, C. A. Smolders, W. P. M. van Swaaij, and G. F. Versteeg, *AIChE J.* **38** (1992) 887.
53. V. T. Zaspalis, W. van Praag, K. Keizer, J. G. van Ommen, J. R. H. Ross, and A. J. Burggraaf, *Appl. Catal.* **74** (1991) 249.

54. P. Cini, and M. P. Harold, *AIChE J.* **37** (1991) 997.
55. O. Shinji, M. Misono, and Y. Oneda, *Bull. Chem. Soc. Japan* **55** (1982) 2760.
56. Y. M. Sun and S. J. Khang, *Ind. Eng. Chem. Res.* **27** (1988) 1136.
57. R. R. Zhao, R. Govind, and N. Itoh, *Sep. Sci. Tech.* **25** (1990) 1473.
58. N. Itoh, *AIChE J.* **33** (1987) 1576.
59. A. M. Champagnie, T. T. Tsotsis, R. G. Minet, and I. A. Webster, *Chem. Eng. Sci.* **45** (1990) 2423.
60. M. P. Harold, V. T. Zaspalis, K. Keizer, and A. J. Burggraaf, *Chem. Eng. Sci.* **48** (1993) 2705.
61. S. Agarwalla, S. and C. R. F. Lund, *J. Membr. Sci.* **70** (1992) 129.
62. L. A. Bernstein and C. R. F. Lund, *J. Membr. Sci.* **77** (1993) 155.
63. Y. Teraoka, H. M. Zhang, S. Furukawa, and N. Yamazoe, *Chem. Letters* 1743 (1985).
64. H. Kruidhof, H. J. M. Bouwmeister, R. H. E. v. Doorn, and A. J. Burggraaf, *Solid State Ionic* **63** (1993) 816.
65. E. A. Hazbun, U.S. Patent 4,287,071 (1989).
66. A. Bielanski and J. Haber, *Oxygen in Catalysis* (Marcel Dekker, New York, 1991).
67. R. Di Cosimo, J. D. Burrington, and R. K. Grasselli, *J. Catal.* **102** (1986) 234.
68. K. Keizer, V. T. Zaspalis, R. de Lange, M. P. Harold, and A. J. Burggraaf, in *Membrane Processes in Separation and Purification*, eds. J. Crespo and K. Boddeker (Kluwer Academic, Dordrecht, 1994).
69. S. Viswanathan and R. Aris, SIAM-AMS Proc. **8** (1974) 99.
70. K. Takeuchi and Y. Uraguchi, *J. Chem. Eng.* Japan **10** (1977) 455.
71. B. K. Cho, R. Aris, and R. W. Carr, *Proc. Roy. Soc.* **A383** (1982) 147.
72. D. Altshuller, *Chem. Eng. Commun.* **19** (1983) 363.
73. T. Petroulas, R. Aris, and R. W. Carr, *Chem. Eng. Sci.* **40** (1985) 2233.
74. K. Takeuchi, T. Miyauchi, and Y. Uraguchi, *J. Chem. Eng. Japan* **11** (1978) 539.
75. B. B. Fish and R. W. Carr, *Chem. Eng. Sci.* **44** (1989) 1773.
76. A. L. Tonkovich, R. W. Carr, and R. Aris, *Science* **262** (1993) 221.

# Epilogue: Future Prospects

The principles and applications of oxidation catalysis remains a subject of enormous industrial and academic interest. Progress continues to be made in our understanding of fundamental redox events occurring between organic molecules and metal (oxide) surfaces or (oxo)metal centers in homogeneous solution. In the former case, for example, sophisticated spectroscopic techniques, such as SEM, TEM, EXAFS, etc. provide valuable information regarding the composition, structure and microscopic distribution of catalytically active species on surfaces. Polyoxometalates and giant metal clusters are interesting in this context as they constitute intermediate stages between monomeric (oxo)metal centers and metal (oxide) surfaces, respectively.

Traditional interdisciplinary barriers separating heterogeneous, homogeneous and biocatalysis are gradually disappearing. Hopefully this course has contributed to a further stimulation of this process. In this context we note that a future course on this subject should include some discussion of oxidation catalysis by metalloenzymes in order to promote a further cross-fertilization of ideas.

Looking to the future, we note that the potentially attractive 'dream' reactions, such as direct catalytic epoxidation of propylene with dioxygen and methane to methanol conversion, remain extremely elusive goals. Nevertheless they will continue to be the focus of considerable research effort in the future and there is still a definite need for new approaches to the activation of dioxygen.

We expect that catalytic oxidations, particularly in the liquid phase, will find broad applications in fine chemicals manufacture, using dioxygen, hydrogen peroxide and alkyl hydroperoxides as primary oxidants. In this context redox molecular sieves and related solid catalysts are likely to find widespread application in liquid phase oxidations. We also expect that significant advances will be made in the area of asymmetric catalytic oxidation using biomimetic catalysts or even (modified) biocatalysts themselves.

Finally, a prerequisite for optimum performance in catalytic processes is a proper choice of reactor configuration. Consequently, we predict further advances in the application of catalytic reactor engineering to oxidation processes. Indeed, further optimization of liquid phase oxidations employing solid catalysts may largely depend on the choice of the optimum reactor type, e.g. fixed-bed and monolithic reactors.

# Index